Environmental Sampling and Analysis for Metals

Environmental Sampling and Analysis for Metals

Maria Csuros • Csaba Csuros

With contributions by
Laszlo Gy. Szabo

LEWIS PUBLISHERS

A CRC Press Company

Boca Raton London New York Washington, D.C.

Library of Congress Cataloging-in-Publication Data

Csuros, Maria.
Environmental sampling and analysis for metals / by Maria Csuros and Csaba Csuros.
p. cm.
Includes bibliographical references and index.
ISBN 1-56670-572-X (alk. paper)
1. Metals—Analysis. 2. Environmental chemistry. 3. Environmental monitoring I. Csuros, Csaba II. Title.

TD196.M4 C78 2002
628.5′2—dc21

2002019440
CIP

Direct all inquiries to CRC Press LLC, 2000 N.W. Corporate Blvd., Boca Raton, Florida 33431.

Visit the CRC Press Web site at www.crcpress.com

Lewis Publishers is an imprint of CRC Press LLC

International Standard Book Number 1-56670-572-X
Library of Congress Card Number 2002019440
1 2 3 4 5 6 7 8 9 0

To all who are so far from us, but always close to our hearts, our sons Geza and Zoltan, and our grandchildren Aaron, Andrew, Daniel, Jordan, and Sebastian.

Preface

Monitoring the environment for metals has become a topic of considerable importance, not only to those industries emitting heavy metals but also to surveillance agencies and other organizations assessing the impact of metals on the environment. Our goal is to provide a comprehensive and easy-to-read text for anyone working in the environmental analytical chemistry arena and to provide essential information to consultants and regulators about analytical and quality control procedures helpful in their evaluation and decision-making procedures. The book is also useful for technicians in their everyday chores. It not only provides a guide for analyzing metals in environmental samples but is useful as a supplementary information source for more general environmental studies and a variety of job-related training programs. In addition, college and university students taking chemical or environmental laboratory courses will find the book easy to use and understand. It will also be helpful to graduate students and chemists seeking information on laboratory practice.

The book provides a detailed introduction to metals and their toxicity and includes sample collection, preservation, correct storage, holding time, preparation for analysis, theory of analytical methods and instrumentation, step-by-step analytical procedures, complete QA/QC requirements, data validation, calculation of analytical results, reporting format, and standards with maximum contaminant levels. The book contains both theoretical and practical applications in metals analysis of environmental samples and incorporates the latest in analytical techniques, instrumentation, and regulations. The appendices provide instant information on a wide array of topics.

This book is part of the "Environmental Sampling and Analysis for Laboratory Technicians" series, and should prove valuable as a practical handbook for students in environmental education and special training programs and for environmental chemists in everyday chores. In addition, the text will help students, chemists, and others understand analytical reports.

Preface

Acknowledgments

We owe a great debt and gratitude to all the people who participated in making this book possible. We are honored to thank Attila Borhidy, head of the Hungarian Academy of Science. He has been responsible for the development of our scientific relationships with the outstanding staff of the University of Pecs, Hungary. Special thanks to Laszlo Szabo, chair of the Biology Department of the University of Pecs, for his helpful comments and support. His contribution to this text has made this book more valuable.

We are pleased to express our gratitude to our son Geza and his wife for their patience in reviewing and correcting the text. Our appreciation goes also to Sandor Barta for proofreading and improving the writing; Lenke Babarci for her patience and hard work in the preparation of the figures and tables; and Barna Csuros, retired library director, for the literature he sent for our review and for his always available helping hand.

We also gratefully acknowledge the support of the outstanding editorial and production staff of CRC Press/Lewis Publishers.

Warm words of thanks to our sons, Geza and Zoltan, and our grandchildren, Aaron, Andrew, Daniel, Jordan, and Sebastian, for their love, encouragement, and cheerful spirit.

To all of you, thank you!

Biographies

Maria Csuros is an environmental research chemist, currently affiliated with the University of Pecs, Hungary. She previously worked as supervisor of the water department for the Environmental and Public Health Laboratory, Hungary, and has served in diverse teaching and research capacities in the United States. Ms. Csuros designed and developed an environmental science program with a focus on sampling and analysis for the Pensacola Junior College, Florida. She received her Ph.D. in environmental chemistry from the Janus Pannonius University, Pecs, Hungary. In addition to this book, Ms. Csuros has authored or co-authored *Environmental Sampling and Analysis for Technicians*, 1994; *Environmental Sampling and Analysis Laboratory Manual*, 1996; and *Microbiological Examination of Water and Wastewater*, 1999, all published by CRC Press/Lewis.

Csaba Csuros teaches advanced microbiology courses and conducts research on the serological diagnosis of parasitic diseases (ascariasis, echinococcosis, and filariasis) at the University of Pecs, Hungary. He previously worked in several capacities in medical and public health testing laboratories, and as a professor of microbiology, anatomy, and physiology in U.S. universities; he has received the excellence in teaching award. Mr Csuros earned his Ph.D. in microbiology from the Jozsef Attila University, Hungary. He has written numerous papers in the field of microbiology and co-authored *Microbiological Examination of Water and Wastewater*, 1999, CRC Press/Lewis.

Laszlo Gy. Szabo is chair of the Botany Department and teaches graduate-level plant physiology and ecological phytochemistry at the University of Pecs, Hungary. He received a Ph.D. and D.Sc. in plant physiology (phytochemistry) from the Hungarian Academy of Science, Budapest, Hungary, and a M.S. degree in pharmacy from Semmelweis University, also in Budapest. Mr. Szabo has written several monographs published by the University of Pecs and Academic Publishers, Budapest. He is vice president of the Medicinal Plant Section, Hungarian Pharmaceutical Society; a member of the International Allelopathy Society, Federation of European Societies of Plant Physiology, and the Botanical Committee of the Hungarian Academy of Sciences; and serves on the editorial board of *The Cultural Flora of Hungary.*

List of Figures and Tables

FIGURES

TABLES

Table of Contents

Chapter 1 Introduction to Metals

Chapter 2 Discussion of Metallic Elements

Chapter 3 Toxicity of Metals

Chapter 4 STANDARDS RELATED TO METALLIC POLLUTANTS

Chapter 5 FUNDAMENTALS OF SPECTROSCOPY

Chapter 6 MOLECULAR SPECTROPHOTOMETRY

Chapter 7 ATOMIC ABSORPTION SPECTROMETRY

Chapter 8 DIRECT ASPIRATION OR FLAME ATOMIC ABSORPTION SPECTROMETRY (FAAS)

Chapter 9 GRAPHITE FURNACE ATOMIC ABSORPTION SPECTROMETRY

Chapter 10 COLD-VAPOR ATOMIC ABSORPTION SPECTROMETRY

Chapter 11 HYDRIDE-GENERATION ATOMIC ABSORPTION TECHNIQUE

Chapter 12 INDUCTIVELY COUPLED PLASMA ATOMIC EMISSION SPECTROSCOPY

Chapter 13 QUALITY CONTROL IN METALS ANALYSIS

Chapter 14 SAMPLE COLLECTION FOR METALS ANALYSIS

Chapter 15 SAMPLE PREPARATION FOR METALS ANALYSIS

Chapter 16 CONVERTING RAW DATA INTO REPORTABLE FORM

Chapter 17 REPORTING ANALYTICAL DATA

Chapter 18 SELECTED METHODS FOR DETERMINATION OF METALS IN ENVIRONMENTAL SAMPLES

Chapter 19 Laboratory Safety Rules

APPENDICES

1 Introduction to Metals

1.1 INTRODUCTION TO ELEMENTS

1.1.1 MATTER

Matter is anything that has mass and occupies space. Matter is found in many different forms, and every year thousands of new types of matter are synthesized. Matter is grouped into two major classes: pure substances and mixtures. *Pure substances* are subdivided into elements and compounds. *Elements* are pure substances that cannot be decomposed by chemical changes. *Compounds* are pure substances that can be decomposed chemically.

1.1.2 Elements

Elements are the basic units of matter. At present, 109 elements have been identified; 92 occur in nature, and the rest are synthetic. At 25°C, 97 elements are solids, 2 are liquids, and 11 are gases.

1.1.2.1 Element Symbols

Element symbols are usually derived from the first one or two letters of the element's name. Twelve of the elements have one-letter symbols that correspond to the first letter of the element's name: hydrogen (H), boron (B), carbon (C), nitrogen (N), oxygen (O), fluorine (F), phosphorus (P), sulfur (S), vanadium (V), yttrium (Y), iodine (I), and uranium (U). Other elements are designated by one-letter symbols but do not correspond to the first letter of their English-language names. K is the symbol for potassium, which is the first letter of its Latin name, *kalium*, which means ashes. Similarly, the symbol for tungsten is W, derived from the Latin word *wolframate*. Most of the remaining elements have been assigned two-letter symbols. The first letter is always an uppercase letter, and the second, a lowercase letter. For instance, the symbol for cobalt is Co, not CO. Some symbols are made up of the first two letters of the element's English-language name, others consist of two letters from a Latin word, and the remainder combine the first letter of the element with some other letter in the name.

1.1.2.2 Element Names

The origins of element names are diverse, including geographical locations, the names of great scientists, mythological gods, and astronomical bodies. Table 1.1 lists the names and symbols of elements derived from Latin words, and Table 1.2 contains the origins of selected elements. The symbol of an element in the periodic table is accompanied by a whole number and a decimal number. To understand the significance of these numbers, a short review of atomic structure is necessary.

TABLE 1.1
Names and Symbols of Elements Derived from Latin Words

Name	Symbol	Latin Word
Antimony	Sb	Stibium
Gold	Au	Aurum
Lead	Pb	Plumbum
Mercury	Hg	Hydrargyrum
Potassium	K	Kalium
Silver	Ag	Argentum
Tin	Sn	Stannum
Tungsten	W	Wolframe

TABLE 1.2
Origins of Selected Element Names

Element	Symbol	Origin of Name
		Location
Americium	Am	America
Berkelium	Bk	Berkeley, CA
Californium	Cf	California
Europeum	Eu	Europe
Francium	Fc	France
Germanium	Ge	Germany
Polobnium	Po	Poland
Stroncium	Sc	Strontia, Scotland
		Scientist
Curium	Cm	Marie and Pierre Curie
Einsteinium	Es	Albert Einstein
Fermium	Fm	Enrico Fermi
Lawrencium	Lr	Ernest O. Lawrence
Mendelevium	Md	Dmitri Mendeleev
Nobelium	No	Alfred Nobel
		God or Astronomical Body
Helium	He	Greek "helios" (sun)
Niobium	Ni	Niobe, daughter of Tantalus
Neptunium	Np	Neptune
Palladium	Pd	Asteroid called Pallas
Plutonium	Pu	Pluto
Selenium	Se	Greek *selene* (moon)
Thorium	Th	Thor
Uranium	U	Uranus

1.1.3 Atoms

Atoms are the smallest particles that retain the chemical properties of elements. In other words, an atom is the smallest unit of an element, and each element is composed of similar atoms. Atoms are extremely small; for example, 1 g of carbon (C) contains 5×10^{22} C atoms.

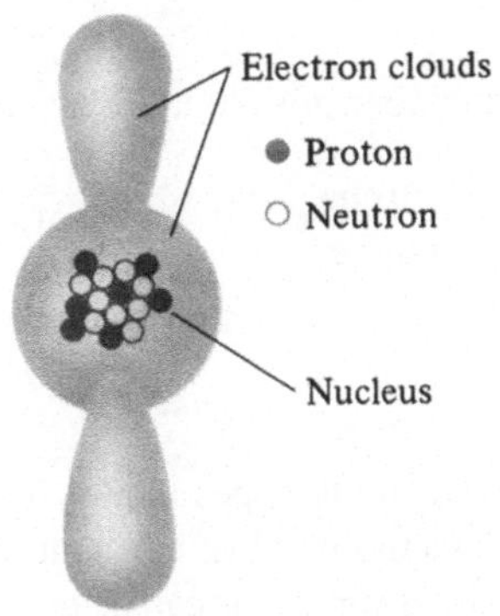

FIGURE 1.1 Typical atom. Protons and neutrons make up the nucleus; electron "clouds" surround the nucleus.

1.1.3.1 Subatomic Particles

Atoms are composed of three fundamental particles — *protons, electrons*, and *neutrons*. (Figure 1.1 illustrates a typical atom.) Particles are characterized by mass and electric charge. Protons and neutrons have approximately the same mass, 1.67×10^{-24} g. The mass of an electron is only 9.11×10^{-28} g, about 1/1837th of a proton or a neutron; in other words, 1837 electrons are needed to equal the mass of one proton. Electrons are negatively charged (1–). Protons possess the same charge as electrons, but it is positive (1+). Neutrons have no charge and are thus electrically neutral. Because an atom is electrically neutral, the number of protons should be equal to the number of electrons. Masses of subatomic particles are frequently expressed using a relative unit, known as a *unified atomic mass unit* (u): $1 \text{ u} = 1.6606 \times 10^{-24}$ g. Properties of subatomic particles are presented in Table 1.3.

Protons and neutrons are located in a very small region of the atom, called the *nucleus*, and are surrounded by electrons. Electrons are located outside of the nucleus in quantized *energy levels*. Each energy level is divided into smaller regions called *sublevels*, and each sublevel is divided into *orbitals*, the location of the electrons. The exact definition of the orbital is a volume of space where there is a specific probability of encountering electrons.

According to the *Heisenberg's uncertainty principle*, it is impossible to accurately determine the exact position and velocity of an electron. Each orbital contains a maximum of two electrons, which spin in opposite directions. Each energy level contains a precise number of sublevels and the sublevels contain a precise number of orbitals; thus, each sublevel contains a specified number of electrons.

1.1.3.2 Atomic Number

The atomic number of an atom equals the number of protons (+) in the nucleus of an atom, because in a neutral atom the number of protons (+) should be equal to the number of electrons (-). Therefore,

$$\text{Atomic number} = \text{number of protons } (+) = \text{number of electrons } (-)$$

TABLE 1.3
Properties of Subatomic Particles

Particle	Symbol	Mass (g)	Mass (u)	Relative Charge
Proton	p^+	1.6726×10^{-24}	1.007276	1+
Neutron	n^o	1.6749×10^{-24}	1.008666	0
Electron	e^{-1}	9.1096×10^{-28}	0.00054861	–

1.1.3.3 Mass Number

The mass number of an atom equals the total number of protons (+) and neutrons (o) in the nucleus:

Mass number = number of protons + number of neutrons
Number of neutrons (n^o) = mass number – atomic number

1.1.3.4 Atomic Mass and Atomic Weight

The atomic mass of an element is the average mass of its naturally occurring isotopes relative to the mass of C^{612}. Most elements are found in nature as mixtures of isotopes in a more or less constant ratio. For some elements, these ratios vary slightly, but for most purposes the slight variations can be ignored.

The atomic weight of an element is a weighted average of the combined mass of the isotopes. The mass of an isotope is approximately the same as its mass number. Some elements — for example, gold, fluorine, and aluminum — occur naturally as a single isotope. The atomic weights of these elements are, of course, close to whole numbers (gold, 196.7; fluorine, 18.998; and aluminum, 26.98). The atomic masses of the elements are decimal numbers. The elements and their atomic numbers and atomic masses are listed in Table 1.4.

1.1.4 Isotopes

Isotopes are atoms with the same number of protons but a different number of neutrons in their nuclei; that is, they have the same atomic number but different mass numbers. A large percentage of the elements are composed of mixtures of different isotopes. For example, three isotopes of uranium occur naturally: U^{92}_{234} contains 142 neutrons, U^{92}_{235} contains 143 neutrons, and the third isotope, U^{92}_{238}, has 146 neutrons. Mass spectroscopy is used to measure relative atomic mass and isotopes. (See Appendix A for a description of mass spectrophotometer operations.)

1.2 PERIODIC TABLE OF ELEMENTS

The periodic table of elements is a tabular arrangement of elements in rows and columns, highlighting the regular repetition of properties of the element. In 1869, Russian chemist Dmitri Mendeleev and German chemist Lothar Meyer, working independently, made similar discoveries. They found that when they arranged the elements in order of atomic weight, they could place them in horizontal rows, one row under the other, so that the elements in any one vertical column have similar properties. In the early part of the twentieth century, scientists demonstrated that the elements are characterized by respective atomic numbers. A modern version of the periodic table, with the elements arranged by atomic numbers, is shown in Table 1.5.

The basic structure of the periodic table is its division into rows and columns, or periods and groups. A *period* consists of the elements in any one horizontal row of the periodic table, and a *group* consists of the elements in any one column of the periodic table. The groups are numbered with Roman numerals, and A's and B's are common. In Europe, a similar convention has been used but the A's and B's are interchanged in some columns. To eliminate this confusion, the International Union of Pure and Applied Chemistry (IUPAC) recommended that the groups be numbered consecutively from 1 to 18. Each period is numbered consecutively from 1 to 7. The periodic table of elements is probably the most important table in chemistry. A modern version of the periodic table, with the elements arranged by atomic numbers and the group numbers by traditional and IUPAC conventions, is presented in Table 1.5.

TABLE 1.4
Table of Elements with Atomic Numbers and Atomic Masses

Name	Symbol	Atomic Number	Atomic Mass
Actinium	Ac	89	227.0278[a]
Aluminum	Al	13	26.98154
Americium	Am	95	243[a]
Antimony	Sb	51	121.75
Argon	Ar	18	39.948
Arsenic	As	33	74.9216
Astatine	At	85	210[a]
Barium	Ba	56	137.33
Berkelium	Bk	97	247[a]
Beryllium	Be	4	9.01218
Bismuth	Bi	83	208.9804
Boron	B	5	10.81
Bromine	Br	35	79.904
Cadmium	Cd	48	112.41
Calcium	Ca	20	40.08
Californium	Cf	98	251[a]
Carbon	C	6	12.011
Cerium	Ce	58	140.12
Cesium	Cs	55	132.9054
Chlorine	Cl	17	35.453
Chromium	Cr	24	51.996
Cobalt	Co	27	58.9332
Copper	Cu	29	63.546
Curium	Cm	96	247[a]
Dysprosium	Dy	66	162.50
Einsteinium	Es	99	252[a]
Erbium	Er	68	167.26
Europium	Eu	63	151.96
Femium	Fm	100	257[a]
Fluorine	F	9	18.998403
Francium	Fr	87	223[a]
Gadolinium	Gd	64	157.25
Gallium	Ga	31	69.72
Germanium	Ge	32	72.59
Gold	Au	79	196.9665
Hafnium	Hf	72	178.49
Helium	He	2	4.00260
Holmium	Ho	67	164.9304
Hydrogen	H	1	1.0079
Indium	In	49	114.82
Iodine	I	53	126.9045
Iridium	Ir	77	192.22
Iron	Fe	26	55.847
Krypton	Kr	36	83.80
Lanthanum	La	57	138.9055
Lawrencium	Lr	103	260[a]
Lead	Pb	82	207.2
Lithium	Li	3	6.941
Lutetium	Lu	71	174.967
Magnesium	Mg	12	24.305
Manganese	Mn	25	54.9380
Mendelevium	Md	101	258[a]
Mercury	Hg	80	200.59
Molybdenum	Mo	42	95.94
Neodymium	Nd	60	144.24
Neon	Ne	10	20.179
Neptunium	Np	93	237.0482
Nickel	Ni	28	58.70
Niobium	Nb	41	92.9064
Nitrogen	N	7	14.0067
Nobelium	No	102	259[a]
Osmium	Os	76	190.2
Oxygen	O	8	15.9994
Palladium	Pd	46	106.4
Phosphorus	P	15	30.97376
Platinum	Pt	78	195.09
Plutonium	Pu	94	244[a]
Polonium	Po	94	209[a]
Potassium	K	19	39.0983
Praeseodymium	Pr	59	140.9077
Promethium	Pm	61	145[a]
Protactinium	Pa	91	231.0359
Radium	Ra	88	226.0254
Radon	Rn	86	222[a]
Rhenium	Re	75	186.207
Rhodium	Rh	45	102.9055
Rubidium	Rb	37	85.4678
Ruthenium	Ru	44	101.07
Samarium	Sm	62	150.4
Scandium	Sc	21	44.9559
Selenium	Se	34	78.96
Silicon	Si	14	28.0855
Silver	Ag	47	107.868
Sodium	Na	11	22.98977
Strontium	Sr	38	87.62
Sulfur	S	16	32.06
Tantalum	Ta	73	180.9479
Technetium	Tc	43	98[a]
Tellurium	Te	52	127.60
Terbium	Tb	65	158.9254
Thallium	Tl	81	204.37
Thorium	Th	90	232.0381
Thulium	Tm	69	168.9342
Tin	Sn	50	118.69
Titanium	Ti	22	47.90
Tungsten	W	74	183.85
Unnilhexium	Unh	106	263[a]
Unnilpentium	Unp	105	262[a]
Unnilquadium	Unq	104	261[a]
Uranium	U	92	238.029
Vanadium	V	23	50.9415
Xenon	Xe	54	131.30
Ytterbium	Yb	70	173.04
Yttrium	Y	39	88.9059
Zinc	Zn	30	65.38
Zirconium	Zr	40	91.22

[a] Not naturally occurring.

TABLE 1.5
Periodic Table of the Elements

Key: 6 — Atomic number; C — Symbol; 12.011 — Atomic mass

Atomic masses are based on ^{12}C. Atomic masses in parentheses are for the most stable isotope.

Periods \ Groups	IA(1)	IIA(2)	IIIB (3)	IVB (4)	VB (5)	VIB (6)	VIIB (7)	VIIIB (8)	VIIIB (9)	VIIIB (10)	IB (11)	IIB (12)	IIIA (13)	IVA (14)	VA (15)	VIA (16)	VIIA (17)	VIIIA (18)
1	1 H 1.0079																	2 He 4.00260
2	3 Li 6.941	4 Be 9.01218											5 B 10.81	6 C 12.011	7 N 14.0067	8 O 15.9994	9 F 18.998403	10 Ne 20.179
3	11 Na 22.98977	12 Mg 24.305											13 Al 26.98154	14 Si 28.0855	15 P 30.97376	16 S 32.06	17 Cl 35.453	18 Ar 39.948
4	19 K 39.0983	20 Ca 40.08	21 Sc 44.9559	22 Ti 47.90	23 V 50.9415	24 Cr 51.996	25 Mn 54.9380	26 Fe 55.847	27 Co 58.9332	28 Ni 58.70	29 Cu 63.546	30 Zn 65.38	31 Ga 69.72	32 Ge 72.59	33 As 74.9216	34 Se 78.96	35 Br 79.904	36 Kr 83.80
5	37 Rb 85.4678	38 Sr 87.62	39 Y 88.9059	40 Zr 91.22	41 Nb 92.9064	42 Mo 95.94	43 Tc (98)	44 Ru 101.07	45 Rh 102.9055	46 Pd 106.4	47 Ag 107.868	48 Cd 112.41	49 In 114.82	50 Sn 118.69	51 Sb 121.75	52 Te 127.60	53 I 126.9045	54 Xe 131.30
6	55 Cs 132.9054	56 Ba 137.33	57 La 138.9055 *	72 Hf 178.49	73 Ta 180.9479	74 W 183.85	75 Re 186.207	76 Os 190.2	77 Ir 192.22	78 Pt 195.09	79 Au 196.9665	80 Hg 200.59	81 Tl 204.37	82 Pb 207.2	83 Bi 208.9804	84 Po (209)	85 At (210)	86 Rn (222)
7	87 Fr (223)	88 Ra 226.0254	89 Ac 227.0278 †	104 Unq (261)	105 Unp (262)	106 Unh (263)	107 Uns (262)	108	109 Une (266)									

* Lanthanide series

58 Ce 140.12	59 Pr 140.9077	60 Nd 144.24	61 Pm (145)	62 Sm 150.4	63 Eu 151.96	64 Gd 157.25	65 Tb 158.9254	66 Dy 162.50	67 Ho 164.9304	68 Er 167.26	69 Tm 168.9342	70 Yb 173.04	71 Lu 174.967

† Actinide series

90 Th 232.0381	91 Pa 231.0359	92 U 238.029	93 Np 237.0482	94 Pu (244)	95 Am (243)	96 Cm (247)	97 Bk (247)	98 Cf (251)	99 Es (252)	100 Fm (257)	101 Md (258)	102 No (259)	103 Lr (260)

The IA–VIIIA groups, or 1, 2, 13, 14, 15, 16, 17, and 18, are called the *main groups* or *representative elements*. The B groups, or 3 to 12, are called the *transition elements*. The two rows of elements at the bottom are called the *inner-transition elements*, where the first row is referred to as the *lanthanides* with atomic numbers 58 to 71, and the second row is known as the *actinides* with atomic numbers 90 to 103.

The elements in a group have similar properties:

- Elements in group IA (1), except hydrogen (H), are called alkali metals.
- Elements in group IIA (2) are called alkaline earth metals.
- Group B elements (1–12) are the transition elements. This group contains the most common metallic elements.
- Group IIIA (13) lacks a unique name and is often called the aluminum or boron-aluminum group.
- Groups IVA (14) and VA (15) are designated as the carbon and nitrogen groups, respectively.
- For group VIA (16), an old name, the chalcogens, is used.
- Group VIIA (17) is known as the halogens.
- Group VIIIA (18) contains the noble gases.

The periodic table with group names is shown in Table 1.6.

Each block contains a single element name and the element's *atomic number* (whole number) and *atomic weight* (decimal number), as illustrated in Figure 1.2.

The *group numbers* of the representative elements in the periodic table, IA (1) through VIIIA (18), indicate the number of electrons in the outer energy level, called the valence shell. The *period number* is equal to the number of the outer energy level.

Group VIIIA (18) elements exist as gases, which consist of uncombined atoms (e.g., neon, Ne). The outer energy levels of gases contain eight electrons. For a long time these elements were considered chemically inert because no compounds were known. Then, in the early 1960s, several compounds of xenon were formed. At present, compounds for krypton and radon are also known.

The elements in Group VIIIA are known as *noble gases* because of their relative poor reactivity. The tendency of atoms in molecules to have eight electrons on valence shells is known as the *octet rule*. The number of electrons that must be lost or gained in order for an atom to have the eight-electron configuration on the outer energy level is called *valence*. When an atom loses electrons it becomes a positively charged ion, or *cation*, and when an atom gains electrons it becomes a negatively charged ion, or *anion*.

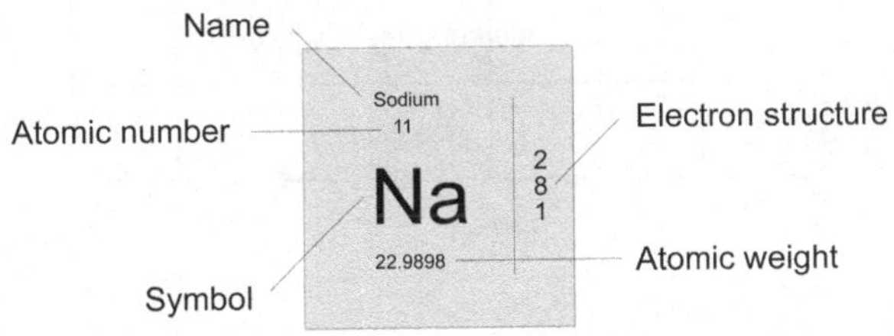

FIGURE 1.2 Each block in the periodic table contains information on one element.

TABLE 1.6
Periodic Table with Names of Chemical Groups

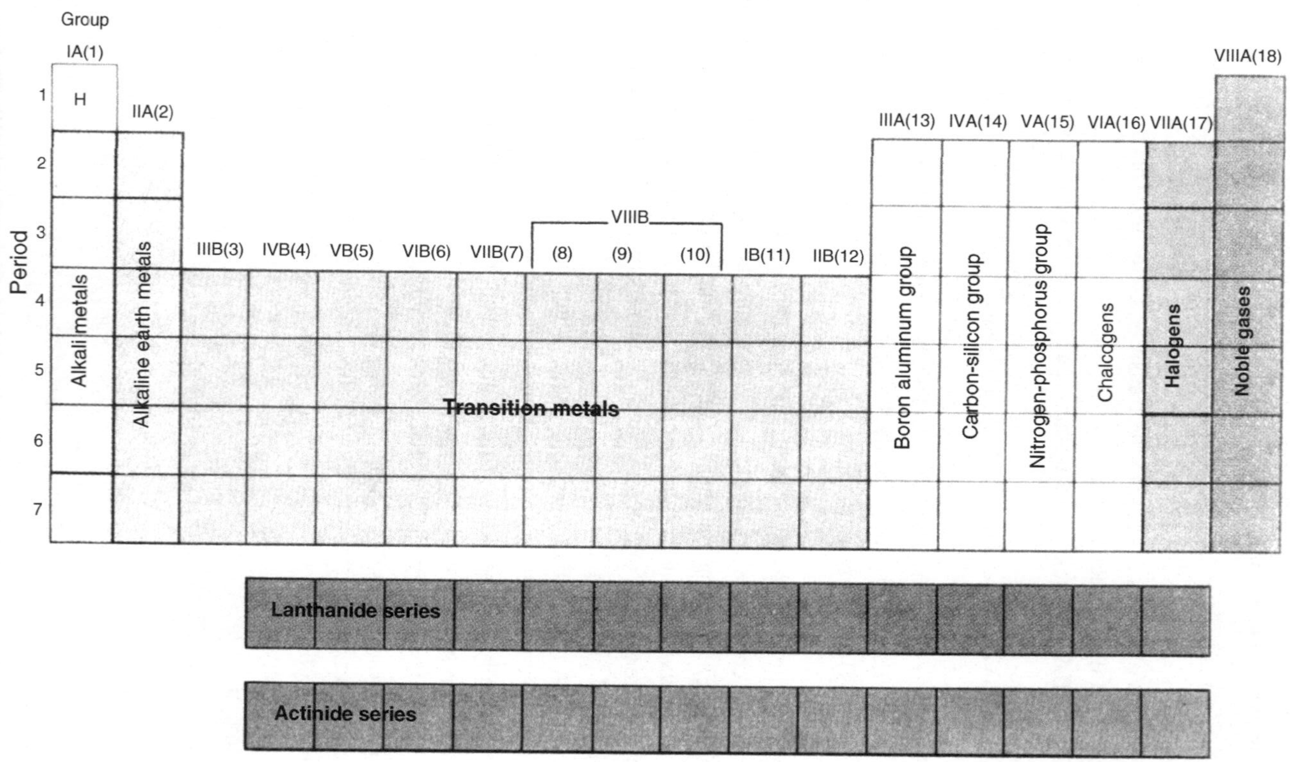

TABLE 1.7
Metals, Nonmetals, and Metalloids Located on the Periodic Table

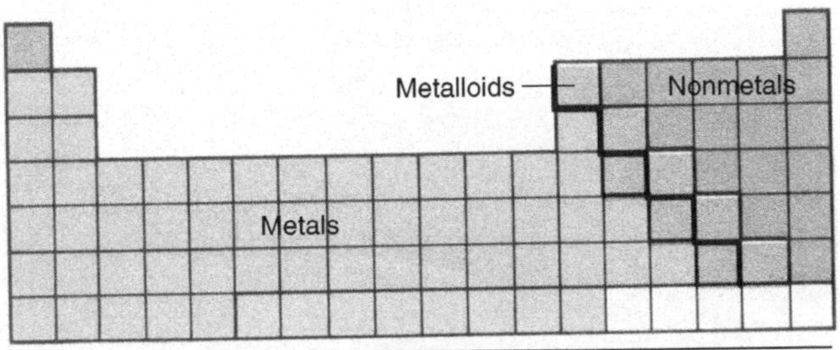

1.3 PROPERTIES OF METALS, NONMETALS, AND METALLOIDS

The elements of the periodic table are divided by a heavy "staircase" line into *metals* on the left and *nonmetals* on the right. Most of the elements bordering the staircase line in the periodic table are *metalloids, or semimetals*, as shown in Table 1.7.

1.3.1 METALS

Metals are substances that have a characteristic luster or shine and are good conductors of heat and electricity; except for mercury (Hg), the metallic elements are solids at room temperature. They are more or less malleable (can be hammered or rolled into thin sheets) and ductile (can be drawn into wire). For example, the production of sheet steel for automobiles and household appliances depends on the malleability of iron and steel, and the manufacture of electrical wire is based on the ductility of copper.

Mercury's low melting point (–39°C) and fairly high boiling point (357°C) make it useful as a fluid in thermometers. Most of the other metals have much higher melting points. Tungsten (W) has the highest melting point of any metal (3400°C), which explains its use as a filament in electric light bulbs.

An important physical property of metals is hardness. Some metals, such as iron and chromium, are very hard, but others, such as copper and lead, are rather soft. The alkali metals are so soft that they can be cut with a knife.

Chemically, metals tend to lose electrons to form positive ions. The special properties of metal result from delocalized bonding, in which bonding electrons are spread over a number of atoms. A very simple picture of a metal depicts an array of (+) ions surrounded by a "sea" of valence electrons (–) that are free to move over the entire metal crystal. The *electron sea model* is presented in Figure 1.3.

The hardness and malleability of metals are explained by the strong electrostatic attraction among positive nuclei and negative electrons; the cations can be easily moved as the metal is hammered into sheet or pulled into wire. Electrical conductivity results from the delocalization of outer electrons; when the metal is connected to a source of electric current, the electrons easily move away from the negative side of the electric source and toward the positive side, forming an electric current in the metal.

1.3.2 NONMETALS

A nonmetal is an element that does not exhibit the characteristics of a metal. Most of the nonmetals are gases (e.g., chlorine, Cl_2, and oxygen, O_2), or solids (e.g., sulfur, S, and phosphorus, P). The solid nonmetals are usually hard, brittle substances. Bromine is the only liquid nonmetal. Table 1.8 shows the differences between metals and nonmetals.

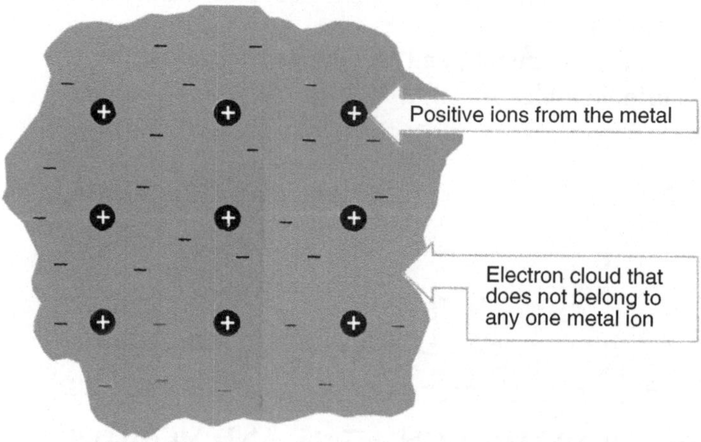

FIGURE 1.3 Electron sea model.

1.3.3 METALLOIDS OR SEMIMETALS

A metalloid or semimetal is an element having both metallic and nonmetallic properties. In most respects, metalloids behave as nonmetals, both chemically and physically. However, in their most important physical property, electrical conductivity, they somewhat resemble metals. Metalloids tend to be semiconductors; they conduct electricity but not nearly so well as metals. These elements, such as silicon (Si) and germanium (Ge), are good *semiconductors* — when pure, they are poor conductors of electricity at room temperature, but moderately good conductors at higher temperatures. The electrical conductivity of a semiconductor is greatly enhanced by adding small amounts of certain elements to it, a process known as *doping*.

Doped semiconductors have useful properties in the manufacture of solid-state electronic devices. Silicon is a basic material of the solid-state electronic industry. Television receivers, microcomputers, and other electronic equipment employ miniature electrical circuits built on silicon chips (see Appendix B).

TABLE 1.8
Characteristics of Metals and Nonmetals

Metals	Nonmetals
Physical Properties	
Good conductors of electricity	Poor conductors of electricity
Ductile	Not ductile
Malleable, lustrous	Not malleable
Solids	Solids, liquids, or gases
High melting point	Low melting point
Good conductors of heat	Poor conductors of heat
Chemical Properties	
React with acids	Do not react with acids
Form basic oxides that react with acids	Form acidic oxides that react with bases
Form cations	Form anions
Form ionic halides	Form covalent halides

1.4 EARLY HISTORY OF METAL USE

The special properties of metals played an important role in the development of human society. In the Copper Age, humans discovered that copper (Cu), found on the surface of the Earth, could be hammered into sheets, which were then used in the manufacture of numerous useful artifacts. Later it was discovered that rocks containing copper (Cu) and tin (Sn) compounds yielded bronze (the first manufactured alloy), so around 4000 BC the Bronze Age began. The first raw iron (Fe) was found in meteorites (the first name of iron was "metal from heaven"), and later, around 2500 BC, iron was smelted from ores, and hence the Iron Age began. Around 100 BC in India, the first steel (90–95% iron and 5–10% carbon) objects appeared. Metallurgy arose from these beginnings. *Metallurgy* is the scientific study of the production of metals from ores and the manufacture of alloys with various useful properties.

1.5 SOURCES OF METALS AND THEIR COMPOUNDS

Metals occur in nature in many different forms. Most are found in compounds, either in the Earth's crust or in the ocean, although some of the less reactive metals are found in the uncombined state (e.g., gold).

Localized deposits of certain metal compounds are called *ores*. An ore is simply a mineral deposit that has a desirable component in a sufficiently high concentration to make its extraction economical. For example, magnesium is found in the mineral called *olivine* (Mg_2SiO_4) with a 30% magnesium (Mg) content. Magnesium concentration in seawater is only 0.3%, but the principal source of magnesium is seawater because it is much more economical to extract it from seawater. The two most abundant metals in seawater are sodium (Na) and magnesium (Mg). Separation of Mg from seawater takes advantage of the low solubility of magnesium hydroxide, $Mg(OH)_2$. Another potential source of metals from the sea is the mining of *manganese nodules* from the ocean floor. Manganese (Mn) nodules are lumps about the size of an orange that contain significant amounts of Mn (about 25%) and iron (Fe) (about 15%). Metal extraction from the oceans is a recent phenomenon. As was mentioned above, metallurgists study the production of metals from their sources, including mining, separation, and preparation for use.

1.6 SOURCES OF METAL POLLUTION

1.6.1 Metal Pollution from Mining and Processing Ores

Digging a mine, removing ore from it, and extraction and processing of the minerals sometimes cause environmental damage. For example, mining operations can destroy habitat, farmland, and homes; produce soil erosion; and pollute waterways via toxic drainage. Emission of toxic materials from smelters — arsenic (As), selenium (Se), lead (Pb), cadmium (Cd), and sulfur oxides, among others — causes serious air pollution. Surface mining produces about eight times as much waste as underground mining, but deep mining can produce even worse problems, such as earthquakes. When underground mines cave in, not only do they kill miners but they also cause subsidence of the surface, forming holes into which roads and houses may collapse. As near-surface minerals are depleted, miners have to dig deeper to find the mineral. A study by the National Academy of Science predicted that copper (Cu) mining operations in the year 2000 would produce three times as much waste per ton of copper output compared to the same activities in 1978.

Exposure of pyrite (FeS) and other sulfide minerals to atmospheric oxygen and moisture results in oxidation of this mineral and the formation of *acid-mine drainage water.* The release of acid-mine drainage from active and abandoned mines, particularly coal mines, has been widely associated with serious water quality problems. It dissolves toxic elements from tailings and soils and carries them into waterways and even groundwater. Water quality problems involve relatively high levels of metals such as iron (Fe), manganese (Mn), zinc (Zn), copper (Cu), nickel (Ni), and cobalt (Co).

Ore processing, smelting, and refining operations can cause deposition of large quantities of trace metals, such as lead (Pb), zinc (Zn), copper (Cu), arsenic (As), and silver (Ag), into drainage basins or direct discharge into aquatic environments.

1.6.2 Other Sources of Metal Pollution

1.6.2.1 Domestic Wastewater Effluents

Domestic wastewater effluents contain large amounts of trace metals from metabolic waste products, corrosion of water pipes — copper (Cu), lead (Pb), zinc (Zn), and cadmium (Cd), and household products, such as detergents — iron (Fe), manganese (Mn), chromium (Cr), nickel (Ni), cobalt (Co), zinc (Zn), boron (B), and arsenic (As). Wastewater treatment usually removes less than 50% of the metal content of the influent, leaving the effluent with significant metal loading. The sludge resulting from wastewater treatment is also rich in metals. Domestic wastewater and the dumping of domestic and industrial sludge are the major artificial sources of cadmium (Cd), chromium (Cr), copper (Cu), iron (Fe), lead (Pb), and mercury (Hg) pollution.

1.6.2.2 Stormwater Runoff

Stormwater runoff from urbanized areas is a significant source of metal pollution in the receiving waters. Metal composition of urban runoff water is dependent on many factors, such as city planning, traffic, road construction, land use, and the physical characteristics and climatology of the watershed.

1.6.2.3 Industrial Wastes and Discharges

Metals and their concentrations in industrial wastes and discharges are specific and depend on the profile of a specific industry.

1.6.2.4 Sanitary Landfills

The metal contents and average concentrations of sanitary-landfill leachates are Cu (5 ppm), Zn (50 ppm), Pb (0.3 ppm), and Hg (60 ppb).

1.6.2.5 Agricultural Runoff

The metal content of agricultural runoff originates in sediments and soils saturated by animal and plant residues, fertilizers, specific herbicides and fungicides, and use of sewage and sludge as plant nutrients.

1.6.2.6 Fossil Fuel Combustion

Fossil fuel combustion is a major source of airborne metal contamination of natural waters.

2 Discussion of Metallic Elements

2.1 REPRESENTATIVE ELEMENTS

As discussed in Chapter 1, in the traditional numbering system of the periodic table, the A group elements are called *main groups* or *representative elements*. Only a few metallic elements occur in nature as free metals. All seven metallic elements known to the ancients (gold, silver, copper, iron, lead, mercury, and tin) have been found in the metallic state. Metals are too reactive chemically to be found in quantity as metallic elements. Except for gold, the metallic elements are obtained principally from their naturally occurring solid compounds or ores. A major source of metals and their compounds is the Earth's crust.

Minerals are naturally occurring inorganic substances or solid solutions with a definite crystalline structure. Thus, a mineral might be a definite chemical substance, or it might be a homogeneous solid mixture. *Rock* is a naturally occurring solid material composed of one or more minerals. An *ore* is a rock or mineral from which a metal or nonmetal can be economically produced. Representative metal groups are listed below.

Group IA (1):	lithium (Li), sodium (Na), potassium (K), rubidium (Rb), cesium (Cs), francium (Fr)
Group IIA (2):	beryllium (Be), magnesium (Mg), calcium (Ca), strontium (Sr), barium (Ba), radium (Ra)
Group IIIA (3):	aluminum (Al), gallium (Ga), indium (In), thallium (Tl)
Group IVA (4):	tin (Sn), lead (Pb)
Group VA (5):	bismuth (Bi)

2.1.1 GROUP IA (1): ALKALI METALS

Alkali metals are soft and the most reactive of all metals; they are never found as free elements in nature, as they always occur in compounds. The pH of their aqueous solution is alkaline. All alkali metals are typically metallic in character, with a bright luster and high thermal and electrical conductivity. They have low densities because they have large atoms; large atoms lead to small ratios of mass per volume (density = mass/volume). When ions of an alkali metal are added to a flame, the resulting brilliant colors are characteristic of the element's atomic spectrum. For example, sodium salts are bright yellow, potassium salts impart a pale violet color to the flame, and lithium salts give a beautiful, deep-red color. All alkali metal salts are water soluble.

2.1.1.1 Lithium (Li)

Lithium is a soft, very rare metal. The source of lithium metal is the ore *spodumene* ($LiAl(SiO_3)_2$), a lithium aluminum-silicate mineral. In recent years, the commercial importance of lithium has risen markedly. Lithium is used in the production of low-density aluminum alloys for aircraft construction, and batteries with lithium metal anodes are also common. Advantages of *lithium batteries* compared to other battery cells include relatively high voltages (about 3.0 V vs. 1.5 V) and typically more electrical energy per mass of reactant, because of lithium's higher voltages and low atomic weight. *Lithium hydroxide* (LiOH) is used to remove carbon dioxide from the air in spacecraft and submarines. *Lithium-6 deuteride* is reportedly the fuel used in nuclear fusion bombs. The Li^+ ion is used in the treatment of mental disorders; for example, lithium carbonate (Li_2CO_3) for treatment of manic depression. Other lithium compounds are used in the preparation of antihistamines and other pharmaceuticals.

2.1.1.2 Sodium (Na)

Sodium is the most familiar alkali metal. Sodium compounds are of enormous economic importance. Common table salt (*sodium chloride*) has been an important article of commerce since prehistoric times. Salt was of such importance in the Roman Empire that a specific allowance of salt was part of soldiers' pay. The word "salary" derives from the Latin *salarium* (salt) for this salt allowance. Major industrial uses of sodium compounds include the manufacturing of glassware, detergents, paper, and textiles. *Soda ash* (*sodium carbonate*, Na_2CO_3) is widely used in water treatment, such as for softening and increasing pH levels. It is also used in organic synthesis, sodium lamps, and photoelectric cells. Household bleach is a 5% solution of *sodium hypochlorite* (NaOCl). An everyday household chemical is *sodium bicarbonate* (baking soda, $NaHCO_3$). Sodium has shown promise as a coolant in certain kinds of nuclear reactors. It has a low melting point and a reasonably high boiling point, and it conducts heat well. Sodium can be pumped through the reactor, where it readily picks up heat, and then pumped through a heat exchanger, where the heat is removed.

Sodium is a natural constituent of water, but its concentration increases with pollution. Sodium salts are extremely soluble in water and, when the element leaches from soil or is discharged into streams by industrial waste processes, it remains in solution. Long-term excessive sodium consumption is responsible for high blood pressure, and consumption of drinking water with high sodium content can be harmful to people with cardiac, circulatory, and renal diseases. In contrast, insufficient replacement of salt leached from the body as a result of sweating will lead to salt depletion, characterized by fatigue, nausea, giddiness, vomiting, and exhaustion. *Sodium sulfate decahydrate* ($Na_2SO_4.10 H_2O$), known as *Glauber salt*, is used as a laxative. Therefore, water containing a high level of sodium sulfate is not recommended for drinking. The American Heart Association recommends a sodium level of less than 20 mg/l for drinking water. Excess sodium concentrations (over 2000 mg/l) in water used by animals for drinking may also be toxic.

Irrigation water with a high sodium level can cause a displacement of exchangeable cations (Ca^{2+}, Mg^{2+}) followed by replacement of the cations by Na. The ratio of Na^+ ions to total cation contents can be used for assessing the suitability of water for irrigation. The ability of water to expel calcium and magnesium by sodium can be estimated by calculating the *sodium absorption ratio* (SAR). Calculation and acceptance criteria are discussed in Section 4.4. With a few exceptions (e.g., seaweed), sodium ions tend to be toxic to plants.

2.1.1.3 Potassium (K)

Potassium, which has properties similar to sodium, is used in organic synthesis in the glass and chemical industries. Both sodium and potassium ions are important in animal metabolism, but potassium ions are far more important than sodium ions in plants and are therefore used extensively as fertilizers. The normal daily intake from food is about 1.6 to 6.0 g. Daily natural potassium intake (1.6–6.0 g) contributes to cardiovascular function, although excessive intake causes *hyperkalemia*, which may cause cardiac arrest. Normal potassium levels in drinking water do not constitute a threat to human health. Consequently, primary and secondary maximum contaminant levels (MCLs) are not available.

The physiological functions of sodium and potassium are essential in all living organisms. The ions of these two elements do not create large and stable complexes with other organic molecules, but they do function in ionic forms. Ion concentrations inside and outside cells are not in equilibrium — potassium ion concentration is greater inside the cell, whereas sodium ions are more concentrated outside the cell (see Figure 2.1). This *asymmetric concentration* is one of the most important energy savers in living organisms and plays an important role in nerve stimulation and muscle function and their physiological functions.

2.1.1.4 Rubidium (Rb) and Cesium (Cs)

Rubidium and cesium are rare and have little commercial importance. The name rubidium is derived from the Latin *rubidus*, which means dark red. The name cesium derived from the Latin *caesius*, which means sky blue. Cesium and rubidium were discovered by Bunsen and Kirchhoff in 1860 and 1861, respectively.

2.1.1.5 Francium (Fr)

Francium has a fleeting existence because all of its isotopes are radioactive and have a very short half-life.

2.1.2 Group IIA (2): Alkaline Earth Metals

Alkaline earth metals are almost as reactive as the group IA metals; therefore, they always occur in compounds. If we compare an alkaline earth metal with an alkali metal in the same period, the

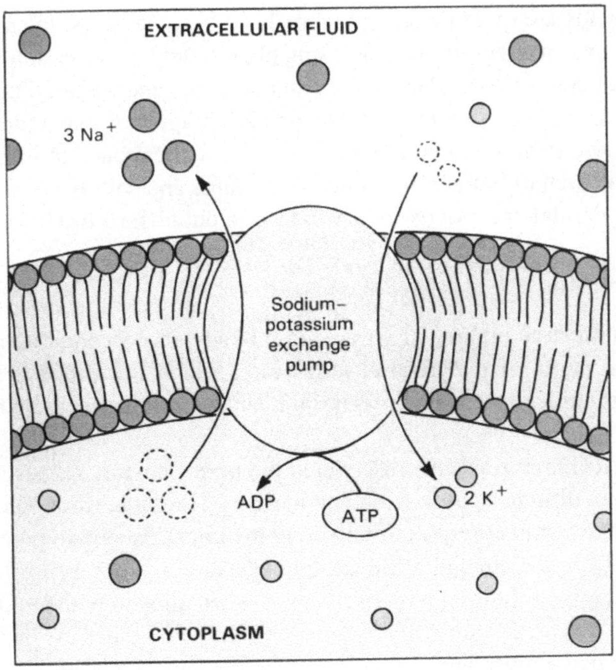

FIGURE 2.1 Sodium–potassium exchange pump. The operation of this pump is an example of active transport, because it depends on energy provided by ATP. For each ATP molecule converted to ADP, this ion pump carries three Na^+ ions out of the cell and two K^+ ions into the cell.

alkaline earth metal is less reactive and harder. For example, lithium is a soft metal, whereas beryllium is hard enough to scratch. The most abundant alkaline earth metals are calcium and magnesium. The most common ions in seawater are Mg^{2+} and Ca^{2+}. Marine organisms take calcium ions from the water to make their calcium carbonate ($CaCO_3$) shells. Underground brine also contains a large concentration of these elements. These metals are found in mineral deposits in the Earth's crust, such as *limestone* (calcium carbonate, $CaCO_3$) and *dolomite* (mixed calcium and magnesium carbonate, $CaCO_3.MgCO_3$). Another important calcium mineral is *gypsum* ($CaSO_4.2H_2O$). Calcium and magnesium are discussed in more detail later.

Like the alkali metals, certain alkaline earth metals give characteristic colors when added to a flame. Calcium salts produce an orange-red color; strontium salts, bright red; and barium salts, yellow-green. These colors are intense enough to serve as *flame tests*. Like alkali metal salts, salts of these metals are used in coloring fireworks displays.

2.1.2.1 Beryllium (Be)

Beryllium is found in the mineral *beryl* ($Be_3Al_2(SiO_3)_6$). Beryl minerals are *emerald* and *aquamarine* and, when cut and polished, they make beautiful gemstones. Beryllium is a very light metal with excellent thermal conductivity and a high melting point, and most of its uses are based on these properties. Because of its low density, excellent thermal conductivity, and elasticity, beryllium is used in high-precision instruments. It is used to make x-ray tube windows, because it is the most transparent mineral to x-rays. This metal is also used in alloys with copper and bronze to give them hardness. Hammers and wrenches made from Be/Cu alloys do not produce sparks when struck against steel and, therefore, can be used in flammable environments. Beryllium absorbs neutrons, which are particles given off in nuclear reactions; consequently, it is used in nuclear power plants and nuclear weapons.

Beryllium compounds are quite toxic, and some have become air pollutants due to combustion emissions, cigarette smoke, and beryllium processing plants. Only its water-soluble salts (sulfates and fluorides) have acute effects, causing dermatitis, conjunctivitis, and, through inhalation, irritation of the respiratory tract. Chronic exposure to beryllium and its compounds may produce *berylliosis,* a frequently fatal *pulmonary granulomatosis*. The toxic effect may be related to inhibition of enzyme activities. There is a small quantity of beryllium in water source and soil. Because the concentration of beryllium in water is minimal, it is not necessary to issue a public health standard.

2.1.2.2 Magnesium (Mg)

Magnesium is the lightest structural metal; its use is limited by its cost and flammability. The metal's name comes from the name of the mineral *magnesite*, which in turn is believed to stem from Magnesia, a site in northern Greece where magnesium and other minerals have been mined since ancient times.

The British chemist Humphrey Davy discovered the pure element magnesium in 1808. He electrolyzed a moist mixture of magnesium oxide and mercury(II) oxide, from which he obtained magnesium amalgam (an alloy of magnesium dissolved in mercury). To obtain pure magnesium, he distilled off the mercury from the amalgam. Because magnesium has a very low density (1.74 g/cm^3) and moderate strength, it is useful as a *structural metal* when alloyed with aluminum. In flashbulbs, a thin magnesium wire is heated electrically by a battery; the heat ignites the metal, which burns very quickly in the pure oxygen atmosphere.

Magnesium is also used in antacids, the cathartic *milk of magnesia* ($Mg(OH)_2$), and *Epsom salts,* $MgSO_4.7H_2O$. Magnesium, together with calcium, contributes to water hardness. New users of drinking water high in magnesium salts may initially experience a cathartic effect, but usually become tolerant. Magnesium is essential for neuromuscular conduction and is involved in many enzyme functions.

The major commercial sources of magnesium are seawater and minerals. It is nontoxic for humans, except in large doses. Magnesium does not constitute a public health hazard; before toxic levels occur in drinking water, the taste cannot be tolerated.

2.1.2.3 Calcium (Ca)

Calcium is a common element that is present in the Earth's crust as silicates, which weather to release a free calcium ion, Ca^{2+}. The ion is about as abundant in seawater as the magnesium ion. Corals are marine organisms that grow in colonies; their calcium carbonate ($CaCO_3$) skeletons eventually form enormous *coral reefs* in warm waters, such as the Bahamas and Florida Keys. Deposits of *limestone* (mostly $CaCO_3$) formed in earlier times as sediments of seashells and coral and by the precipitation of $CaCO_3$ from seawater.

Gypsum, hydrated calcium sulfate ($CaSO_4.2H_2O$), is another important mineral of calcium. When heated moderately, it loses some water and the formula changes to $(CaSO_4)_2.H_2O$ or $CaSO_4.1/2H_2O$; the water content changes to half of the original quantity. This partially dehydrated form of gypsum is called *plaster of Paris.* (Early sources were mines in the Paris Basin, France.) When ground to a fine powder and mixed with water to form a paste, it hardens within just a few minutes. This property designated its uses, such as covering the interior walls of buildings, plasterboard, and plaster casts. The fine-grained crystalline form of the mineral is called *alabaster*. It is a soft stone, easily carved by sculptors; when highly polished, alabaster takes on a beautiful appearance. *Calcium chloride* ($CaCl_2$) has a special high affinity to moisture. Calcium chloride can be purchased in hardware stores for use in removing moisture from places with high humidity such as damp basements.

Calcium oxide (CaO) is among the top ten industrial chemicals. Calcium oxide is known commercially as *quicklime*, or simply lime. Calcium oxide reacts exothermally with water to produce calcium hydroxide ($Ca(OH)_2$), commercially called *slaked lime*. Calcium hydroxide solutions react with gaseous carbon dioxide (CO_2,) to form calcium carbonate ($CaCO_3$). An important use of this reaction and the formation of the precipitated calcium carbonate is as a *filler* in the manufacture of paper. (The purpose of the filler is to improve the paper's characteristics, such as brightness and ink absorption.) Large amounts of quicklime (CaO) and slaked lime ($Ca(OH)_2$) are used to soften municipal water supplies.

Numerous calcium compounds have therapeutic uses, such as antispasmodic, diuretic, and antacid (e.g., *Tums*) preparations and treatment of low-calcium tetany. As discussed in Section 2.5.4, calcium is essential for healthy bones and teeth. *Hypercalcemia* (excess calcium) occurs in vitamin D poisoning in infants, hyperparathyroidism, sarcoidosis, and malignancy. Calcium toxicity can result in anorexia, nausea, vomiting, dehydration, lethargy, coma, and death. Excessive calcium levels in drinking water may relate to the formation of kidney and bladder stones. Calcium concentration in water is related to water hardness. High sodium and low calcium intake contributes to the development of high blood pressure.

2.1.2.4 Strontium (Sr) and Barium (Ba)

Strontium and barium have few commercial uses as metals, other than as reducing agents in specialized metallurgical operations, and are thus produced in small quantities. One of the important uses of barium sulfate ($BaSO_4$) is in obtaining x-ray photographs of the digestive tract. A patient drinks a suspension of barium sulfate in water and then the x-ray photograph is taken. The path of the patient's digestive tract is clearly visible on the film because $BaSO_4$ is opaque to x-rays. Even though the barium ion (Ba^{2+}), like most heavy metal ions, is very toxic to humans, barium sulfate is safe, because its solubility is so low and Ba^{2+} ions are barely absorbed by the body.

Other uses of barium sulfate are based on its whiteness; it is used as a whitener in photographic papers and as a filler in paper and polymeric fibers. The source of barium pollution is from mining industries (coal), combustion (aviation and diesel fuel), and the mud resulting from oil well drilling. Acute exposure to barium results in gastrointestinal, cardiac, and neuromuscular effects. Its maximum contaminant level (MCL) in drinking water is 5 mg/l.

2.1.3 Group IIIA (13) Metals

The Group IIIA elements clearly show the trend of increasing metallic characteristics when moving downward in the column of elements in the periodic table. *Boron* (B), at the top of the column, is a metalloid, and its chemistry is typical of nonmetals. The rest of the elements in the column are metals.

2.1.3.1 Aluminum (Al)

Aluminum is the third most abundant element, and the most abundant metal in the Earth's crust. It occurs primarily in aluminum silicate minerals. The weathering of these rocks results in aluminum-containing *clay*. Further weathering of the clay yields *bauxite*, the chief ore of aluminum. Bauxite contains aluminum in the form of hydrated oxide ($Al_2O_3.xH_2O$).

Aluminum always exists as the Al^{3+} ion. Aluminum has many uses, ranging from aluminum foil to airplane construction. Its structural uses — building construction, electrical wiring and cables, packaging and containers — are based on its low weight and moderate strength. Other interesting uses of aluminum include drain cleaners, which consist mostly of NaOH along with small bits of aluminum metal. When sprinkled into a clogged drain, the bubbles caused by the release of hydrogen gas cause a stirring effect in the clogged drain.

A thin layer of aluminum is used to reflect light in large visible-light telescopes. *Dur-aluminum*, a solution of aluminum, manganese, and calcium, is used in the construction of buildings, boats, and airplanes. Another alloy of aluminum is *alnico*, a mnemonic for aluminum, nickel, and cobalt. Because the world supply of copper is diminishing, aluminum now replaces copper as the electrical conductor in wires and cables. Pure aluminum, when heated in air at a high temperature, is totally converted to aluminum oxide (Al_2O_3) or *alumina*. It is used as a carrier or support for many heterogeneous catalysts required for chemical processes, including those used in the production of gasoline. *Aluminum oxide* is used in the manufacture of *ceramics*. The word "ceramics" derives from the Greek *kerimikos*, which means "of pottery," referring to objects made by firing clay.

When aluminum oxide is fused (melted) at a high temperature, it forms *corundum*, one of the hardest materials known. Corundum is used as an abrasive for grinding tools. The presence of impurities results in various colors and produces gem-quality corundum. If the impurities in the corundum structure are chromium oxides, then the crystal has a red color and is called *ruby*. Synthetic rubies, for example, contain about 2.5% chromium oxide (Cr_2O_3). Ruby is used in fine instrument bearings (*jewel bearings*) and in making lasers (see Appendix C). If the impurities are cobalt and titanium, then the crystal is blue and it is called *sapphire*. If the impurities are iron oxides, the crystal is called *oriental topaz*. *Amethyst* results when manganese oxide is the impurity in corundum.

When aluminum combines with iron(III) oxide, it releases a tremendous amount of energy, enough that the resulting iron becomes molten. This reaction is known as the *thermite reaction*. Because temperatures in excess of 3000°C are obtained, metals are welded using the thermite reaction.

Important aluminum compounds include *aluminum hydroxide* ($Al(OH)_3$), which is an ingredient in antacids. *Potassium aluminum sulfate* ($KAl(SO_4)_2.12H_2O$), commonly called *alum*, is used as an additive to neutralize base components of soils. *Aluminum chloride* ($AlCl_3$) is frequently used as a catalyst in laboratory syntheses and as an intermediate in a procedure for isolating aluminum from bauxite. *Aluminum sulfate* ($Al_2(SO_4)_3$) is used to make paper water resistant. Aluminum sulfate is also used in water treatment plants, where it is added to the water along with lime (CaO). The CaO reacts with water to make the solution alkaline. Gelatinous aluminum hydroxide will precipitate, thereby removing suspended solids and certain bacteria. Aluminum compounds are also used to prevent *hyperphosphatemia* in renal disease, and as *antidotes*. Until recently, aluminum was considered nontoxic. Because *Alzheimer's disease* patients have a high aluminum content in certain brain cells, research is now focused on high aluminum intake as a possible causal factor. High aluminum intake originates from packaging, aluminum cooking vessels, aluminum foil, and aluminum-containing antacids.

2.1.3.2 Gallium (Ga), Indium (In), and Thallium (Tl)

These metals have +1 and +3 oxidation states. *Gallium* has a melting point of only 29.8°C, so human body temperature (37°C) is high enough to cause the metal to melt in the palm of your hand. *Thallium* compounds are highly toxic; for humans, doses of 14 mg/kg and above are fatal. Thallium is used mostly in electrical and electronic applications. Previously used in rodenticides, fungicides, and in cosmetics, these products are now banned.

2.1.4 Group IVA (14) Metals

The two metallic elements in this column are tin (Sn) and lead (Pb). Both metals were known in ancient times.

2.1.4.1 Tin (Sn)

Tin is a relatively rare element, ranking 50th or so in abundance in the Earth's crust. The element occurs in localized deposits of the tin ore *cassiterite* (SnO_2). Sn refers to its original name, *stannum*. Elemental tin occurs in three allotropic forms. The most common is called *white tin*, the shiny tin coating over steel. If tin is kept for long periods below 13.2°C, the white tin gradually changes to *gray tin*, a powdery, nonmetallic form. Therefore, when tin objects are kept at low temperatures for long periods, lumps develop on the surface. The phenomenon is called *tin sickness* or tin disease; historically, it was thought to be caused by an organism. For instance, during a cold winter in the 1850s, the tin pipes of some church organs in Russia and other parts of Europe began crumbling from tin disease. Tin disease is simply the transition from white tin to gray tin. The third allotropic form is *brittle tin*, and its properties reflect its name. Tin is not found naturally in environmental samples; therefore, its presence always indicates industrial pollution. The level of tin in drinking water systems is negligible.

Tin(IV) oxide (SnO_2) is used to give glass a transparent, electricity-conducting surface. *Bis-(tributyltin)oxide* is used in wood treatments to prevent rot. It has also been used in antifouling paints that are applied to boat hulls to prevent the growth of marine organisms such as barnacles. However, its high toxicity to all forms of marine life has led to a ban on its use for this purpose. Tin is used to make tinplate, which is steel (iron alloy) sheeting with a thin coating of tin. Tinplate is used for food containers ("tin cans"). *Tin(II) chloride* ($SnCl_2$) is used as a reducing agent in the preparation of dyes and other organic compounds. An excellent reducing agent, $SnCl_2$ is used in the preparation of dyes and other organic compounds. *Tin(IV) chloride* ($SnCl_4$) is a liquid; it freezes at −33°C. A tin coating protects iron from reacting with air and food acids. Tin is also used to make numerous alloys, including *solder*, a low-melting alloy of tin and lead, and *bronze*, an alloy of copper and tin.

2.1.4.2 Lead (Pb)

Lead occurs in the form of *lead sulfide* (PbS), known as *galena*. The Latin word for lead is *plumbum*, thus its symbol, Pb. The word "plumber" comes from the early use of lead water pipes and pipe joints. Lead is a very heavy, soft, highly malleable, bluish-gray metal and exists in +2 and +4 oxidation states, although lead(II) compounds are the more common.

In lead storage batteries, the cathode is *lead(II) oxide* (PbO, called *litharge*), which is packed into a lead metal grid (PbO is a reddish-yellow solid). When the battery is charged, the PbO is oxidized to *lead(IV) oxide* (PbO_2 is a dark brown powder). The metal is used to make batteries and solder and to manufacture *tetraethyllead* ($(C_2H_5)_4Pb$), a gasoline octane booster. The use of lead-containing additives in gasoline has been phased out in many countries (but not all) because of environmental hazards.

Lead is toxic to the nervous system and children are especially susceptible to its effects. It is readily absorbed from the intestinal tract and deposited in the central nervous system. The first lead water pipes were used in ancient Rome by upper-class citizens; their children drank the water throughout childhood and thus were at high risk of lead toxicity. This fact may explain the bizarre behavior of certain notorious Roman emperors and the fall of the Roman Empire. In recent years, exposure to lead toxicity has become widespread. Sources are lead-containing paint, air, soil, dust, food, and drinking water. The presence of lead in the body is indicated by lead blood levels, expressed as micrograms of lead per deciliter of blood (μg/dl). Blood lead levels of 10 μg/dl and higher may contribute to learning disabilities, nervous system damage, and stunted growth. Many children suffered lead poisoning from ingestion of lead-based paints. Lead-based paint was used inside many homes until Congress passed the Lead-Poisoning Prevention Act in 1971. Lead is encountered in air, soil, and water. The concentration of lead in natural waters has been reported to be as high as 0.4 to 0.8 mg/l, mostly from natural sources, such as galena deposits. High contamination levels may be caused by industrial and mining pollution sources. High levels of lead in drinking water are mostly the result of corrosion products from lead service pipes, solders, and household plumbing. According to a survey by the Environmental Protection Agency, infants dependent on formula may receive more than 85% of their blood lead levels from drinking water. Lead as a corrosion product in drinking water is associated with copper. Copper is needed for good health, and in low levels it has a beneficial effect, but in high concentrations it is toxic, causing diarrhea and vomiting. The maximum contaminant level (MCL) established for lead in drinking water is 0.02 mg/l, but the maximum contaminant level goal (MCLG) for lead is zero, and for copper, 1.3 mg/l.

2.1.5 Group VA (15) Metals

2.1.5.1 Bismuth (Bi)

The only metallic element in group VA is bismuth. It is one of the few substances that expand slightly at freezing. This property makes bismuth ideal to use for *castings* because it expands to fill all details of the mold. The other principal use of bismuth is in making alloys with unusually low melting points. For example, *Wood's metal*, an alloy, contains 50% bismuth, 25% lead, 12.5% tin, and 12.5% cadmium. The alloy melts when dipped into boiling water (melting point is 70°C).

2.2 TRANSITION METALS

2.2.1 General Discussion

The transition elements or metals are elements normally placed in the body of the periodic table, the B groups. The *inner transition elements* are located in the long row, usually found just below the main body of the table. Elements in the first row are called *lanthanides* because they follow lanthanum. Elements in the second row are called *actinides* because they follow actinium. The lanthanides and actinides are rare elements (see Sections 1.2 and 2.2.2).

Many of the transition elements have properties in common. One of the most important characteristics of the transition metals is the occurrence of multiple oxidation states. The oxidation state of the metal is expressed by using special nomenclature for these elements. In the *stock system*, the full name of the metal is followed by its oxidation number (valence) in Roman numerals enclosed in parentheses. The old nomenclature system assigned names to metals in a different way. The ending "-ic" designates the higher oxidation states, while "-ous" identifies the lower oxidation state of the metal. The names of metals with multiple oxidation states are listed in Table 2.1.

Another property of transition elements is the tendency of ions to combine with neutral molecules or anions to form complex ions, or *chelates*. The number of *complexes* formed by the transition metals is enormous, and their study is a major part of chemistry. (Chelate formation and its importance in medicine are discussed in Section 3.2.) Many compounds and complexes of the transition metals have beautiful colors, because the transition metal in the complex ion can absorb visible light of specific wavelengths. For instance, all chromium compounds are colored; in fact, chromium gets its name from the Greek *chroma*, which means color.

Many of the atoms and ions of the transition elements contain unpaired electrons. Substances with unpaired electrons are attracted to a magnetic field and are said to be *paramagnetic*. The attraction tends to be weak, however, because the constant movement and collision between the individual atomic-sized magnets prevent large numbers of them from becoming aligned with the external magnetic field. The magnetic property we often associate with iron is its strong attraction to the magnetic field. In reality, iron is one of three elements (iron, cobalt, and nickel) that exhibit this strong magnetism, called *ferromagnetism*. Ferromagnetism is about 1 million times stronger than paramagnetism. Ferromagnetism is a property specific to the solid state. Alloys with ferromagnetic properties have been manufactured, such as *alnico magnets* — alloys of iron, aluminum, nickel, and cobalt. Manganese is paramagnetic, but by adding copper to manganese a ferromagnetic alloy is formed.

Transition metals have many uses. For instance, iron is used for steel; copper for electrical wiring and water pipes; titanium for paint; silver for photographic paper; manganese, chromium, vanadium, and cobalt as additives to steel; and platinum for industrial and automotive catalysts. Transition metal ions also play a vital role in living organisms. For example, iron complexes provide the transport and storage of oxygen, molybdenum and iron compounds are catalysts in nitrogen fixation, zinc is found

TABLE 2.1
Metals with Multiple Oxidation States

Metal	Oxidation	Stock Name	Old Name
Copper	+1	Copper(I)	Cuprous
	+2	Copper(II)	Cupric
Mercury	+1	Mercury(I)	Mercurous
	+2	Mercury(II)	Mercuric
Iron	+2	Iron(I)	Ferrous
	+3	Iron(III)	Ferric
Chromium	+2	Chromium(II)	Chromous
	+3	Chromium(III)	Chromic
Manganese	+2	Manganese(II)	
Manganous			
	+3	Manganese(III)	Manganic
Cobalt	+2	Cobalt(II)	Cobaltous
	+3	Cobalt(III)	Cobaltic
Tin	+2	Tin(II)	Stannous
	+4	Tin(IV)	Stannic
Lead	+2	Lead(II)	Plumbous
	+4	Lead(IV)	Plumbic
Titanium	+3	Titanium(III)	Titanous
	+4	Titanium(IV)	Titanic

Note: Mercury(I) is a diatomic molecule; that is, it exists in pairs as Hg^{22+}. Whatever the notation style of mercury(I), it indicates a pair of mercury ions.

in more than 150 biomolecules in humans, copper and iron play a crucial role in the respiratory cycle, and cobalt is found in essential biomolecules such as vitamin B_{12}.

The transition metals behave as typical metals, possessing metallic luster and relatively high electrical and thermal conductivities. Silver is the best conductor of heat and electrical current. However, copper is a close second, which explains copper's wide use in electrical systems. In spite of these metals' many similarities, their properties vary considerably. For example, tungsten has a melting point of 3400°C and is used for filaments in light bulbs, and mercury is a liquid at 25°C. Some transition metals, such as iron and titanium, are hard and strong and are thus very useful structural materials. Others, such as copper, gold, and silver, are relatively soft. Chemical properties also vary significantly. Some react readily with oxygen to form oxides. These metals, such as chromium, nickel, and cobalt, form oxides that adhere tightly to the metallic surface, protecting the metal from further oxidation. Others, such as iron, form oxides that scale off, exposing the metal to further corrosion. Noble metals, such as gold, silver, platinum, and palladium, do not form oxides. An introduction to some of these important metals and their specific properties follows.

2.2.1.1 Scandium (Sc)

Scandium's atomic number is 21. Scandium is a rare element that exists in compounds, mainly in the +3 oxidation state. This metal is not widely used because of its rarity, high reactivity, and high cost. It is found in some electronic devices, such as high-density lamps.

2.2.1.2 Titanium (Ti)

Titanium is widely distributed in the Earth's crust. Because of its relatively low density and high strength, titanium is an excellent structural material, especially in jet engines where light weight and stability at high temperatures are required. It is used also in manufacturing racing bicycles. Its resistance to chemical reactions makes it useful material for pipes, pumps, and reaction vessels in the chemical industry. *Titanium(IV) oxide* (TiO_2) is used as the white pigment in papers, paints, linoleum, plastics, synthetic fibers, and cosmetics. Titanium is found in several minerals; one of the most important is *rutile* (TiO_2). *Titanium tetrachloride* ($TiCl_4$) is a clear, colorless, volatile liquid with a boiling point of only 136°C and whose vapors react almost instantly with moist air to form a dense smoke of TiO_2. The reaction was once used by the U.S. Navy to create smoke screens during naval battles.

2.2.1.3 Vanadium (V)

Vanadium is widely spread in the Earth's crust. A gray, relatively soft metal, it is found in various minerals. It is used mostly in alloys with other metals, such as *vanadium steel* (80% vanadium), a hard steel used in engine parts and axles. *Vanadium(V) oxide*, (V_2O_5, vanadium pentoxide), is used as an industrial catalyst. Vanadium salts have low oral toxicity and medium toxicity via inhalation. Vanadium is possibly a protective agent against atherosclerosis.

2.2.1.4 Chromium (Cr)

Although very rare, chromium is a very important industrial metal. It is a grayish-white crystalline, very hard metal, with high resistance to corrosion. Chromium maintains a bright surface by developing a tough *invisible oxide coating*. These properties make it an excellent decorative and protective coating for other metals, such as brass, bronze, and steel. *Chrome plate* is deposited electrolytically on automobile parts such as bumpers.

Large amounts of chromium are used to produce alloys, such as *stainless steel*, which contains about 18% chromium, 8% nickel, and small amounts of manganese, carbon, phosphorus, sulfur and silicon, all combined with iron. *Nichrome*, an alloy of chromium and nickel, is often used as a wire-heating element in devices such as toasters.

The many colorful compounds of this element are a fascinating feature of chromium chemistry. The common oxidation states of chromium compounds are +2, +3, and +6. The color of the *chromium(III)* species depends on anions in solution that can form complexes with Cr^{3}. The ion is frequently green. *Chromium(VI) oxide* (CrO_3, also called chromium trioxide), is a red crystalline compound. It precipitates when concentrated sulfuric acid is added to concentrated solutions of a dichromate salt. Red chromium(VI) oxide (CrO_3) dissolves in water to give a strong, acidic, red-orange solution; when made basic, the solution turns yellow. CrO_3, the anhydride of chromic acid (H_2CrO_4), is a highly poisonous red-orange compound. At a higher pH, two other forms predominate, the yellow *chromate ion* (CrO_4^{2-}) and the red-orange *dichromate ion* ($Cr_2O_7^{2-}$).

A mixture of chromium(VI) oxide and concentrated sulfuric acid, commonly called cleaning solution, is a powerful oxidizing medium that can remove organic materials from analytical glassware, yielding a very clean surface. Commercial substitutes for dichromate-sulfuric acid, such as *Nichromix*, do not contain chromium and hence are safer to use. One of the principal uses of chromium compounds is in pigments for coloring paints, cements, and plasters. The Cr^{2+} ion is a powerful *reducing agent* in aqueous solution; therefore, it is used to remove traces of oxygen from other gases by bubbling through a Cr^{2+} solution. The Cr^{6+} ions are excellent *oxidizing agents. Zinc yellow pigment* ($ZnCrO_4$, *zinc chromate*) is used as a corrosion inhibitor on aluminum and magnesium aircraft parts. Cr^{3+} (trivalent) chromium may be essential in human nutrition, but Cr^{6+} (hexavalent) is highly toxic. Among other health problems, intake of hexavalent chromium can cause hemorrhaging in the liver, kidneys, and respiratory organs. Workers exposed to hexavalent chromium have developed dermatitis and ulceration and perforation of the nasal septum. Gastric cancers, presumably from excessive inhalation of dust containing chromium, have also been reported.

2.2.1.5 Manganese (Mn)

Manganese is found in many minerals as oxides, silicates, and carbonates. One interesting source of manganese is manganese nodules found in the ocean floor. These roughly spherical "rocks" contain a mixture of manganese and iron oxides as well as smaller amounts of other metals, such as cobalt, nickel, and copper. Apparently the nodules were formed at least partly from the action of marine organisms (see Section 1.5).

Manganese is a very brittle metallic element resembling iron, but harder, and is complicated by the existence of six oxidation states from +1 to + 7, although +2 and +7 are the most common. Manganese(II) forms an extensive series of salts with all of the common anions. Manganese(VII) is found in the purple-colored *permanganate ion* (MnO_4^-). Manganese is principally used in iron alloys, dry cells, and oxidizing chemicals, as *potassium permanganate* ($KMnO_4$). The metal is also used as a steel additive and in the preparation of other alloys, such as *manganese bronze* (a copper–manganese alloy) and *manganin* (an alloy of copper, manganese, and nickel, whose electrical resistance changes slightly with temperature).

Manganese toxicity to humans has been shown only on exposure to high levels in the air. Inhalation of large doses of manganese compounds, especially the higher oxides, can be lethal. Inhalation of manganese fumes causes *manganese pneumonia*, which can be fatal. Chronic manganese toxicity is well known in miners, mill workers, and others exposed to high concentrations of manganese-laden dust and fumes, and drinkers of well water containing excessive manganese (often in mining villages). The usual symptoms involve the central nervous system. Characteristic *manganese psychosis* involves inappropriate laughter, euphoria, impulsiveness, and insomnia, followed by overwhelming somnolence. These symptoms may be accompanied by headache, leg cramps, and sexual excitement, followed by lethargy. In the final stage, speech disturbance, masklike facial expression, general clumsiness, and micrography (very minute writing) are characteristic. Although patients may become totally disabled, the syndrome is not lethal.

2.2.1.6 Iron (Fe)

Iron is the most abundant heavy metal. Its chief ores are the red-orange *hematite* (Fe_2O_3) and the black *magnetite* (Fe_3O_4). Iron contains both the +2 and +3 oxidation states. Iron and its carbon alloy, steel, constitute the backbone of modern industrial society. It is a white, lustrous, not particularly hard metal that is very reactive toward oxidizing agents. For example, in moist air iron is rapidly oxidized to form *rust*, a hydrated oxide, whose formula is usually given as $Fe_2O_3.xH_2O$ (Figure 2.2). Rust does not adhere well to the metal, but instead falls away, exposing fresh iron to attack. One way to prevent rusting is to coat the iron with another metal such as tin. Another way to prevent corrosion is called *cathodic protection*, which involves placing the iron in contact with another metal that is more easily oxidized. This causes iron to react as a cathode (the electrode at which reduction occurs during an electrochemical change) and the other metal to be the anode (the electrode at which oxidation occurs during an electrochemical change). If corrosion occurs, the iron is protected from oxidation because it is cathodic and the other metal reacts instead. Zinc is most often used to provide cathodic protection to other metals.

Corrosion protection is illustrated in Figure 2.3. Steel objects that must withstand weather are often coated with a layer of zinc, a process called *galvanizing*. Iron is also quite reactive to nonoxidizing acids, such as hydrochloric acid (HCl) and sulfuric acid (H_2SO_4). Iron does not react with concentrated nitric acid (HNO_3). Instead, because its surface becomes quite unreactive, the iron is said to have been made passive. The chemistry of iron mainly involves its +2 and +3 oxidation states. Iron(II) salts are generally light green, and iron(III) salt solutions usually range from yellow to brown.

Iron ions form many complex ions. Iron is the central metal in the *hemoglobin* molecule, and iron is used in the therapy of iron-deficiency anemia. Iron and its compounds are used as pigments, magnetic tapes, catalysts, disinfectants, tanning solutions, and fuel additives. Iron is an essential mineral, but toxic in high doses.

Iron content of environmental samples is mostly attributed to feeding aquifers, corrosion from pipes, leachate from acid mine drainage, and iron-product industrial wastes. *Ferrous* (Fe^{2+}) and *ferric* (Fe^{3+}) iron are soluble in water, but ferrous iron is easily oxidized to ferric hydroxide, which is not soluble in water and thus flocculates and settles. High iron concentration in water can cause staining of laundry and porcelain and a bittersweet astringent taste. To prevent the formation of black iron

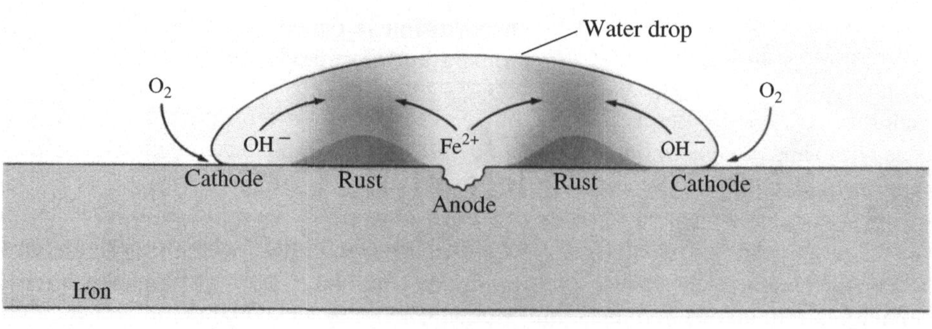

FIGURE 2.2 Electrochemical process involved in rusting of iron. Shown here is a single drop of water containing ions from a voltaic cell in which iron is oxidized to an iron(II) ion at the center of the drop. Hydroxide ions and iron(II) ions migrate together and react to form iron(II) hydroxide. Iron(II) hydroxide is oxidized to iron(III) hydroxide by more O_2 that dissolves at the surface of the drop. Iron(III) hydroxide precipitates and settles to form rust on the surface of the iron.

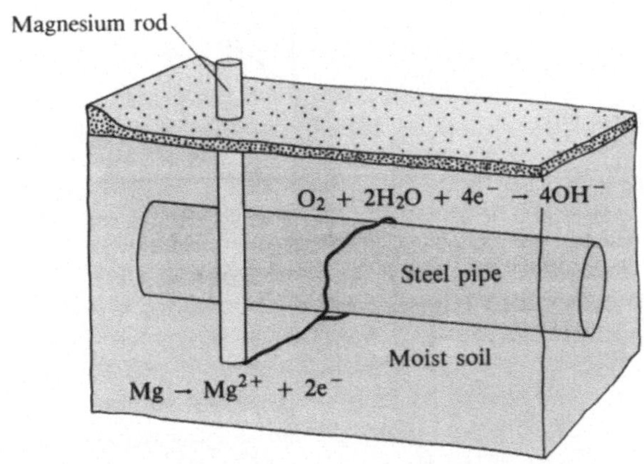

FIGURE 2.3 Rust prevention: cathodic protection of a buried steel pipe. Iron in the steel becomes the cathode in an iron–magnesium voltaic cell. Magnesium rather than iron is oxidized.

deposits and iron bacterial growth, oxygen in the water should be higher than 2 mg/l and the free-chlorine residual concentration should be higher than 0.2 mg/l. Maintaining a pH above 7.2 in the distribution system also helps to avoid high levels of iron deposition.

2.2.1.7 Cobalt (Co)

Cobalt is relatively rare and is found in ores such as *smaltite* ($CoAs_2$) and *cobaltite* (CoAsS). Cobalt is a hard, bluish-white metal that is used mainly in alloys, such as stainless steel and *stellite* (an alloy of iron, copper, and tungsten), which is used in surgical instruments. Cobalt is also used to prepare the alloy *alnico*, which forms powerful magnets. Aqueous solutions of cobalt(II) salts are characteristically rose colored. Cobalt salts, the usual oxidation states II and III, are used to give a brilliant blue color to glass, tiles, and pottery. *Anhydrous cobalt(II) chloride* ($CoCl_2$) is used in water quality testing and as a heat-sensitive ink. Artificially produced cobalt-60 is used as a radioactive tracer and cancer treatment agent. Cobalt is a part of vitamin B_{12} (*cyanocobalamin*) and is considered an essential nutrient, but concentrations higher than 1 mg/kg of body weight are regarded as a health hazard. The formula for vitamin B_{12} appears in Figure 2.4. Cobalt exhibits toxic effects on the heart, kidneys, and thyroid gland. Consumption of large quantities of coffee or beer may lead to high concentrations of cobalt. Cobalt toxicity resulting in heart failure (about 40% mortality) has been reported among heavy beer drinkers who had consumed products containing cobalt additives used as a foam stabilizer.

2.2.1.8 Nickel (Ni)

This element ranks 24th in abundance in the Earth's crust. Nickel metal is a silver-white, malleable, ductile substance with high electric and thermal conductivity. Because it is quite resistant to corrosion, nickel is often used in plating more active metals. Nickel and chromium are the chief additives to iron in making stainless steel. Combined with copper, nickel produces a hard, strong, corrosion-resistant alloy called *monel*. Because boats operate in the corrosive environment of seawater, monel is used in the manufacture of boat propeller shafts.

Nickel is also used as a *catalyst* for the hydrogenation of organic compounds that contain double bonds. Nickel in compounds is almost exclusively in the +2 oxidation state. Aqueous solutions of nickel(II) salts have a characteristic emerald-green color. Nickel and its compounds have little toxicity. *Nickel itch* or contact dermatitis is the most commonly seen reaction to nickel compounds,

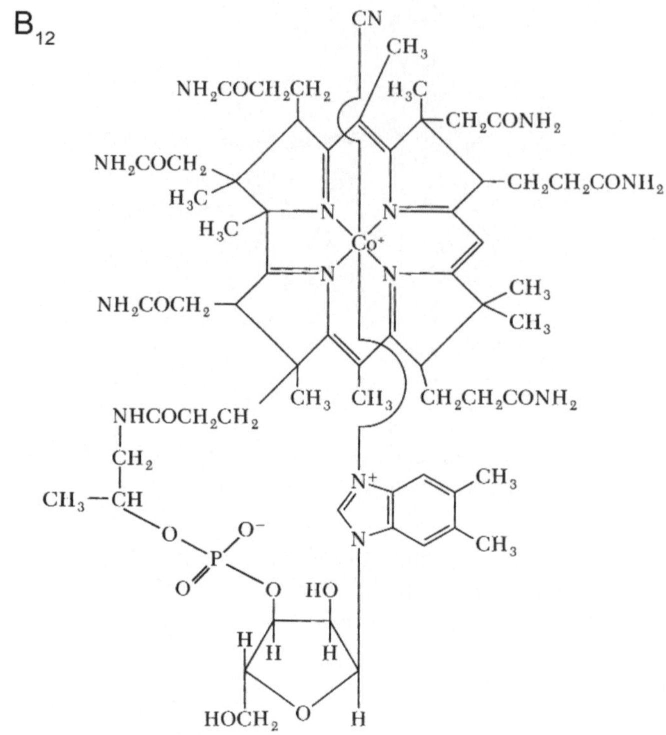

FIGURE 2.4 Formula of vitamin B_{12}.

especially in women, resulting from use of nickel in costume jewelry, especially earrings. Chronic exposure to nickel causes cancer in the respiratory tract and the lungs.

2.2.1.9 Copper, Silver, and Gold

Copper, silver, and gold are often called the "coinage metals" because they have been used for that purpose since ancient times. They can be found in nature as free metals, a reflection of their stability.

Copper (Cu) is widely distributed in nature in ores containing sulfides, arsenides, chlorides, and carbonates. A reddish-brown, malleable, ductile metal, copper is valued for its high electrical conductivity and resistance to corrosion. It is used in plumbing and electrical applications. The reddish-colored metal oxidizes slowly in air; when CO_2 is also present, its surface becomes coated with a green film of $Cu_2(OH)_2CO_3$. The outer surface of the Statue of Liberty is made of copper, and this compound gives the statue its green color. Copper has long been used in the United States to make pennies, but since 1981 new pennies have been made from zinc with a thin copper coating. Copper principally exists in the +2 oxidation state, but compounds containing copper(I) ion are also known. *Copper(II) oxide* (CuO) is black, and *copper(I) oxide* (Cu_2O) is red. Usually, *copper(II) compounds* have a characteristic bright blue color.

Although trace amounts of copper are essential for life, copper in large amounts is quite toxic. For example, copper salts are used to kill bacteria, fungi, and algae, and paints containing copper are used on ship hulls to prevent fouling by marine organisms. Copper is essential to human nutrition because it plays a major role in enzyme functions.

Silver (Ag) has the highest thermal and electrical conductivity of any metal. Its value as a coinage metal, however, makes it too expensive to be used often as an electrical conductor. Silver has a high luster, and, when polished, reflects light very well. This makes it valuable for jewelry and for the reflective coating on mirrors. Silver is soft and usually alloyed with copper. *Sterling silver*, for example, contains 7.5% copper, and silver used for jewelry often contains as much as 20% copper. Silver is even more difficult to oxidize than copper. Metallic silver is not attacked by oxygen in the air, but it does tarnish in air by reacting with oxygen and traces of hydrogen sulfide, H_2S (formed in nature by decomposing vegetation). The *black tarnish* deposit is silver sulfide (Ag_2S). Similar reactions occur if silver utensils are left in contact with sulfur-containing foods, such as eggs and mustard.

One of silver's most important applications is in photography. *Silver salts* tend to be unstable and sensitive to light. *Silver iodide* is used to "seed" clouds to bring on rain. The most important oxidation state of silver is +1.

The major problem in humans arising from overexposure to silver is called *argyria*, which is characterized by blue-gray coloration of the skin, mucous membranes, and internal organs. According to a report by the World Health Organization in 1987, a continuous daily dose of 0.4 mg of silver intake may produce argyria.

Gold (Au) is valuable as bullion and as a decorative metal in jewelry and other artifacts. This element is also used occasionally to plate electrical contacts because of its low chemical reactivity. Pure gold is very soft and it is particularly ductile and malleable. Gold leaf is made by pounding gold into very thin sheets. Gold is so unreactive that even concentrated nitric acid (HNO_3) fails to attack it. A special solution, called *aqua regia*, dissolves gold slowly. (Aqua regia consists of one part concentrated HNO_3 and three parts concentrated HCl.) Gold is found as a free element in nature because its compounds are so unstable.

2.2.1.10 Zinc (Zn)

This metal is mainly refined from *sphalerite* ((ZnFe)S), which often occurs in galena (PbS). Zinc is a white, lustrous, very active metal that behaves as an excellent reducing agent and tarnishes rapidly. Because of zinc's excellent reactivity, its surface quickly acquires a film of a basic carbonate, $Zn_2(OH)_2CO_3$; this coating protects the metal below from further oxidation. About 90% of the zinc produced is used for galvanizing steel. (See detailed discussion of galvanization and cathodic protection against corrosion in Section 2.2.1.6.) The automotive industry has used galvanized steel to make rustproof automobile bodies.

Zinc exists in the +2 oxidation state, and its salts are colorless. Zinc compounds are used in many applications. *Zinc oxide* (ZnO), a white powder, is used in various creams, such as sunscreens, and to make quick-setting dental cements. *Zinc sulfide* (ZnS) can be used to prepare phosphor substances that glow when bathed in ultraviolet light or the high-energy electrons of cathode rays. Such phosphors are used on the inner surface of television picture tubes and the CRT displays of computer monitors and in devices for detecting atomic radiation. Zinc is also used in dry batteries.

Zinc is an essential trace element in human nutrition. High concentrations of zinc are found in the male reproduction system, muscles, kidneys, liver, pancreas, and the thyroid and other endocrine glands. Zinc is also an important component of enzymes. Excessive zinc intake may inhibit copper absorption and lead to copper deficiency. Acidic beverages packaged in galvanized containers may produce toxic zinc concentration levels, causing nausea, vomiting, stomach cramps, and diarrhea.

2.2.1.11 Yttrium (Y)

The yttrium metals include *terbium* (Te), a lanthanide (at. no. 65); *erbium* (Er), another lanthanide (at. no. 68); *ytterbium* (Yb), yet another lanthanide (at. no. 70); and *yttrium* (Y), a transition metal (at. no. 39). These metals are all related to ores found in Ytterby, a small town near Stockholm.

Yttrium–aluminum garnets ($Y_3Al_2O_{15}$), commonly referred as YAGs, are used in lasers (see Appendix C) and electronic equipment (microwave filters) and as synthetic gems.

2.2.1.12 Zirconium (Zr) and Hafnium (Hf)

Zirconium and hafnium occur together in nature because their ions are the same size and have the same charge. These similarities make it difficult to separate them from each other.

Zirconium and zirconium oxide (ZrO) are highly resistant to high temperatures. Their primary use has been in spacecraft that must reenter the atmosphere. Hafnium is named after the Latin term for Copenhagen. This element was originally found in samples that had been mistakenly identified as pure zirconium, as well as in zirconium ores.

2.2.1.13 Niobium (Nb) and Tantalum (Ta)

Niobium and tantalum were "tantalizingly" difficult to separate, and thus named after the mythological Tantalus and his daughter Niobe. Both are transition elements, with the atomic numbers of 41 and 73, respectively. Niobium steel is used in atomic reactors because it has sufficient strength to handle high temperatures over long periods of time.

2.2.1.14 Molybdenum (Mo)

Molybdenum is a lustrous, silver-white, metallic element, mostly used in alloys, and is particularly valuable in enhancing the quality of stainless steel. Molybdenum is also used in nuclear energy production, electrical products, and glass and ceramics. Molybdenum is an essential trace mineral in the meats of ruminants and in plants. Deficiencies are unknown in humans; apparently practically any diet supplies sufficient amounts to carry out this element's roles in enzyme functions.

2.2.1.15 Tungsten (W)

This symbol refers to its Latin name, *wolframate*. The metal is prepared from *tungsten(VI) oxide*, a canary-yellow compound obtained from the processing of tungsten ore. One of the most important uses of tungsten metal is the production of filaments for incandescent light bulbs. This usage depends on the fact that tungsten has the highest melting point (3410°C) and highest boiling point (5900°C) of any metal. To be useful, the incandescent filament in a light bulb must not melt and should not vaporize excessively. The tungsten metal filament does slowly vaporize, and the condensed metal often appears as a black coating on the inside surface of a burned-out bulb. In a light bulb, a coiled wire of tungsten becomes white hot when an electric current flows through it. The wire is enclosed in a glass bulb containing gases that do not react with the tungsten, such as nitrogen and argon. The gases carry the heat away from the wire, which would otherwise overheat and boil away.

Cobalt, chromium, and tungsten form the alloy stellite, which retains its hardness even when hot. This characteristic makes stellite useful for high-speed cutting tools used to machine steel. In *interstitial carbides*, carbon atoms occupy spaces or interstices within the lattice of metal atoms, which results in a material with many characteristics of a metal, such as conductivity and luster. An industrial example is *tungsten carbide* (WC), which is used to make high-speed cutting tools because it is exceptionally hard and chemically stable even as the tool becomes very hot during use.

2.2.1.16 Technetium (Tc)

Technetium is a transition metal with an atomic number of 43. It has no isotopes. The nucleus of every technetium isotope is radioactive and decays or disintegrates, producing an isotope of another element. Because of its nuclear instability, technetium is not found naturally on Earth. Nevertheless, it is produced commercially in kilogram quantities from other elements by nuclear fission, a process in which nuclei are transformed. Technetium derives its name from the Greek word *tekhnetos*, meaning artificial. Technetium was the first new element produced in the laboratory from another element. It was discovered in 1938 by Carlo Pierrer and Emilio Segre when the element molybdenum was bombarded with deuterons (nuclei of hydrogen, each consisting of one proton and one neutron).

Technetium is one of the principal isotopes used in medical diagnostics based on radioactivity. A compound of technetium is injected into a vein, where it concentrates in certain organs. The energy emitted by technetium nuclei is detected by special equipment and provides an image of the organs.

2.2.1.17 Ruthenium (Ru), Osmium (Os), Rhodium (Rh), Iridium (Ir), Palladium (Pd), and Platinum (Pt)

These metals are collectively known as *platinum metals.* The six elements following technetium (Tc), element 43, and rhenium (Re), element 75, are similar and occur together in various combinations in nature.

2.2.1.18 Cadmium (Cd)

Cadmium is less abundant than zinc and is usually found as an impurity in zinc ores. The free metal is soft and moderately active. Its chief use is as a protective coating on other metals, including metals exposed to an alkaline environment, and for making nickel-cadmium batteries.

Cadmium compounds are quite toxic; if absorbed by the body they can cause high blood pressure, heart disease, and even death. Acute overexposure to cadmium fumes may cause pulmonary damage, while chronic exposure is associated with renal tube damage and an increased risk of prostate cancer. The high level of cadmium in cigarette smoke contributes to air pollution. Cadmium may contaminate water supplies from mining, industrial operations, and leachate from landfill. It also may enter water distribution systems through corrosion of galvanized pipes.

2.2.1.19 Mercury (Hg)

Mercury is a heavy, silver-white liquid metal. Its symbol corresponds to the Latin *hydrargyrum*, which means quick silver. Its chief ore is *cinnabar* or mercury sulfide (HgS). Mercury is liquid at room temperature; it freezes at –38.9°C and boils at 357°C. This large and convenient liquid temperature range accounts for mercury's use as the fluid in thermometers. A useful property of mercury is its ability to dissolve many other metals to form solutions called *amalgams.* A silver amalgam used in teeth fillings for many years is no longer used because of the highly toxic effects of mercury.

Mercury is a less-active metal than zinc or cadmium. In compounds, mercury occurs in two oxidation states, +1 and +2. *Mercury(I) chloride* (Hg_2Cl_2), also known as calomel, is very insoluble in water. Its low solubility permitted its uses as an antiseptic and treatment for syphilis before the discovery of penicillin. The body retains very little mercury because so little Hg_2Cl_2 is able to dissolve.

Mercury(II) chloride ($HgCl_2$) is water soluble and highly poisonous. The addition of H_2S to a solution containing mercury(II) chloride produces a black precipitate of HgS. When heated, its crystal structure changes and becomes a brilliant red substance, called *vermilion.*

Because mercury is absorbed by lung tissue, mercury vapor is hazardous, especially when heated. Mercury is a nervous system toxin, causing tremors, ataxia (uncoordinated muscle movements), irritability, slurred speech, psychiatric disorders, blindness, and death. (Thus, when thermometers break inside infant incubators, the spilled mercury vapor can leak into the heating unit, causing a severe hazard to infants.) *Mercuric nitrate* ($Hg(NO_3)_2$) was once used in the manufacture of felt for hats. Workers often developed severe mercury poisoning, an affliction that leads to central nervous system disorders, loss of hair and teeth, loss of memory, and tremors or "hatter's shakes" (hence the term, "mad as a hatter"). In the 1950s, an outbreak of mercury poisoning from contaminated seafood in Minamata Bay, Japan, raised awareness of the mercury hazard. The main sources of mercury pollution are industrial wastes and incinerators, power plants, laboratories, and even hospitals.

In streams and lakes, inorganic mercury is converted by bacteria into two organic forms: *dimethyl mercury* and *methyl mercury*. Dimethyl mercury is very volatile and evaporates quickly, but methyl mercury remains in the bottom sediment and is slowly released into the water, where it enters organisms in the food chain and is biologically magnified. Freshwater fish are particularly at risk, especially near paper plants where *mercuric chloride* ($HgCl_2$) is used as a bleach for paper and then discharged into the water. Organic mercury compounds continue to be used as fungicides in seeds for crop planting.

2.2.2 Inner Transition Elements

2.2.2.1 Lanthanides

The elements from *lanthanum* (La, at. no. 57) through *lutetium* (Lu, at. no. 71) are collectively called the lanthanides, or the *rare earth elements*. To the ancient Greeks, metal oxides were known as "earths." Because these elements were first found in rare minerals as oxides, they became known as the rare earth elements. Although often difficult to isolate, many of the rare earth metals are not particularly rare. *Cerium* (Ce, at. no. 58) is the most abundant rare earth element; *thulium* (Tm, at. no. 69) and *promethium* (Pm, at. no. 61) are the least abundant.

All lanthanides are shiny, silvery, reactive elements. Most readily tarnish in air by the formation of oxides, although *gadolinium* (Gd, at. no. 64) and *lutetium* (Lu, at. no. 71) are quite stable. Some form white oxides and colorless ions in aqueous solutions, while others have colored ions and oxides. The pure metals range in density from 6.2 g/cm^3 for lanthanum to 9.8 g/cm^3 for lutetium, and their melting points all fall between about 800 and 1600°C. The principal use of lanthanide compounds is in petroleum-cracking catalysts. The glass and metallurgy industries also consume lanthanide compounds. In some alloys, rare earths are used to impart desirable properties and in others to react with sand to remove undesirable impurities.

Praseodymium (Pr, at. no. 59) and *neodymium* (Nd, at. no. 60) are added to the glass in welders' goggles to absorb the bright yellow light of the sodium flame. *Cerium oxide* is effective in polishing camera and eyeglass lenses. Pure *neodymium oxide* is added to glass to produce a beautiful purple color. A mixed *oxide of europium* and *yttrium* (Eu_2O_3 and Y_2O_3) produces a brilliant red phosphor that is used in color television screens. To mention just one more application of a lanthanide, *yttrium–aluminum garnets* (YAGs) are used in electronic equipment (e.g., microwave filters) and as synthetic gems.

2.2.2.2 Actinides

The elements from *actinium* (Ac, at. no. 89) through *lawrencium* (Lr, at. no. 103) are collectively called the actinides. All actinides are radioactive.

Elements with atomic numbers greater than 92 (the at. no. of uranium is 92) are called the transuranium elements, the naturally occurring elements of greatest atomic number. In 1940, E.M. McMillan and P.H. Abelson, at the University of California, Berkeley, discovered the first transuranium element. They produced an isotope of element 93, which they named neptunium. The next transuranium element to be discovered was plutonium (at. no. 94). The next two transuranium elements were americium (at no. 95) and curium (at. no. 96). Transuranium elements have a number of commercial uses. For instance, plutonium-238 isotope has been used as a power source for space satellites, navigation buoys, and heart pacemakers. Americium-241 is used in home smoke detectors.

2.3 METALLOIDS

2.3.1 Group IVA (14)

2.3.1.1 Silicon (Si)

Silicon is a representative metalloid; it is a brittle, shiny, black-gray solid that appears to be metallic but is not. Structurally, silicon resembles dismount (a pure form of carbon).

Silicon is extremely hard, is capable of scratching glass, melts at 1414°C, and boils at 2327°C. Swedish chemist Jons Jacob Berzelius discovered silicon in 1823. Silicon is the second-most abundant element in the Earth's crust (oxygen is the most abundant).

Silicon is an element, and silicone is a complex compound. Quartz, sand, agate, jasper, and opal are silicon oxides. In many compounds, silicon is combined chemically with both oxygen and metals. Common examples include talc, mica, asbestos, beryl, and feldspar. Silicon compounds are commercially important, especially the group of compounds known as silicates. Clay, cement, and glass are silicates. When Si is combined with C, the resulting compound is silicon carbide, a very hard compound that has many industrial uses. Very pure Si is used in the production of transistors and integrated circuits.

2.3.2 Group VA (15)

2.3.2.1 Arsenic (As)

Arsenic is silvery white, very brittle, and semimetallic. It is toxic to humans, especially the trivalent compounds. In low doses, arsenic is used as a medication to enhance growth. At low intake levels, arsenic can accumulate in the body over time. Arsenic is used in bronzing, pyrotechnics, dye manufacturing, insecticides, and pharmaceuticals.

An arsenic compound, *gallium arsenide* (GaAs), has fascinating and useful properties. Because GaAs can convert electricity directly into laser beams of coherent light, it is used in light-emitting diodes. These diodes are used in audio disc players and visual display devices. Like silicon, gallium arsenide is a semiconductor (see Section 1.2 and Appendix B), but because it is more expensive than silicon, it is not used the manufacture of computer chips. However, GaAs conducts an electrical current more rapidly than silicon at the same or lower power, producing less waste heat. When manufacturers seek to make chips for computers running at speeds in excess of 100 million instructions per second, GaAs will be needed.

Groundwater may contain arsenic in high concentrations originating from geological materials. Sources of arsenic pollution are industrial wastes, arsenic-containing pesticides, and smelting operations.

2.3.2.2 Antimony (Sb)

Antimony is a brittle, crystalline, solid semimetal. It is a poor electricity conductor. The symbol, Sb, derives from the Latin word *stibium*. Chemically and biologically, antimony resembles arsenic. It is used in alloys, and certain compounds are being used for fireproofing textiles, in ceramics and glassware, and as an antiparasitic drug. Antimony and arsenic toxicity symptoms are similar.

2.4 HEAVY METALS

Although the term "heavy metal" has become entrenched in the literature of environmental pollution, use of the term in this and other contexts has caused a great deal of confusion. One of the most common definitions of "heavy metal" is a metal with a density greater than 5 g/cm^3 (i.e., specific gravity

> 5). Although relatively clear and unambiguous, this definition causes confusion because it is based on a rather arbitrarily chosen physical parameter and consequently includes elements with very different chemical parameters. According to other definitions focused on chemical parameters, these elements are classified as class A, class B, and borderline elements.

2.5 METALLIC SUBSTANCES ESSENTIAL TO LIFE

Minerals, including some metals, constitute about 4% of total body weight and are concentrated most heavily in the skeleton. Minerals known to perform functions essential to life include potassium, sodium, magnesium, calcium, manganese, cobalt, copper, selenium, zinc, chromium, chloride, iodine, and phosphorus.

Other minerals, such as aluminum, silicon, arsenic, and nickel are present in the body, but their exact functions have not yet been determined. Calcium and phosphorus form part of the structure of bone, but because minerals do not form long-chain compounds they are otherwise poor building materials. Their chief role is to help regulate body processes. Calcium, iron, magnesium, and manganese are constituents of some coenzymes. Magnesium also serves as a catalyst for the conversion of *ADP* (adenosine diphosphate) to *ATP* (adenosine triphosphate). Without these minerals, metabolism halts and the body dies. Generally, the body uses mineral ions rather than nonionized forms. Some minerals, such as chlorine, are toxic or even fatal in the nonionized form.

2.5.1 Most Important Metals in Human Metabolism

2.5.1.1 Calcium (Ca)

Calcium is the most abundant cation in the body. It is important to the formation of bones and teeth, blood clotting, normal muscle and nerve activity, and glycogen metabolism and synthesis, and it helps prevent hypertension. Vitamin D and lactose help improve calcium absorption by the body. Oxalic acid, found in some leafy green vegetables (notably spinach), somewhat reduces the absorption of calcium from those foods.

The recommended daily amount (RDA) for adults is 1200 mg, dropping to 800 mg after age 25. Sources are dairy products, leafy green vegetables, egg yolks, shellfish, broccoli, canned sardines and salmon, some types of tofu, and some fortified cereals. In megadoses (ten or more times the RDA), calcium depresses nerve function and causes drowsiness, extreme lethargy, calcium deposits, and kidney stones. *Hypercalcemia* (elevated blood calcium concentrations) occurs in diseases such as hyperparathyroidism, sarcoidosis, malignancy, and vitamin D poisoning. Sudden death may occur if calcium levels remain above 160 mg/l. Calcium toxicity signs and symptoms include anorexia, nausea, vomiting, dehydration, lethargy, coma, and death. Kidney damage and kidney stones may develop in hypercalcemia, and the condition may be associated with congenital heart disease. Excessive calcium levels in drinking water may be related to the formation of kidney or bladder stones, but there is no toxicity concern in these cases.

Calcium deficits may cause muscle tetany, osteomalacia, osteoporosis, retarded growth, and rickets in children. According to a recent survey of studies on various drinking water parameters, high sodium and low calcium intake have been implicated as factors in the development of high blood pressure.

2.5.1.2 Iron (Fe)

Iron accounts for 66% of *hemoglobin*. The hemoglobin in red blood cells carries oxygen (O_2) to cells throughout the body. Hemoglobin is a very large molecule and has four iron (Fe) atoms. Each of these four atoms is embedded in a part of hemoglobin called *heme*. The iron atom is in the center. The structural formula of heme is illustrated in Figure 2.5. Every hemoglobin molecule has four heme units, each containing one Fe atom. When hemoglobin picks up O_2 in the lungs, each O_2 molecule bonds to one of the Fe atoms.

The bonding ability of Fe in hemoglobin is not restricted to O_2. Many other substances can bond with Fe in hemoglobin, such as the poison carbon monoxide (CO). CO is poisonous because the bond it forms with Fe is stronger than the O_2 bond. When a person breathes in CO, the hemoglobin combines with this molecule rather than with O_2. The cells, deprived of O_2, can no longer function, and the person dies.

Only 2 to 10% of dietary iron is absorbed, because of the mucosal barrier. Heme iron, the type found in meat and other animal products, is better absorbed by the body than nonheme iron, the type found in foods derived from plants. Consuming a food high in vitamin C enhances the absorption of iron. The body loses iron in menstrual flow, shed hair, sloughed skin, and mucosal cells. The recommended daily amount (RDA) for males is 10 mg; for females, 18 mg. Normal plasma levels are 1290 μg/l in men and 1100 μg/l in women. The best sources of iron are meat, liver, shellfish, egg yolks, dried fruits, nuts, legumes, and molasses. Iron is found in virtually every food, with higher concentrations in animal tissues than in plants. Generally, men consume about 16 mg/d, and women, about 12 mg/d. Inhalation of urban air contributes about 27 μg/d to total intake.

Megadoses of iron cause hemochromatosis (inherited condition of iron excess), damage to the liver (cirrhosis and liver cancer), cardiac disorders, and diabetes. Large amounts of stored iron are associated with an increased risk of cancer because iron serves as a nutrient for cancer cells. Signs of toxicity are caused by free iron that appears after the carrier is saturated. The first sign of acute toxicity is vomiting, followed by gastrointestinal bleeding, lethargy, restlessness, and perhaps gray cyanosis. If the patient survives for 3 or 4 days, complete recovery follows rapidly.

Chronic excessive iron intake can lead to *hemosiderosis* (a generalized increased iron content) or *hemochromatosis* (specific histological site of hemosiderosis), possibly accompanied by fibrosis. This condition is relatively benign but may be accompanied by glucose metabolism or exacerbation of existing cardiac disease. Chronic inhalation of iron fumes leads to mottling of the lungs, a siderosis that is considered benign, nonfibrotic, and not favorable to tubercle bacilli.

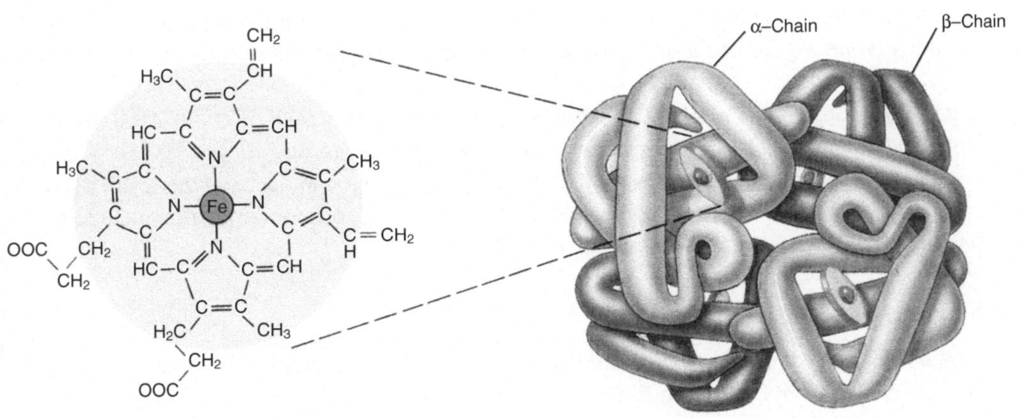

FIGURE 2.5 Hemoglobin structure. Hemoglobin consists of four globular protein subunits. Each subunit contains a single molecule of heme, a porphyrin ring surrounding a single ion of iron.

Inadequate iron intake causes iron-deficiency anemia, pallor, lethargy, flatulence, anorexia, paresthesia, impaired cognitive performance in children, inability to maintain body temperature, and reduced production of phagocytic white blood cells (and thus reduced immune system response).

2.5.1.3 Copper (Cu)

Copper is required along with iron for synthesis of hemoglobin. It is also a component of the enzyme necessary for melanin pigment formation. Humans ingest copper in food and water. Concentrations in food vary widely from less than 10 to more than 25,000 μg/100 calories, and the maximum contaminant level (MCL) in drinking water is 1.0 mg/l.

The RDA is 2 to 3 mg, and the average blood level is 1 mg/l. Rich sources are oysters, liver, kidney, nuts, dried legumes, and potatoes. The average copper content of drinking water is 0.61 to 250 μg/l, and this amount has increased over time due to pipe corrosion and chlorination. Undesirable taste and odor are often perceived at levels higher than 1 mg Cu/l. Copper is actively absorbed in the stomach and duodenum.

Acute exposure overdose causes an immediate metallic taste, followed by epigastric burning, nausea, vomiting, and diarrhea. Symptoms include ulcers and other damage to the gastrointestinal tract, jaundice, and suppression of urine production. Fatal cases often include secondary effects, such as hypertension, shock, and coma. Some cases of copper overdose have been the result of consuming large amounts of acidic foods (e.g., fruit juices and carbonated beverages) in copper-lined containers or dispensed through machines with copper components..

Inhaled dust and fumes cause irritation of the respiratory tract. Chronic exposure may produce *metal fume fever*, an influenza-like syndrome that lasts a day or so.

The role of copper in human metabolism involves the turnover of copper-containing enzymes. Two inherited diseases disrupt these enzymes. *Menke's disease*, apparently an inability to absorb copper, produces copper deficiency. *Wilson's disease* is the opposite, leading to excessive accumulation of copper.

2.5.1.4 Sodium (Na)

Of all sodium in the body, 50% is found in extracellular fluid, 40% in bone salts, and 10% in cells. Sodium is also part of the bicarbonate buffer system and strongly affects distribution of water through osmosis, thus the acid–base balance of blood. Sodium is necessary in neuromuscular function, as it is essential for transport of glucose and other nutrients. Absorption is rapid and almost complete. The hormone aldosterone regulates the metabolism of sodium. Excretion occurs mainly through urination.

The RDA for sodium has not been established, although daily intake of about 2500 mg is typical. Sources include table salt (1 tablespoon = 2000 mg), cured meats, and cheese. Excess sodium intake causes hypertension and edema. Sodium deficiency is rare but can occur as the result of, for example, excessive vomiting, diarrhea, and sweating. Symptoms of sodium deficiency include nausea, abdominal and muscle cramping, and convulsions.

2.5.1.5 Potassium (K)

Potassium, a principal cation in intracellular fluids, plays a role in the transmission of nerve impulses and in muscle contraction. Potassium is necessary for proper cardiovascular function, as it helps regulate blood pressure and water balance in cells. There is some evidence that a high potassium diet may reduce the risk of hypertension and stroke. The body maintains a high concentration of K^+ ions inside the cells even though K^+ concentration outside the cells is low. The reverse is true for Na^+. To prevent K^+ from diffusing out of cells and to prevent Na^+ from entering the cells, special transport proteins in the cell membranes constantly pump K^+ into the cells and Na^+ out. This pumping requires energy that is supplied by the hydrolysis of ATP (adenosine triphosphate).

The RDA for potassium has not been established. A diet adequate in calories provides an ample amount of about 2500 mg/d. Sources are most foods, especially avocados, bananas, dried apricots, oranges, potato skins, yogurt, meat, poultry, fish, and milk.

Excess potassium usually causes renal failure, severe dehydration, muscular weakness, and cardiac abnormalities. Deficits are rare but may result from severe diarrhea or vomiting, causing muscular weakness, paralysis, nausea, tachycardia, or heart failure. In a condition called *hypoglycemia*, the body's output of insulin is elevated and blood sugar is depleted. The condition may suddenly shift the already small amount of K^+ from the extracellular media into the cells. The general result is inadequate nerve impulses going to the muscles and the extremities. Muscular weakness and numbness in fingers and toes are symptoms of K^+ deficit. Because the heartbeat is also influenced, later symptoms may include tachycardia (fast heartbeat) and, still later, weak pulse and falling blood pressure. Intravenous potassium chloride (KCl) solution is used to prevent a severe K^+ deficit from causing cardiac arrest.

Sweating causes loss of K^+ ions. Hence, strenuous physical activity in warm weather often leads to severe muscle cramping.

2.5.1.6 Magnesium (Mg)

Magnesium is an important constituent of many coenzymes, is vital to many basic metabolic functions, and also aids in bone growth and the function of nerves, bones, and muscles, including heart rhythm regulation. In coastal areas, seawater can penetrate drinking-water wells if the water table becomes depleted. The water in such wells contains higher-than-normal concentrations of magnesium salts. These salts, especially *magnesium sulfate* and *magnesium citrate*, are incompletely absorbed in the intestines. High concentration of these salts in the intestines creates a hypertonic condition relative to neighboring tissues. Consequently, water flows from the tissues to the intestine, diluting the stool and causing diarrhea. At the same time the tissues are dehydrated. This is also the principle used in treating hemorrhoids in a sitz bath. When hemorrhoidal tissue is swollen, a hypertonic solution of magnesium sulfate draws out water and shrinks the tissue. Swollen feet respond to a hypertonic solution when soaked in a hot magnesium sulfate bath.

The RDA for magnesium is 300 to 350 mg. Sources are dairy products, meat, whole-grain cereals, nuts, legumes, leafy green vegetables, bananas, and apricots. Excess magnesium intake causes diarrhea. Deficits cause neuromuscular problems, tremors, muscle weakness, irregular heartbeat, diabetes, hypertension, high cholesterol levels, pregnancy problems, and vascular spasms. Low magnesium intake has been linked to high blood pressure, heart-rhythm abnormalities, and consequently, heart attacks.

2.5.1.7 Zinc (Zn)

Zinc is an important part of many enzymes that are necessary for normal tissue growth and healing of wounds and the sense of taste and appetite. As a part of peptidase, zinc is important in protein digestion. Zinc is also necessary for prostate gland function. Next to iron, zinc is the second most abundant trace mineral in the body.

The RDA is 15 mg. Sources are seafood, meat, cereal grains, legumes, nuts, wheat germ, whole-grain bread, and yeast. Zinc excess may raise cholesterol levels and cause difficulty in walking, slurred speech, hand tremors, involuntary laughter, and a masklike facial expression. Zinc is relatively nontoxic except in extremely high doses. Acidic beverages made in galvanized containers may produce toxic levels of zinc concentration and can cause nausea, vomiting, stomach cramps, and diarrhea. Zinc deficiency may be involved in impaired immunity and learning disabilities and can cause growth retardation and loss of taste and smell. In general, zinc deficiency is rare, but several groups are at risk, such as heavy drinkers (alcohol speeds zinc excretion), athletes (sweating causes significant zinc depletion), and strict vegetarians (fruits and vegetables contain little zinc).

2.5.1.8 Manganese (Mn)

Manganese activates several enzymes necessary for hemoglobin synthesis, growth, reproduction, lactation, bone formation, production and release of insulin, and preventing cell damage. The RDA is 2.5 to 5.0 mg. The best sources are nuts, legumes, whole grains, leafy vegetables, and fruits. Excessive manganese appears to contribute to obsessive behavior and hallucinations and may interfere with iron absorption. The effects of manganese deficit are not known.

2.5.1.9 Cobalt (Co)

Cobalt is a constituent of vitamin B_{12} (see illustration in Figure 2.4) and is needed for *erythropoiesis*, the process in which *erythrocytes* (red blood cells) are formed. Cobalt is found in all cells, with higher concentrations in bone marrow.

The RDA has not been established. Good sources are liver, lean red meats, poultry, fish, and milk. Megadoses may cause goiter and damage to the heart muscle. Deficits (mainly impaired absorption) cause the same symptoms as vitamin B_{12} deficiency, such as pernicious anemia, weight loss, and neurological disorders.

2.5.1.10 Chromium (Cr)

Chromium is necessary for the proper utilization of sugars and other carbohydrates by optimizing the production and effects of insulin. It is widely distributed in the body.

The RDA is 0.05 to 2 mg. Sources include liver, meat, cheese, whole grains, yeast, and wine. The effects of excess chromium are not known. Deficits cause impaired insulin function, hence increased insulin secretion and the risk of adult-onset diabetes mellitus.

2.5.1.11 Selenium (Se)

Selenium is a nonmetal, listed in the VIA (16) periodic group. An antioxidant, it prevents chromosome breakage, certain birth defects, and certain types (e.g., esophageal) of cancer. It is necessary for the beneficial action of vitamin E; if vitamin E in the diet is inadequate, more selenium is required. Besides its cancer-prevention activity, selenium slows down the process of aging and makes heart muscles stronger.

The RDA is 0.05 to 2 mg. Estimated selenium intake is 132 μg/d for an adult man, but in seleniferous areas intake may increase to 0.7 to 7 mg daily. The recommended drinking water standard is 10 μg/l, but the maximum contaminant level goal (MCLG) is 5 μg/l. Selenium dietary supplements are recommended due to its anticarcinogenic effects. Selenium deficiency occurs when the diet contains less than 0.02 to 0.05 ppm Se. Sources are meat, seafood, and cereals. Selenium content of vegetables depends on the concentration of selenium in the soil.

Chronic toxicity has been reported in humans ingesting 1 mg Se/kg/d. Toxic effects include gastrointestinal complaints, jaundice, skin hyperpigmentation, hair loss, dental caries, arthritis, dizziness, and fatigue.

Selenium concentrations in air are high near metallurgical industries. Signs of inhalation exposure are similar to allergenic responses, such as inflammation of mucous membranes and eyes, sneezing, coughing, and frontal headache. Absorption through the skin has not been observed in people who use antidandruff shampoo containing selenium sulfide.

Problems resulting from selenium deficits are not well known. People living in the Keshan province of China suffer from an endemic cardiomyopathy known as *Keshan sickness*, probably due to the very low selenium content of the soil.

2.5.2 Common Plant Nutrients

Of the 18 elemental essential plant nutrients, 15 are minerals. Of the 15 minerals, 11 are metals, including potassium, calcium, magnesium, boron, copper, iron, manganese, molybdenum, sodium, vanadium, and zinc. *Potassium* (K) is needed for enzymatic control of the interchange of sugars, starches, and cellulose. *Calcium* (Ca) and *magnesium* (Mg) are available as Ca^{2+} and Mg^{2+} ions. Chlorophyll requires magnesium; therefore, deficiencies cause chlorosis, or low chlorophyll content. *Iron* (Fe) is also an essential catalyst in chlorophyll formation. Green plants suffering from iron deficiency turn yellow. *Boron* (B) is a trace element and is toxic to most plants in concentrations above a relatively narrow range. See Appendix D for more information on the roles of metals as plant nutrients.

3 Toxicity of Metals

3.1 GENERAL DISCUSSION OF TOXICITY

Toxic substances, or toxins, are chemicals that adversely affect living organisms. *Toxicology* is the study of these effects. Chemical substances exert a wide range of effects, depending on the amount ingested, inhaled, or absorbed.

3.1.1 Toxicytosis

Toxicytosis is the type and intensity of response evoked by a chemical. To determine the response to chemicals, the *toxicologist* administers controlled doses to laboratory test animals and uses the information to approximate the hazards for humans.

3.1.2 Toxic Effects

The *toxic effects* of chemicals are various. Some chemicals interfere with the function of an organ (e.g., kidneys, lung, or liver), and others disrupt the blood-formation mechanism, enzyme activities, the central nervous system, or the immune system. For example, dioxin, an extremely toxic compound, affects DNA and ultimately the immune system.

3.1.3 Acute Effects

Acute effects are symptoms that appear right after exposure. These effects are generally caused by fairly high concentrations of chemicals during a short exposure period.

3.1.4 Chronic Effects

Chronic effects are delayed, but long-lasting, responses to toxic agents. They may occur months to years after exposure and usually persist for years. They are generally the result of low-level exposure over a long period.

3.1.5 Lethal Effects

Lethal effects can be defined as responses that occur when physical or chemical agents interfere with cellular and subcellular processes in the organism to such an extent that death directly follows. Examples are suffocation and interference with movement to obtain food or escape predators.

3.1.6 Sublethal Effects

Sublethal effects disrupt physiological or behavioral activities but do not cause immediate mortality, although death may follow. Examples include interference with feeding, growth retardation, alteration in blood chemistry, changes in the number and type of blood cells, and tumor formation.

3.1.7 Two D's (Dose and Duration)

Toxic effects are determined by the concentration (or dose) of the toxin and the duration of the exposure, known as the two D's. In general, the higher the dose and the longer the exposure, the greater the effect.

3.1.8 LD_{50} (Lethal Dose 50)

The dose that kills half of the test animals is called the LD_{50}, or the lethal dose for 50% of the test animals, and is expressed as milligrams of toxin per kilogram of body weight. The lower the LD_{50}, the more toxic the chemical. For example, a chemical with an LD_{50} of 200 mg per kilogram of body weight is half as toxic as one with an LD_{50} of 100 mg.

3.1.9 Classification of Toxic Substances

Toxic substances can be classified according to the way in which they disrupt body chemistry. Modes of toxic substances are described as corrosive, metabolic, neurotoxic, mutagenic, teratogenic, and carcinogenic.

3.1.9.1 Corrosive Poisons

Corrosive poisons are toxic substances that actually destroy tissues. Examples are *strong acids* and *alkalis* and many *oxidants*, such as those found in laundry products. Examples are sulfuric acid (found in auto batteries), hydrochloric acid (also called muriatic acid, used for cleaning purposes), and sodium hydroxide (used in clearing clogged drains). Some poisons act by undergoing chemical reaction in the body and producing corrosive material. *Phosgene*, the deadly gas used during World War I, is an example. When inhaled, it is hydrolyzed (broken down by water) in the lungs to hydrochloric acid, which causes pulmonary edema (a collection of fluid in the lungs) owing to the dehydrating effect of the strong acid on tissues, so that oxygen cannot be absorbed effectively by the flooded and damaged tissues. Some corrosive poisons destroy tissue by oxidizing it. This type of material includes *ozone* and *nitrogen dioxide*. Selected corrosive poisons and their effects are presented in Table 3.1.

3.1.9.2 Metabolic Poisons

The word "metabolism" derives from the Greek *metabolein*, meaning to change or alter. Metabolic poisons interfere with a vital biochemical mechanism by preventing the proper function of a biochemical mechanism or by completely stopping its activity. For example, *carbon monoxide* (CO) reacts with hemoglobin, making hemoglobin unable to transport oxygen. The *cyanide* ion (CN^-) is the toxic agent in cyanide salts. One of the most rapidly working poisons, the cyanide ion interferes with oxidative enzymes, such as cytochrome oxidase. The mechanism of cyanide poisoning is described in detail in Appendix E.

3.1.9.3 Metal Toxicity

Metal toxicity is the most common of all the metabolic poisons. Metal toxicity is discussed separately in Section 3.2.

TABLE 3.1
Selected Corrosive Poisons

Substance	Formula	Toxic Action	Possible Contact Source
Hydrochloric acid	HCl	Acid hydrolysis	Cleaning products
Sulfuric acid	H_2SO_4	Acid hydrolysis dehydrates tissue, oxidizes tissue	Auto batteries
Phosgene	ClCOCl	Acid hydrolysis	Combustion of chlorine-containing plastics (PVCs)
Sodium hydroxide	NaOH	Base hydrolysis	Caustic soda, drain cleaners
Trisodium phosphate	Na_3PO_4	Base hydrolysis	Detergents, household cleaners
Sodium perborate	$NaBO_3.4H_2O$	Base hydrolysis oxidizing agent	Laundry detergents, denture cleaners
Ozone	O_3	Oxidizing agent	Ambient air, electric motors
Nitrogen dioxide	NO_2	Oxidizing agent	Polluted air, automobile exhaust
Iodine	I_2	Oxidizing agent	Antiseptics
Hypochlorite	OCl^-	Oxidizing agent	Bleach
Peroxide	O_2^{2-}	Oxidizing agent	Bleach, antiseptics
Oxalic acid	$H_2C_2O_4$	Reducing agent	Bleach, tanning solutions, spinach, tea
Sulfite	SO_3^{2-}	Reducing agent	Bleach
Chloramine	NH_2Cl	Oxidizing agent	Produced when ammonia and chlorinated bleach are mixed
Nitrosyl	NOCl	Oxidizing agent	Produced when ammonia and bleach are mixed

3.1.9.4 Neurotoxins

Neurotoxins are metabolic poisons but their actions are limited to the nervous system. Such poisons include *strychnine, curare* (used on darts to bring down game by a group of South American Indians), *atropine, acetylcholine, nicotine, caffeine, codeine,* and *morphine*. Many neurotoxins are useful in medicine. Atropine is used to dilate the pupil of the eye to facilitate examination of its interior and as an antidote for anticholinesterase poisons. Atropine sulfate and other atropine salts are excellent painkillers when applied to the skin. Curare is useful as a muscle relaxant. Nicotine causes stimulation and then depression of the central nervous system. Morphine is the most effective pain reliever known. Codeine in small quantities is an ingredient in cough syrups.

Chemical warfare agents constitute another group of neurotoxins. The Greeks used sulfur dioxide gas during the war between Athens and Sparta. Chemical weapons were used in World War I, including *mustard gas* (dichloroethyl sulfide), *phosgene* (Cl_2CO), *chlorine gas* (Cl_2), *hydrogen cyanide* (HCN), and *tabun* and *sarin* nerve gases. In the 1980s, during the war between Iran and Iraq, chemical agents were also used. Some insecticides, such as *parathion* and *malathion,* also qualify as neurotoxins.

3.1.9.5 Teratogens

Teratogens are chemical agents with toxic effects on reproduction; they are classified as radiation, viral agents, and chemical substances. The study of birth defects caused by chemical agents is called *teratology* (*terat* is a Greek word for "monster"). The *thalidomide* disaster is a good example of a teratogen. Thalidomide was used as a tranquilizer and sleeping pill. Many pregnant women who took the drug gave birth to babies with deformities, such as missing arms and fingers. In 1961,

TABLE 3.2
Teratogenic Substances and Effects on Fetuses of Selected Species

Substance	Species	Effects on Fetus
Metals		
Arsenic	Mice, hamsters	Increase in males born with eye defects, renal damage
Cadmium	Mice, rats	Miscarriage
Cobalt	Chickens	Eye and lower extremity defects
Gallium	Hamsters	Spinal defects
Lead	Humans, chickens	Low birth weight, brain damage, stillbirth, early- and late-pregnancy death
Lithium	Primates	Heart defects
Mercury	Humans	Minamata disease (Japan)
	Mice	Fetal death, cleft palate
	Rats	Brain damage
Thallium	Chickens	Growth retardation, miscarriage
Zinc	Hamsters	Miscarriage
Organic compounds		
DES (diethyl-stilbestrol)	Humans	Uterine anomalies
Caffeine (15 cups/d equivalent)	Rats	Skeletal defects, growth retardation
PCBs (polychlorinated biphenyls)	Chickens	Central nervous system and eye defects
	Humans	Growth retardation, stillbirth

thalidomide was taken off the market and has not been sold since. Chemicals with teratogenic effects are listed in Table 3.2.

3.1.9.6 Mutagens

Mutagens are chemical substances that alter the structures of *deoxyribonucleic acid* (DNA), which contains the organism's genes and chromosomes, and cause abnormalities in offspring. In other words, a *mutagen* is a chemical that can change the hereditary pattern of a cell and *mutation* is an error in the copying of the base sequence of DNA resulting in a change in heredity.

Every embryo formed by sexual reproduction inherits genes from the parent sperm and egg cells. The transmission of the hereditary information from one generation to the next takes place in the *chromosomes* of cell nuclei. Each species has a different number of chromosomes in cell nuclei. *Genes*, located inside the chromosomes, contain the information that determines external characteristics (red hair, blue eyes, etc.) and internal characteristics (blood group, hereditary diseases, etc.). The genes that carry inheritable traits lie in sequence along the chromosomes.

Chemical analysis shows that nuclei are largely made up of special basic proteins called *histones* and a compound called *nucleic acids*. Only the nucleic acid, DNA, carries hereditary information. Genes, then, are located in DNA. (See Appendix F for components of nucleic acids.)

In the early 1980s, Bruce Ames and colleagues at the University of California, Berkeley, developed a simple test (*Ames test*) that identifies chemicals capable of causing mutations in sensitive strains of bacteria. In this test, the analyst uses a bacterial strain, such as *Salmonella*, which feeds on the amino acid *histidine*. When the bacteria are grown in a medium that does not contain histidine, very few survive. If a mutagen is added to the medium, however, some of the bacteria may undergo mutations that can live without a supply of histidine. The mutated bacteria multiply and show up as

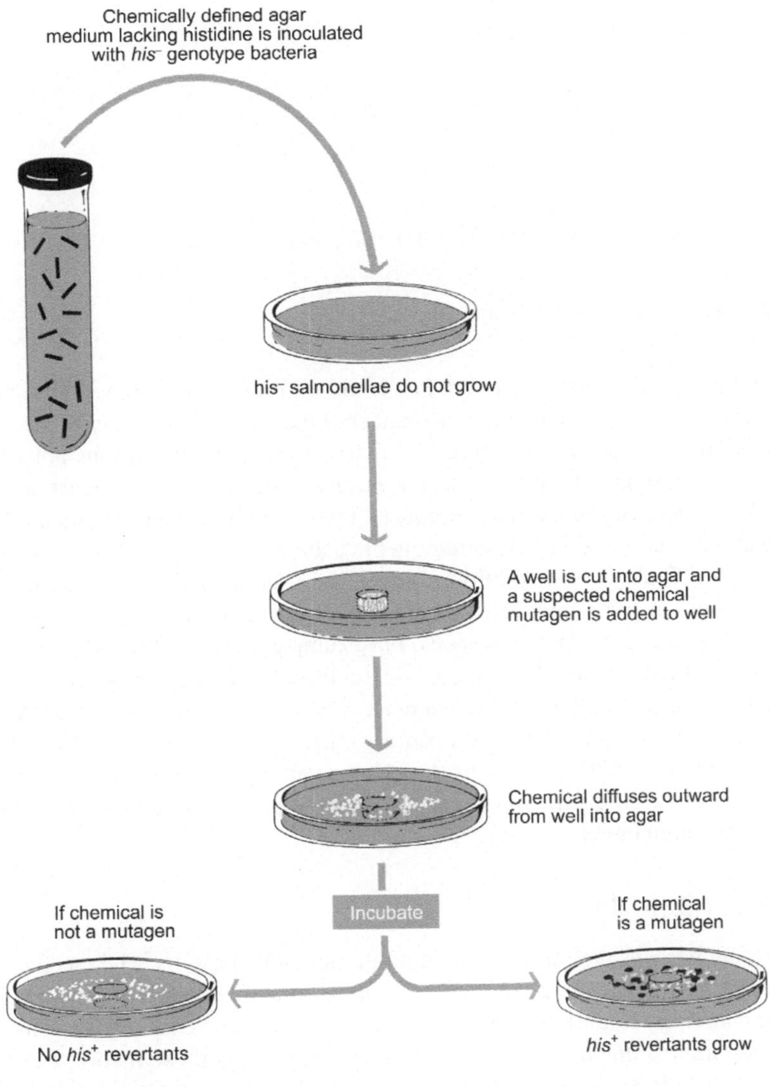

FIGURE 3.1 Ames test for detecting chemical mutagen.

a heavy growth of bacteria colonies. With such a simple test, many chemicals can be tested for mutagenic activity. Mutagenic chemicals can then be further tested in animals to determine whether they are also carcinogens. The Ames test is illustrated in Figure 3.1.

3.1.9.7 Carcinogens

Carcinogens are chemicals that cause cancer, an abnormal growth condition in an organism. The rate of cell growth in cancerous tissue differs from the rate in normal tissue. Cancerous cells spread to other tissues and show partial or complete loss of specialized functions. Almost all human cancers caused by chemicals have a long induction period, which makes it extremely difficult for researchers to obtain meaningful interpretation of exposure data.

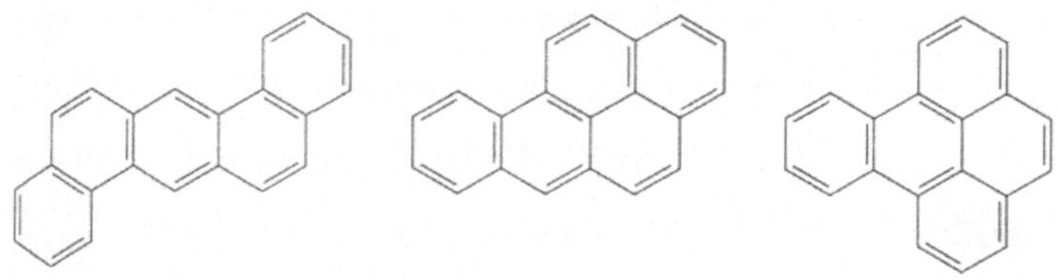

Dibenz[a,h]anthracene Benzo[a]pyrene Benzo[e]pyrene

FIGURE 3.2 Carcinogenic aromatic hydrocarbons.

About 200 years ago, London surgeon Percivall Pott found that chimney sweeps (boys employed to clean chimneys) were especially prone to cancer of the scrotum and other parts of the body. Today, it is known that these cancers were caused by fused aromatic hydrocarbons present in the chimney soot. Carcinogenic aromatic hydrocarbons have at least four rings and at least one angular junction (see Figure 3.2). These carcinogens are produced by automobile exhausts and are found in cigarette smoke. Researchers have verified the carcinogenic behavior of a large number of chemicals, some of which are listed in Table 3.3. In addition to industrial chemicals that are known to contaminate air and drinking water, our everyday diets contain a great variety of natural carcinogens. Some of these chemicals are also mutagens and teratogens. For example, celery contains *isoimpinellin* — a member of the chemical family called *psoralens* — at a level of 100 μg/100 g. This level increases 100-fold if the celery is diseased. Psoralens, when activated by sunlight, damage DNA. Oil of bergamot, which is found in citrus fruits, contains a psoralen that was once used by a French manufacturer of suntan oil. Sunlight caused the psoralens to enhance tanning. Black pepper contains small amounts of *safflere*, a known carcinogen. Oil of mustard and horseradish contain *allyl isothiocyanate*, which is mutagenic and carcinogenic.

3.2 METAL TOXICITY

Heavy metals are perhaps the most common of all metabolic poisons. The mechanism of metal toxicity is different from other metabolic poisons. Metal toxicity can affect enzymes, the cellular proteins that regulate many important chemical reactions. Heavy metals are toxic primarily because they react with and inhibit *sulfhydryl* (SH) *enzyme* systems, such as those involved in the production of cellular energy. Figure 3.3 illustrates the reaction of a heavy metal with glutathione. The metal replaces the hydrogen in two sulfhydryl groups on adjacent molecules and the strong bond effectively eliminates the two glutathione molecules from further reaction.

TABLE 3.3
Selected Inorganic Chemicals Carcinogenic to Humans

Compound	Use or Source	Site Affected
Arsenic and compounds	Insecticides, alloys	Skin, lungs, liver
Asbestos	Brake linings, insulation	Respiratory tract
Beryllium	Alloy with copper	Bone, lungs
Cadmium	Metal plating	Kidneys, lungs
Chromium	Metal plating	Lungs
Nickel	Metal plating	Lungs, sinuses

$$2\ \text{Glutathione} + \text{metal ion}\ M^{2+} \longrightarrow M\,(\text{glutathione})_2 + 2H^+$$

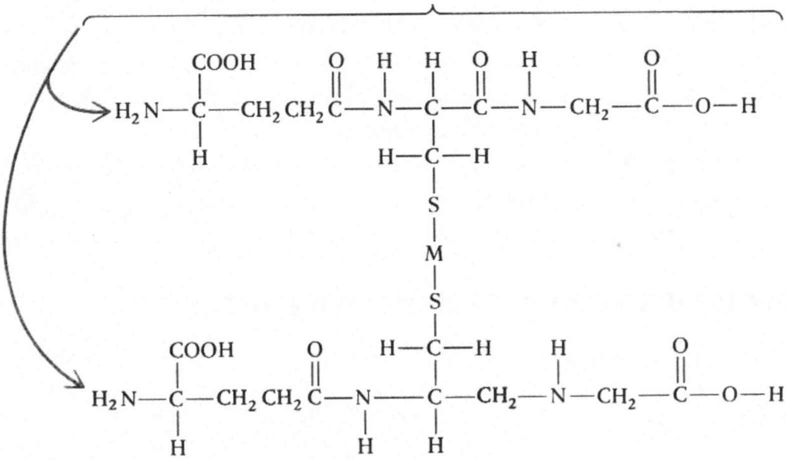

Glutathione-metal complex

FIGURE 3.3 Glutathione reaction with a metal. (From *World of Chemistry*, 1st ed., by M.D. Joesten, D.O. Johnston, J.T. Netterville, J.L. Wood © 1990. Reprinted with permission of Brooks/Cole, an imprint of the Wadsworth Group, a division of Thomson Learning. Fax 800 730-2215.)

A disturbance in enzymatic activity can seriously alter the functioning of the organ or tissue. As examples, mercury and arsenic both bind to certain enzymes, thereby blocking their activity. Lead binds to the thiol (SH–) chemical group in the enzymes and consequently reduces the body's ability to synthesize enzymes necessary for respiration. The addition of chelating agents is used to eliminate such metal poisoning. Transition metals are known for their ability to form many complex ions — substances in which a metal cation is surrounded by and bounded to one or more other ions or molecules. Complexes are often called chelates (from the Greek *chele*, meaning "claw") because a chelating agent encases an atom or ion like a crab grasps food. In the same way a chelating agent envelops a metal ion, and when the metal ion is tied up, the sulfhydryl groups are freed and the enzyme again functions normally. For example, an effective chelating agent for removing lead from the human body is *ethylenediamine-tetraacetic acid* (EDTA). The calcium disodium salt of EDTA is used in the treatment of lead poisoning because EDTA by itself would remove too much of the blood serum's calcium. In solution, EDTA has a greater tendency to complex with lead (Pb^{2+}) than with calcium (Ca^{2+}). As a result, the calcium is released and the lead is tied up in the complex, as seen in Figure 3.4. The lead chelate is then excreted in the urine.

FIGURE 3.4 Structure of chelate formed when the anion of the EDTA envelops a Pb^{2+} ion. (From *World of Chemistry*, 1st ed., by M.D. Joesten, D.O. Johnston, J.T. Netterville, J.L. Wood © 1990. Reprinted with permission of Brooks/Cole, an imprint of the Wadsworth Group, a division of Thomson Learning. Fax 800 730-2215.)

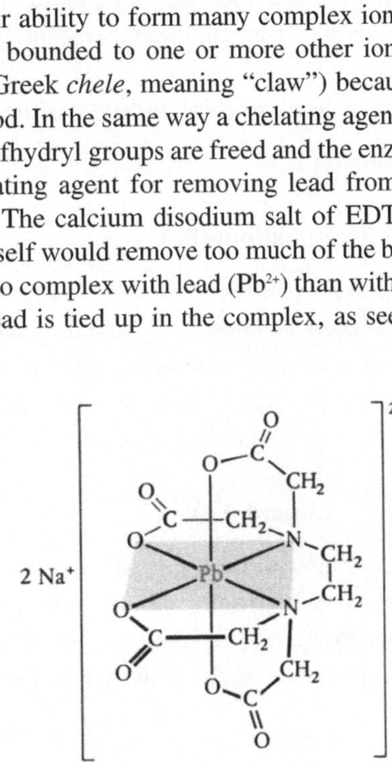

Metals can form lipid-soluble organo-metallic ions, involving Hg, As, Sn, Tl, and Pb, capable of penetrating biological membranes and accumulating within cells. Some metals in metallo-proteins exhibit oxidation-reduction activity, such as Cu^{2+} to Cu^{+}, which can alter structural or functional integrity. Certain metals displace others in biomolecules. For example, when Zn^{2+} is replaced by Ni^{2+} or Be^{2+} to Mg^{2+} in enzymes, the enzymes are deactivated. In addition, the replacement of Ca^{2+} with other metals in membrane proteins causes functional disorders.

Because heavy metals are elements, they cannot be broken down, either chemically or by decomposer organisms. The only ways to dispose of them are to dilute them to levels at which they are no longer toxic or to treat them with chemicals that convert them into less toxic compounds.

3.3 TOXIC EFFECTS OF SELECTED REPRESENTATIVE METALS

3.3.1 Group IA (1): Alkali Metals

3.3.1.1 Lithium (Li)

Lithium is widely found in plant and animal tissues. *Daily intake* has been estimated at 2 mg/d. Therapeutic doses of lithium (used as an antidepressant) range from 90 to 1800 mg/d. When patients are first dosed with lithium carbonate, they often experience nausea, vomiting, and abdominal pain about an hour after each dose, but these symptoms soon disappear.

Chronic toxicity usually affects the gastrointestinal tract, nervous system, and kidneys. Additional symptoms of acute toxicity include increased thirst, excessive salivation, and diarrhea. Chronic toxicity effects include tremors (especially of the hands), muscular weakness, ataxia, giddiness, drowsiness, muscular hyperirritability and fasciculation, lethargy, stupor, and, in extreme cases, coma and seizures. Renal symptoms include polyuria, elevation of nonprotein nitrogen, and, in the terminal stages, oliguria. An increase in a rare cardiac defect, *Ebstein's anomaly*, has been reported in children of women dosed therapeutically with lithium.

3.3.1.2 Sodium (Na) and Potassium (K)

See Section 2.5.

3.3.1.3 Rubidium (Rb)

Rubidium is present in the body in larger than trace metal amounts and can replace potassium in certain processes, but the body's requirement of this metal is not known. It functions similarly to potassium in altering heart muscle contractions and can alter behavior and manic-depressive states, but its metabolic function is not understood. All animal tissues contain 20 to 40 ppm (mg/kg) of this metal. The toxicity of rubidium appears to be relatively low.

3.3.1.4 Cesium (Cs)

Cesium is able to substitute for potassium to some extent. For example, cesium partially protects the kidneys and heart in potassium-deficiency conditions, and it concentrates in erythrocytes, as does potassium. Almost half the average daily intake of about 10 mg/d derives from food (red meats, eggs, and dairy products). Its toxicity is not known.

3.3.2 Group IIA (2): Alkaline Earth Metals

3.3.2.1 Beryllium (Be)

Beryllium inhibits a number of enzymes. A small intake of beryllium from water and soil (via food) occurs, estimated at 100 μg/d. Airborne beryllium is the result of coal combustion, cigarette smoke, and, in a few areas, beryllium-processing plants. The toxic effects of beryllium are ascribed to damage of lysosomes, which release cell-destroying enzymes. Chronic exposure to beryllium and its compounds can produce a frequently fatal pulmonary granulomatosis called *berylliosis*. Major signs and symptoms include pneumonitis with accompanying cough, chest pain, and general weakness and often pulmonary dysfunction. The first symptom is shortness of breath.

3.3.2.2 Magnesium (Mg) and Calcium (Ca)

See Section 2.5.

3.3.2.3 Strontium (Sr)

Strontium substitutes for calcium in many normal mechanisms, often with no apparent ill effects. Strontium is concentrated in the skeleton. Dietary strontium intake ranges from 0.98 to 2.2 mg/d for adults, about one third of which is from milk. Acute strontium toxicity causes death from respiratory failure, but most strontium compounds have a low toxicity. Evidence of chronic effects is negligible.

3.3.2.4 Barium (Ba)

Barium is absorbed through the lungs and the gastrointestinal tract and, once absorbed, accumulates in the bones. Small proportions of barium accompany calcium in virtually every foodstuff. It is estimated that the average daily intake is 1.33 mg. The national interim primary drinking water standard is 1 mg/l. Barium is commonly found in urban ambient air, because barium compounds are used as diesel fuel, smoke suppressants, and automotive lubricants.

The soluble salt of barium causes toxicity. Soluble salts are irritants to skin and mucous membranes, and the barium dispersant in automotive lubricants is a mild eye irritant. Barium compounds (nitrate, sulfide, carbonate, and chloride) have been involved in accidental and suicidal poisonings. Signs are nausea, vomiting, colic, and diarrhea, followed by myocardial and general muscular stimulation with tingling of the extremities. Severe cases continue to loss of tendon reflexes, heart fibrillation, general muscular paralysis, and death from respiratory arrest. A fatal dose of barium chloride ($BaCl_2$) for a human is 0.8 to 0.9 g (0.55–0.6 g Ba).

Chronic exposure to barium causes a benign pneumoconiosis, known as *baritosis* (numerous evenly distributed nodules in the lungs), which has occurred in workers exposed to finely ground barium sulfate ($BaSO_4$). Baritosis nodules usually disappear after cessation of exposure, but bronchial irritation may persist. Barium is mutually antagonistic to all muscular depressants.

3.3.3 Group IIIA (13): Boron–Aluminum (B–Al)

3.3.3.1 Aluminum (Al)

Aluminum is found in all human tissues, but is most concentrated in the lungs, presumably from inhaled air. Oral doses of aluminum induce phosphorus depletion syndrome and deplete red blood cell ATP (adenosine triphosphate). Unprocessed foods contain aluminum in very small quantities, although some vegetables and fruits may contain up to 150 mg/kg. Total daily intake is estimated at about 80 mg.

Aluminum compounds are used for storing and processing food (e.g., baking powder, cooking vessels, and metal foil). Inhalation of aluminum compounds has been used in the prevention of silicosis. Aluminum compounds are also used to prevent *hyperphosphatemia* in renal disease. High aluminum intake originates from packaging, aluminum cooking vessels, aluminum foil, and aluminum-containing antacids.

Aluminum is generally considered nontoxic. Because *Alzheimer's disease* patients have a high aluminum content in certain brain cells, research is now focused on high aluminum intake as a possible causal factor. In patients with this disease, the nerve fibers in the cerebral cortex are entangled, and some of the nerve endings degenerate and form plaque. The brain becomes smaller, and part of the cortex atrophies.

3.3.3.2 Gallium (Ga)

Gallium is chiefly deposited in bone tissue and is relatively immobile. Human exposure to gallium has included the use of radioactive plus stable gallium in therapeutic doses, so reported toxicity may be due to radioactivity. Signs of toxicity include dermatitis, gastrointestinal disturbances, and bone marrow loss.

3.3.3.3 Indium (In)

The daily human intake from food is estimated at less than 8 μg. Indium is the lowest-volume metal used by the body. Drinking water is unlikely to be the major source of human exposure. However, indium might be expected to leach from galvanized iron pipes. No drinking water concentrations have been reported. Fish and shellfish containing bioconcentrated indium from contaminated waters can lead to human oral exposure. Lead-smelting emissions can produce elevated indium levels in ambient air. Soluble and colloidal indium compounds are generally more toxic than insoluble noncolloidals.

3.3.3.4 Thallium (Tl)

Thallium at low concentration as Tl^+ has an affinity for certain enzymes and an activating ability ten times that of K^+. Thallium salts inhibit several enzymes that play major roles in bone formation. Toxic doses adversely affect protein synthesis and cause disaggregation of ribosomes. Consumption is about 0.5 ton/year and is not well defined. Biota in thallium-contaminated areas currently have thallium levels (>3 ppm) that could be high enough to cause toxic symptoms in mammals if their entire diet derives from the contaminated biota. Accidental poisonings have occurred from use of thallium rodenticides, but their use has been banned. The use of thallium acetate as a cosmetic depilatory around 1930, as well as its use for about 50 years as a therapeutic epilant in the treatment of fungal scalp infections, was often accompanied by severe poisoning and fatalities.

Dermal exposure to thallium may occur while handling thallium preparations used in laboratory analyses. After acute poisoning, the kidneys — especially the renal medulla — contain the highest thallium concentrations. In the final stages of fatal poisoning, thallium appears in all organs and tissue concentrations tend to equalize. For humans, doses of 14 mg/kg and above are fatal. In mammals, toxic effects are usually delayed for 12 to 48 h.

Symptoms include gastrointestinal discomfort, pain and paralysis in the extremities, high blood pressure, optic nerve dystrophy and blindness, psychic excitement (10 days after poisoning), liver and kidney damage, and hair loss. In the absence of known association of the patient with possible sources, diagnosis of thallium poisoning is difficult. The usual cause of death is respiratory arrest, the end result of pneumonia and general respiratory depression. Other deaths from thallium poisoning have been attributed to cardiac failure, dehydration, and progressive impairment of the brain and vagus nerve. In still other cases, severe parenchymatous changes were found, including fatty heart and liver (degenerative changes due to fat deposits in cells), lung edema, meningeal congestion, renal damage, gastroenteritis, and widespread degeneration of the nerve cells and axons in the brain.

Few of the reported human and mammalian studies of thallium toxicity provide conclusions about the dangers of very low chronic intake (10 to 20 µg/d). In humans, *alopecia* is the hallmark of long-term thallium poisoning, with hair loss beginning within about 10 days and epilation being complete within a month. However, alopecia does not always occur, even after severe poisoning.

Selenium- and sulfur-containing compounds may offer some protection against thallium toxicity. The only proven antidote to date is *Prussian blue* (potassium ferrihexacyano-ferrate(II)).

3.3.4 Group IVA (14): Carbon

3.3.4.1 Tin (Sn)

Daily intake of tin from food ranges from 0.2 to 10 mg. SnF_2 in toothpaste is another oral source. Tin in food may be augmented by tin leached from unlacquered cans. The general population inhales up to 7 µg/d from ambient air. Toxic effects have been seen in people who eat canned food, with accumulations of 250 mg Sn per kg or more. Tin toxicity may cause nausea, vomiting, and diarrhea. Organotins are more toxic than inorganic compounds. Contact with tin compounds may cause skin irritation. The only lethal incident associated with tin compounds mentioned in the literature at the time of this writing was the Stalinon catastrophe in France, in which about 100 people died from ingesting an impure, untested drug preparation that was contaminated with triethyltin.

3.3.4.2 Lead (Pb)

Lead is toxic to the nervous system, and children are especially susceptible to its effects. Lead is readily absorbed through the intestinal tract and deposited in the central nervous system. The first lead water pipes were used in ancient Rome by upper-class citizens; their children drank the water throughout childhood and thus were at high risk of lead toxicity. This fact may explain the bizarre behavior of certain notorious emperors and may have contributed to the fall of the Roman Empire. In recent years, exposure to lead toxicity has become widespread. Sources are lead-containing paint, air, soil, dust, food, and drinking water. The presence of lead in the body is indicated by lead blood levels, expressed as micrograms of lead per deciliter of blood (µg/dl). Blood lead levels of 10 µg/dl and higher may contribute to decreased cognition, nervous system damage, and stunted growth. Many children have suffered lead poisoning after ingestion of lead-based paint. Lead-based paint was used inside many homes until Congress passed the Lead-Poisoning Prevention Act in 1971.

Lead concentrations as high as 0.4 to 0.8 mg/l in natural waters — mostly from natural sources, such as galena deposits — have been reported. High contamination levels may be caused by industrial and mining pollution sources. High levels of lead in drinking water consist mainly of corrosion products from lead service pipes, solders, and household plumbing. Lead as a corrosion product in drinking water is associated with copper. Copper is needed for good health, and at low levels it has a beneficial effect, but in high concentration it is toxic, causing diarrhea and vomiting. The maximum contaminant level (MCL) established for lead in drinking water is 0.02 mg/l, but the maximum contaminant level goal (MCLG) for lead is zero, and for copper is 1.3 mg/l.

Acute toxicity is quite unusual, as lead is a relatively insoluble, cumulative poison. Reported symptoms include fatigue, sleep disturbance, and constipation, followed by colic, anemia, and neuritis. Symptoms of chronic lead poisoning include loss of appetite, metallic taste, constipation, anemia, pallor, malaise, weakness, insomnia, headache, irritability, muscle and joint pains, tremors, encephalopathy, and colic. Some lead-poisoning victims develop weakness in the extensor muscles, known as *wrist drop* or *foot drop*.

Lead interacts with a number of other metals. A typical detoxification treatment involves chelation of the lead with *calcium ethylenediamine tetraacetate* (EDTA) given parenterally. Repeated treatments leach lead out of bone tissue. *Lead arthralgia* (joint pains) is lead-induced gout caused by lead's interference with uric acid excretion by the kidneys. Lead toxicity affects the kidneys and causes tubular dysfunction, or *nephrotoxicity*. Lead is associated with the depression of many endocrine functions, particularly the thyroid and adrenal glands. It causes premature deliveries and spontaneous abortions in humans, as well as chromosome aberrations, but there is no evidence of teratogenic or carcinogenic effects.

3.3.5 Group VA (15): Nitrogen–Phosphorus (N–P)

3.3.5.1 Bismuth (Bi)

Bismuth is not available in food. Drinking water contains an average of 0.01 mg/l. Certain over-the-counter drugs sold for gastrointestinal disturbances contain bismuth compounds (e.g., Pepto-Bismol). Bismuth, as BiOCl, is used as a white pearlescent coloring material in lipsticks and other cosmetics.

3.4 TOXICITY OF SELECTED TRANSITION METALS

3.4.1 Period 4

3.4.1.1 Scandium (Sc)

Scandium intake from food and drinking water is considered negligible. Recently, intake by inhalation found in Italy was 0.04 μg/d. No toxic effects have been found.

3.4.1.2 Titanium (Ti)

Titanium levels in food vary widely; total daily intake has been estimated at between 300 to 600 μg and 100 to 1600 μg. The highest titanium concentrations are found in butter and corn oil. Typically, intake from drinking water is about 2 μg/d, and from inhalation about 2 to 4 μg/d from ambient air. Titanium dioxide is only slightly absorbed. Titanium dioxide and metal are practically inert.

3.4.1.3 Vanadium (V)

Vanadium concentrations in animal and plant foods are very low, probably a few parts per billion (ppb) in wet weight, and are also very low in drinking water. Vanadium in the ambient air results from combustion of coal, crude oils, and undersulfurized heavy-fuel oils; airborne vanadium doubles during cold weather. The highest vanadium concentrations in the body have been found in the lungs. Vanadium salts have intermediate inhalation toxicity and low oral toxicity. Industrial exposures are generally described as acute episodes with relapses and sometimes chronic coughing and bronchitis. Sequelae of acute vanadium intoxication may include chronic respiratory symptoms, but pneumoconiosis, fibrosis, and emphysema do not develop.

Epidemiological investigations have correlated concentrations of vanadium and other metals in ambient air with disease mortality indexes. V, Cd, Zn, Sn, and Ni in the air of 25 localities correlated well with heart disease and nephritis. V, As, and Zn in the air showed a weak association with lung cancer. V was strongly associated with bronchitis, V and Be with pneumonia; V, Be, and Mo correlated with other cancers.

Exposure to vanadium irritates the skin and eyes, and a greenish-black discoloration of the tongue and oral mucosa may occur with a salty or metallic taste. These symptoms disappear 2 to 3 days after cessation of exposure.

3.4.1.4 Chromium (Cr), Manganese (Mn), Iron (Fe), Cobalt (Co), Copper (Cu), and Zinc (Zn)

These elements are discussed in Chapter 2.

3.4.1.5 Nickel (Ni)

Nickel uptake is mostly from food at 300 to 500 μg/d. Airborne nickel derives from combustion of coal and petroleum products. High concentrations are found in the brain, liver, and kidneys. The major toxicity responses to lead and lead compounds are contact dermatitis and allergic sensitization, but generally nickel and nickel compounds have little toxicity. *Nickel itch* usually begins with a sensation of burning and itching, followed by erythema and a nodular eruption, which may progress to pustules or ulcers. Once exposure ends, recovery occurs in about a week. This reaction is found mostly in women who use nickel-plated garter clips (before the panty hose era) and costume jewelry, especially earrings.

Inhalation of *nickel carbonyl* causes pulmonary disturbances and sometimes death. Chronic exposure to nickel dust, nickel compounds, or a combination of compounds causes cancer of the respiratory tract and lungs. Nickel in drinking water also correlates with mortality from oral and intestinal cancer, but not from respiratory cancer.

3.4.2 Period 5

3.4.2.1 Yttrium (Y)

Yttrium is discussed in Section 3.4.3 together with the lanthanides (rare earth metals).

3.4.2.2 Zirconium (Zr)

Zirconium uptake is mostly from meat, poultry, eggs, dairy products, algae, and shellfish. However, plants generally do not translocate zirconium above the roots, so only root crops would be affected. Zirconium in lipstick, nonaerosol deodorants, and poison ivy remedies also contribute to oral intake. Zirconium contamination of superphosphate fertilizers is a possible route into human foodstuffs. Information on zirconium in the ambient air is scarce, but it is likely present because of its high natural background concentration in the soil. Little or no hazard is expected from its emission.

Niobium (Nb), also known as in metallurgical industry as *columbium*, has been found in the diet at levels of 600 μg/d and in drinking water, at about 20 μg/d. Niobium is found in almost every food — meat, cereals, dairy products, fish, vegetables, and fruits — but concentrations are above average in tea, coffee, pepper, and fats. Occupational exposure occurs is ore processing, metal fabrication, and welding. No toxic effects in human have been reported as of this writing.

3.4.2.2 Molybdenum (Mo)

Daily intake of Mo is about 100 to 500 μg. The main dietary contributors are meat, grains, and legumes. Concentrations in drinking water average about 1.4 μg/l (based on the drinking water of 100 large cities). Toxicity is negligible for molybdenum, with the exception of the hexavalent compounds (molybdenum valence ranges from 0 to +6). A gout-like disease has been observed in people living in a region of Armenia characterized by high ambient levels of Mo; dietary intake in this region is 10 to 15 mg/d.

3.4.2.3 Ruthenium (Ru)

No information is available on the toxicity of oral, dermal, or inhalation exposure to Ru. Ruthenium red is a tumor inhibitor, apparently because it interferes with mitochondrial transport of calcium.

3.4.2.4 Rhodium (Rh)

There is no information available about the toxicity of oral, dermal, or inhalation exposure to Rh. The effectiveness of rhodium chloride ($RhCl_3$) as an antiviral chemotherapy has been variously explained by the ability to act as a cobalt antagonist and to form lipid-soluble complexes, which interfere with phospholipid formation by the virus (Browning 1969).

3.4.2.5 Palladium (Pd)

Palladium used in dental alloys is innocuous. Dental uses include pins in porcelain teeth, dental wires, and gold alloys for inlays, crowns, bridges, and partial dentures. Information on dermal or inhalation exposure is not available. The (ineffective) use of colloidal palladium to treat tuberculosis and gout has no adverse effects.

3.4.2.6 Silver (Ag)

Oral exposure mainly occurs through drinking water. Natural waters do not contain toxic levels of silver (over 0.05 mg/l). Silver at 0.1 to 0.2 mg/l was used in the Apollo space program and on Soviet spacecraft to purify drinking water and wastewater. Numerous silver-containing medications and dental silver amalgam fillings have contributed to high levels of oral exposure; however, at present these uses are banned. A few drops of silver nitrate are applied to the conjunctiva of newborn infants to prevent *ophthalmia neonatorum*, a result of gonorrhea transmitted from the mother. Silver salt solutions and ointments are used to treat burns.

Absorption does not occur through contact with the skin. Inhalation from ambient air near mines and smelters and from occupational exposure causes problems. The major problem in humans from overexposure to silver is argyria, characterized by a blue-gray coloration of the skin, mucous membranes, and internal organs. The disease has occurred almost exclusively among workers in the manufacture of silver nitrate ($AgNO_3$). Argyrosis of the conjunctiva has occasionally developed from the use of silver-nitrate-containing hair dyes for coloring the eyebrows and eyelashes.

3.4.2.7 Cadmium (Cd)

Daily consumption of cadmium varies from 17 to 64 μg according to various national estimates. Concentrations in drinking water range from 0.2 to 0.7 μg/l. Airborne cadmium comes primarily from the steel industry and waste incineration, followed by volcanic activity and zinc production. There is no evidence of dermal absorption, and oral absorption is very low. The effects of acute exposure to cadmium include vomiting (15–30 min after ingestion), increased salivation, abdominal pain, and diarrhea. Acute inhalation toxicity is characterized by coughing and tightness in the chest (4–10 h after exposure). Chronic exposure produces a variety of effects on kidneys, lungs, heart, bones, and gonads. Cadmium fumes can damage the olfactory organs. Cadmium toxicity is decreased by the presence of other metals, especially zinc, calcium, copper, iron, and selenium.

3.4.3 Period 6

3.4.3.1 Lanthanum (La) and Lanthanides or Rare Earth Elements: Cerium (Ce), Praseodymium (Pr), Neodymium (Nd), Promethium (Pm), Samarium (Sm), Europium (Eu), Gadolinium (Gd), Terbium (Tb), Dysprosium (Dy), Holmium (Ho), Erbium (Er), Thullum™, Ytterbium (Yb), and Lutetium (Lu)

Yttrium (Y) and the lanthanides are discussed as group because of their chemical and toxicological similarities. The naturally occurring rare elements range from 0.2 to 46.1 ppm for cerium (Ce). No information is available on oral and dermal exposure; inhalation is possible from ambient air. All lanthanides deposit in the bones and liver. No definitive evidence of poisoning has been reported. Lanthanum as La^{3+} counteracts Ca^{2+} binding in heart muscle.

The blood anticoagulant activity of the lanthanides has been studied for the prevention of thrombosis. The anticoagulant effect of the lanthanide salts is counteracted by vitamin K.

3.4.3.2 Hafnium (Hf)

Information on exposure to hafnium is the same as for zirconium (see Section 2.2).

3.4.3.3 Tantalum (Ta)

Tantalum is relatively nontoxic.

3.4.3.4 Tungsten (W)

Tungsten is basically inert. No information is available for oral, dermal, or inhalation exposure. There is no evidence of acute toxicity. Chronic exposure to WC (tungsten carbide) in grinding wheels produces *hard metal disease,* with symptoms of coughing, dyspnea, wheezing, minor radiological abnormalities, and asthma. The disease is primarily attributed to the cobalt content of WC (see Section 2.2).

3.4.3.5 Rhenium (Re)

No information is available for oral, dermal, or inhalation exposure.

3.4.3.6 Osmium (Os)

Inhalation exposure is expected near burning or smelting of copper concentrates. Acute effects of *osmium tetroxide* (OsO^4) in humans include a purulent discharge and eye and respiratory damage; permanent and temporary blindness can develop. Toxic exposure occurs among precious metal workers and histologists who use OsO^4 solution as tissue stain. Some workers have reported contact dermatitis, and some arthritis patients treated with OsO_4 have also developed dermatitis. Injections of OsO_4 solution into knee joints of patients afflicted with rheumatoid arthritis affect chemical synovectomy — the solution destroys the synovial membrane and allows regeneration of a thickened synovial membrane. OsO_4 articular injections are toxic; thus, great care must be taken in handling OsO_4 before injection to avoid inhalation by medical personnel or the patient.

3.4.3.7 Iridium (Ir)

Iridium is chemically similar to rhodium (Rh). No information is available on exposure or toxicity.

3.4.3.8 Platinum (Pt)

Oral exposure is unknown. After inhalation, lung clearance is rather slow, and kidney and bone tissues accumulate platinum. Some people develop allergic dermatitis upon wearing platinum jewelry. *Platinosis*, which resembles an allergic hypersensitivity that improves upon removal from exposure, is an asthma-like condition with dermatitis. A low-grade fibrosis in the lungs may also develop in platinosis. Irritation of the eyes and upper respiratory tract is common, with coughing, tightness of the chest, wheezing, and shortness of breath. In the most severe cases, cyanosis, diaphoresis, feeble pulse, and clammy coldness of the extremities are possible.

3.4.3.9 Gold (Au)

Oral exposure commonly originates from dental inlays and crowns. It is doubtful that these materials contribute any gold to the "body burden" (general metabolism), as attrition and leaching are unlikely. No information is available on inhalation exposure. Dermal irritation and eczema may develop from contact with gold jewelry. Colloidal gold compounds are used to treat arthritis. Contact with or injections of gold compounds may cause chronic contact-sensitivity dermatoses.

3.4.3.10 Mercury (Hg)

Major exposure via oral ingestion occurs through eating fish. Mercury levels are usually below 200 mg/g in fish and below 20 mg/g in other foods. Total intake of inorganic mercury is estimated at less than 10 μg/d. Dermal exposure is minimal, although mercury compounds used for disinfecting diapers may cause mercury poisoning. The main organs infected are the brain and kidneys.

Mercury is poisonous and, to make matters worse, mercury and its salts accumulate in the body. Because mercury is a cumulative poison, even small amounts absorbed over extended periods can lead to serious medical problems. *Acute mercury poisoning* is usually the result of toxic exposure to soluble inorganic salts. After gastrointestinal disturbances (abdominal pain, nausea and vomiting, bloody diarrhea, and shock), stomatitis, and loosening of the teeth, nephritis and hepatitis occur. Death results from ulceration and bleeding in the gastrointestinal tract. Mercury vapors cause erosive bronchitis and bronchiolitis with interstitial pneumonia. *Mercurialism*, or *chronic intoxication* by elemental mercury vapor or mercury salts, is much more common than acute toxicity, due to the cumulative nature of mercury. Symptoms include mental and emotional disturbances (the victim becomes depressed and excitable and irascible, especially when criticized), headache, fatigue, weakness, loss of memory, drowsiness, insomnia, muscular tremor, and general neuralgia. Other symptoms are gingivitis, stomatitis, digestive disturbances, and ocular lesions. Symptoms of toxicity from exposure to mercury salts are similar, but kidney disorders are more frequent. Metallic mercury — the type found in thermometers, sphygmomanometers, and other instruments — is not absorbed by the gastrointestinal tract and therefore is not very hazardous if swallowed. However, it is absorbed by lung tissue. Therefore, mercury vapors are hazardous, especially when heated.

Mercury poisoning has become an acute problem in recent years because of industrial dumping of mercury compounds into streams and lakes. Previously it was believed that mercury, as a heavy metal, would settle to the bottom of lakes and rivers and would be harmlessly buried there, covered by sand. However, certain microorganisms convert mercury metal to *organic mercury compounds*, mainly methylmercury and dimethylmercury. *Dimethylmercury* evaporates quickly from the water, but *methylmercury* remains in the bottom sediments and is slowly released into the water, where it enters organisms in the food chain and is biologically magnified (by buildup of chemical elements or substances in organisms in successively higher trophic levels). Methylmercury is concentrated in fish, and people who eat the contaminated fish can get mercury poisoning.

When mercury was dumped in the Minamata Bay, Japan, ethylmercury and other alkylmercury compounds produced *Minamata disease*, which has the clinical appearance of encephalitis. The earliest signs are gradual decreases in the senses of touch, vision, hearing, and taste; numbness in the fingers, toes, lips, and tongue; and tunnel vision (which develops to complete blindness). Other signs are loss of balance, lack of coordination, and tremor and mood changes, similar to mercurialism. Of the 52 reported cases in Japan, 17 people died and 23 were permanently disabled. Once contaminated foods were removed from the market, the number of mercury-poisoning cases reported dropped drastically.

Metallic mercury is mixed vigorously with silver to form amalgams used in tooth fillings. Significant mercury exposure may result upon opening amalgamators (amalgam-mixing devices). Mercury is also released when dentists carve and shape amalgams. Overall, dentists and dental assistants experience prolonged exposure, compared to patients' intermittent exposure. In the nineteenth century, the felt industry used *mercury(II) nitrate* ($Hg(NO_3)_2$), to stiffen the felt used in making hats. The factory workers frequently developed tremors or *hatter's shakes* and lost hair and teeth; hence, the term "mad hatters." A vivid description of the psychological changes produced by mercury poisoning can be found in Lewis Carroll's *Alice in Wonderland*, specifically the Mad Hatter character.

Mercury is a byproduct of manufacturing vinyl chloride and is also emitted in the aqueous wastes of chemical manufacture, incinerators, power plants, laboratories, and even hospitals. Organic mercury compounds continue to be used as fungicides in seeds for planting crops. In one severe outbreak in New Mexico, a family consumed a pig that had eaten contaminated seeds. Mercury intoxication is treated by chelation (see Section 3.2).

3.4.4 Selected Metals of Period 7, Including Actinides

3.4.4.1 Thorium (Th)

Major oral exposure occurs in medical use of thorium as a radiopaque medium, and some patients have developed tumors. Drinking water contains no thorium. Dermal and inhalation exposures are found mostly in workers handling thorium compounds. The effect of chronic exposure to thorium is radiotoxicity.

3.4.4.2 Uranium (U)

All naturally occurring uranium isotopes are radioactive. Daily intakes in urban areas of various countries have been estimated at 1.0 to 1.5 μg/d. Major food sources of uranium are table salt, vegetables, and cereals. Drinking water is also important source of uranium intake in areas with overlying uranium deposits. Oral toxicity is low, but inhalation is highly toxic. UF_4 is the most toxic by inhalation and U_2O_8 the least toxic. Long-term exposure results in a high radiation hazard. Absorbed uranium deposits in bone tissue, where it is bound in the hydroxyapatite complex, substituting for calcium.

3.5 TOXICITY OF SELECTED METALLOIDS

3.5.1 Boron (B)

Oral intake originates mostly from boric acid and sodium borate used as food preservatives. Boron is an important plant nutrient (especially for tobacco, cabbage, and sugar beets); therefore, most food is rich in boron. Seawater has a relatively high boron concentration, as do algae and sea sponges. Boron participates in the hormonal regulation of calcium and plays an important role in cell division. Boric acid, whether from ingestion or skin absorption, causes nausea, vomiting, diarrhea, abdominal cramps, and erythematous lesions on skin and mucous membranes. High doses cause circulatory

TABLE 3.4
Selected Arsenic-Containing Insecticides

Insecticide	Formula
Lead arsenate	$Pb_3(AsO_4)_2$
Monosodium methanarsenate	CH_3–AsHO–O–O^-.Na^+
Paris green (copper acetoarsenite)	3 CuO.3 As_2O_3.$Cu(C_2H_3O_2)_2$

collapse, tachycardia, cyanosis, delirium, convulsions, and coma. Chronic exposure may cause dry skin, eruptions, and gastric disturbances.

3.5.2 Germanium (Ge)

The daily intake of germanium from food is about 1.5 mg. Because of its presence in coal ash, urban air may contain germanium. This metalloid is nontoxic; toxic effects result only after intake of high doses.

3.5.3 Arsenic (As)

Trace quantities are usually found in food and water, and the highest concentrations occur in seafood. Arsenic is a feed additive for swine and poultry. Dietary intake ranges from 0.15 to 0.40 mg/d. Arsenic is released into the air by coal combustion and the use of arsenic-containing pesticides. Table 3.4 presents selected arsenic-containing insecticides.

During metabolism arsenic is methylated to *methylarsinic acid* (*cacodylic acid*) and *monomethylarsenic acid*, thereby detoxifying arsenic. Acute effects are caused by accidental, suicidal, or homicidal ingestion of large doses. The effects of arsenic overdose are collapse caused by high blood pressure, restlessness, convulsions, coma, and death. Effects of chronic exposure, either environmental or occupational, are carcinogenesis, cardiovascular disease, and neurological disturbances. Effects on the mucous membrane, peripheral nervous system, and gastrointestinal system are quite common.

Selenium is capable of reducing the carcinogenicity of arsenic, perhaps even eliminating it. Arsenic's effect on various enzymes is based on the reaction between the metal and the thiol chemical group in the enzymes and cofactors (such as glutathione). A compound known as *British Anti-Lewisite* (BAL) was developed as an antidote to *Lewisite*, an arsenic-containing poison gas used in World War I. BAL is a chelating agent that bonds to the metal at several sites. The chelation of arsenic by BAL is illustrated in Figure 3.5. With the arsenic or heavy metal tied up, the sulfhydryl

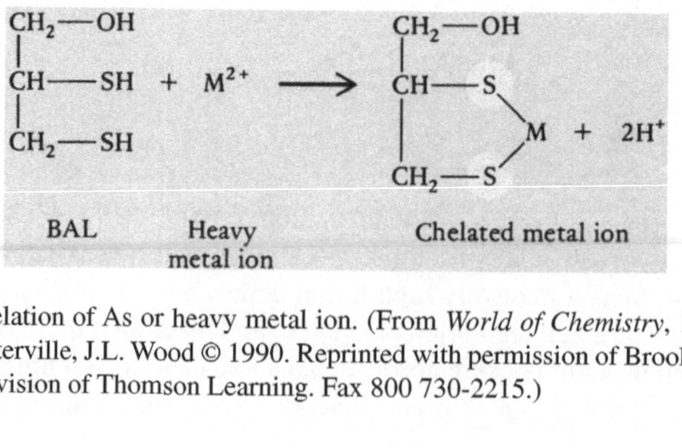

FIGURE 3.5 BAL chelation of As or heavy metal ion. (From *World of Chemistry*, 1st ed., by M.D. Joesten, D.O. Johnston, J.T. Netterville, J.L. Wood © 1990. Reprinted with permission of Brooks/Cole, an imprint of the Wadsworth Group, a division of Thomson Learning. Fax 800 730-2215.)

groups in vital enzymes are freed and can resume their normal functions. BAL is used routinely to treat heavy metal poisoning.

3.5.4 Antimony (Sb)

Antimony belongs to the same periodic group as arsenic and resembles it chemically and biologically. Oral exposure to antimony occurs primarily through foods. Daily intake ranges from 10 to 1250 μg/d. Drinking water also contains antimony in trace quantities; the maximum acceptable limit is 146 μg/l. Concentrations of antimony in the air of large U.S. cities averages 32 ng/m^3. Tobacco contains 0.1 mg/kg antimony by dry weight, so cigarette smoke also contributes to inhalation exposure. Dermal contact may occur from textiles that have been fireproofed by antimony compounds. Medical uses include radioantimony injections and antimony compounds in the treatment of parasitic diseases.

According to the literature (1977), children ingesting about 30 mg/l antimony from contaminated soft drinks suffered vomiting, nausea, and diarrhea. The effect from inhalation exposure is disturbance of the upper respiratory tract. Exposure to heavy antimony fumes causes vomiting, abdominal cramps, and diarrhea. Long-term exposure may participate in the development of gastrointestinal and lung problems and heart disease.

3.5.5 Tellurium (Te)

Tellurium intake occurs mostly from fatty foods, some processed foods, baking powder, and beverages. Tellurium has not been found in drinking water and ambient air. The toxicity of tellurium has been observed in accidental poisoning and exposure of research animals. Effects included garlic breath, digestive disturbances, stunted growth, somnolence, and loss of hair.

4 Standards Related to Metallic Pollutants

The increasing number of the toxic pollutants in the environment has become a major problem. Over the years, many laws have been enacted to protect the environment and human health. The Environmental Protection Agency (EPA) is the federal government regulatory agency charged with managing and enforcing environmental protection legislation issued by Congress. The EPA sets standards for permissible levels of pollutants and continuously updates them. Metals are powerful pollutants, and they are perhaps the most common metabolic poisons. Teratogenic and carcinogenic effects of some metals are also well known. (Metals with teratogenic and carcinogenic effects are listed in Tables 3.2 and 3.3, respectively). Therefore, metals are important components of regulatory standards related to diverse different environmental matrices.

4.1 ENVIRONMENTAL LAW

Environmental law is more than simply a collection of statutes on environmental topics. It can best be described as an interrelated system of statutes, regulations, guidelines, factual conclusions, and case-specific judicial and administrative interpretations. The environmental law system is an organized way of using all aspects of the legal system to minimize, prevent, punish, or remedy the consequences of actions that damage or threaten the environment and public health and safety. The environmental law system, then, includes the Constitution, statutes, regulations, rules of evidence, rules of procedure, judicial interpretations, common law, and, indeed, criminal law, to the extent that these elements are being applied toward environmental ends. In summary, environmental law encompasses all environmental protections that emanate from the following sources:

- Laws, including federal and state statutes and local ordinances
- *Regulations* promulgated by federal, state, and local agencies
- *Court decisions* interpreting laws and regulations
- Common law
- U.S. Constitution and state constitutions
- International treaties

4.1.1 FEDERAL AND STATE ENVIRONMENTAL LAW

Many federal statutes establish regulatory programs under which the states have the opportunity to enact and enforce laws meeting minimum federal criteria to achieve the regulatory objectives established by Congress. States are generally the primary permitting and enforcement authorities and are subject to federal intervention only if they do not enforce effectively or rigorously enough. The laws and interpretations used to apply and enforce federal laws vary considerably from state to state and these variations may not be readily apparent. Many states provide their citizens and environment with protections beyond minimum federal criteria.

4.1.2 Environmental Regulations

Environmental statutes generally empower an administrative agency, such as the EPA, to develop and promulgate regulations. Rule making is a process of adopting regulations. Final regulations are published in the *Federal Register*. The regulations are consolidated annually into the *Code of Federal Regulations* (CFR).

4.1.3 Selected Regulatory Programs

The major federal environmental statutes define most of the substantive compliance obligations of the environmental law system. Programs created by federal statutes are aimed at protection and appropriate management of environmental systems, such as groundwaters, surface waters, and drinking water quality. Examples of federal statutory programs are summarized below.

4.1.3.1 Clean Water Act (CWA)

The CWA controls the discharge of toxic materials into surface streams. The act regulates pollution levels by setting discharge limits and water quality standards. The concept of federal discharge permits was incorporated into the *National Pollutant Discharge Elimination System* (NPDES). The EPA set up 34 industrial categories covering over 130 toxic pollutants that are discharged into surface waters. Entities responsible for discharges of these substances are required to use the *best available technology* (BAT) to achieve discharge limits. Toxic and hazardous wastes discharged directly to a receiving body of water are regulated by *NPDES permits*, whereas materials acceptable to an industrial or municipal sewer system are discharged without a federal permit. The CWA also includes guidelines to protect wetlands from dredge-and-fill activities.

4.1.3.2 Safe Drinking Water Act (SDWA)

The SDWA was established to protect groundwaters and drinking water sources. The EPA established maximum contaminant levels (MCLs) and maximum contaminant level goals (MCLGs) for each contaminant that may affect human health. The SDWA includes over 83 contaminants, grouped as inorganic chemicals, synthetic organic chemicals, and microbiological and radiological contaminants. It also regulates the injection of liquid wastes into underground wells to ensure that disposal methods do not damage the quality of groundwater and groundwater aquifers. Details of this program are discussed later in this chapter.

4.1.3.3 Resource Conservation and Recovery Act (RCRA)

The primary concern of this program is to protect groundwater supplies by creating a management system for hazardous waste, from the time it is generated until it is treated and disposed of. Waste that contains chemicals on EPA's list of toxic chemicals may be deemed hazardous waste.

4.1.3.4 Toxic Substances Control Act (TSCA)

The EPA has the authority to control the manufacture of chemicals. The TSCA bans the manufacture of *polychlorinated biphenyls* (PCBs) and also controls the disposal of these chemical substances (40 CFR, Parts 712–799).

4.1.3.5 Federal Insecticide, Fungicide, and Rodenticide Act (FIFRA)

The FIFRA controls the manufacture and use (i.e., registration process) of pesticides, fungicides, and rodenticides (40 CFR, Parts 162–180). Examples of canceled-registration chemicals include *DDT, kepone,* and *ethylene dibromide* (EDB).

4.1.3.6 Comprehensive Environmental Response, Compensation, and Liability Act (CERCLA, Superfund)

This program is designed to address the problems of cleaning up existing hazardous waste sites. CERCLA provides the EPA with "broad authority for achieving clean-up at hazardous waste sites" and the clean-ups are financed jointly by private industry and the government (*Superfund*). According to CERCLA, substances which "when released into the environment may present substantial danger to the public health or welfare or the environment" are hazardous. CERCLA establishes a list of substances that, when released in sufficient amounts, must be reported to the EPA.

The *Superfund Amendments and Reauthorization Act* (SARA) of 1986 pertains to carcinogen testing and regulations. Section 121 requires that clean-ups at Superfund sites "[a]ssure protection of human health and the environment." SARA provides authority and financing to the EPA to act quickly in the event of hazardous material spills.

Title III, Section 313 of SARA, the *Emergency Planning and Right To Know Act*, requires private-sector and public-sector facilities to report annually to the EPA on the types of hazardous substances they handle and all releases of such compounds into various media (e.g., air and water). Program enforcement is provided by state governments after receiving EPA approval.

4.2 DRINKING WATER STANDARDS

The correct definition of drinking or potable water is water delivered to the consumer that can be safely used for drinking, cooking, and washing. Regulatory agencies establish physical, chemical, bacteriological, and radiological quality standards for potable water. Water supplies in the United States and elsewhere are endangered by the introduction of new chemicals and pollutants every year. Drinking water standards in the United States, established by the EPA, reflect the best scientific and technical judgment available.

The *World Health Organization* (WHO), a U.N. agency dedicated to public health, first issued *Guidelines for Drinking-Water Quality* in 1984–1985 as a basis for developing standards that, if properly implemented, would ensure the safety of drinking water supplies. Although the main purpose of these guidelines is to provide a basis for developing standards, the guidelines are also useful to countries in implementing alternative control procedures where the implementation of drinking water standards is not feasible.

4.2.1 Safe Drinking Water Act (SDWA)

Drinking water quality is protected by laws and regulations that must be enforced. Currently about 200,000 public water systems are regulated under the *Safe Drinking Water Act* (SWDA). The rest of the population is served by private wells not subject to regulation under SDWA. Drinking water risks are the highest priority of public health issues because everyone drinks water and because so many potentially toxic substances can contaminate drinking water. In accordance with the SDWA, the EPA sets standards as close as possible to a level "at which no known or anticipated adverse effects on the health of persons occur and which allows an adequate margin of safety." Systems that fail to meet MCLs must be treated using the BAT. Under the revised SDWA, it will be easier for the EPA to ensure that the states take enforcement actions swiftly and effectively.

The federal SDWA requires a broader appreciation of the "philosophy" of water. Water utility service is distinguished from all other types of utilities in three important ways: (1) water service is the only utility essential for life; (2) unlike other utilities, water is ingested; and (3) the investment in facilities per customer to provide water service far exceeds the comparable cost for other utility services.

The content of water in terms of aesthetics (taste, color, and odor) and health-risk contaminants is the result of natural processes, external pollutants, or byproducts of accepted water treatment methodologies. For example, iron, manganese, and radium naturally occur in some groundwater. Pollutants such as nitrates and pesticides can be found in surface waters and arise from stormwater runoff and drainage. Disinfection byproducts can result from chlorination at a treatment plant pursuant to methodology accepted and mandated for a hundred years. The SDWA places the burden on water utilities to treat water content, regardless of "contamination" source.

On August 5, 1998, the EPA published guidelines on the definition of a public water system under the SDWA. In the same publication, the EPA stated that bottled and packaged water and natural bodies of water that have been altered by humans fall under the jurisdiction of the SDWA.

4.2.2 SDWA Regulations

Drinking water regulations fall into primary and secondary categories. *Primary regulations* are aimed at protecting public health, and define "clean" water. *Secondary regulations* are intended to protect the "public welfare" by offering unenforceable guidelines on the taste, odor, or color of drinking water, among other considerations. Primary and secondary drinking water standards are listed in Table 4.1.

4.2.2.1 Maximum Contaminant Levels (MCLs) and Maximum Contaminant Level Goals (MCLGs)

The MCLs are enforceable standards that must be established as close to respective MCLGs as is feasible. "Feasible" means with the use of the best technology, treatment techniques, and other available means, while taking cost into consideration. The 1986 amendments to the SDWA require the EPA to establish *national primary drinking water regulations* (NPDWRs) for 83 specified contaminants with MCLs and MCLGs. In addition, the EPA must publish a list of contaminants that may require regulation every 5 years, beginning in February 1998. At 5-year intervals, the EPA must determine whether to regulate at least five of the listed unregulated contaminants.

4.2.3 SDWA Amendments

Since 1986, *regulatory impact analyses* have been developed for amending the SDWA. The changes are discussed below.

4.2.3.1 Fluoride Studies

In 1986 and 1990, the EPA requested new toxicological studies about the health effects of fluoride to determine whether the current standard was adequate (*Fed. Reg.*, 51, 11396, April 1986; *Fed. Reg.*, 55, 160, 3 January 1990). Besides the existing 4-mg/l primary standard, the EPA established a secondary standard with an MCL of 2 mg/l. According to study results, the previous 4-mg/l MCL for fluoride is adequate as a primary standard.

4.2.3.2 Volatile Organic Compounds (VOCs) Rule

The VOCs rule that went into effect in 1989 (*Fed. Reg.*, 52, 23690, 8 July 1987; *Fed. Reg.*, 53, 25108, 1 July 1988) established standards for eight compounds. The EPA suggested new regulations, including changes in analytical methods and laboratory certification and the redesign of monitoring programs of unregulated contaminants by using targeted sampling.

4.2.3.3 Surface Water Treatment Rule (SWTR)

Promulgated in 1989, the SWTR is currently in effect. Utilities served by surface water or groundwater under the direct influence of surface water should monitor *disinfectant concentration* and *disinfectant contact time* and, based on summaries of collected data, submit a proposal for the *Disinfectant–Disinfectant By-Products Rule* (*Fed. Reg.*, 54, 27488, 29 June 1989) to set MCLGs for *Giardia,* viruses, and *Legionella.* SWTR also established treatment techniques for surfacewater supply sources and ground water under direct influence of surface water, including filtration and disinfection requirements. In addition, the rule set turbidity standards. Filtration is required unless criteria are met for avoidance (*Fed. Reg.*, 54, 27486–27541, 29 June 1989).

As required under the 1996 SDWA amendments, the *Interim Enhanced Surface Water Treatment Rule* was issued in December 1998. The purpose of the rule is to improve the control of microbial pathogens in drinking water. It is expected that this rule will further reduce the occurrence of *Cryptosporidium, Giardia,* and other waterborne bacteria or viruses in finished drinking water supplies. This rule applies to public water systems that use surface water or ground water under direct influence of surface water and serve at least 10,000 people. The rule also requires primacy states to conduct sanitary surveys for all surfacewater and groundwater systems, regardless of size.

In 2000, the EPA issued its *Long Term 1 Enhanced Surface Water Treatment and Filter Backwash Proposed Rule* (*Fed. Reg.*, 65, 19046, 10 April 2000). The purpose of the proposed rule is to increase protection of finished water from contamination by cryptosporidium and other microbial pathogens. The proposal is intended to extend the rule to small systems serving less than 10,000 people.

4.2.3.4 Groundwater Disinfection Rule

Another proposed rule that has been pending for several years provides for groundwater disinfection. In May 2000, the EPA published its proposed rule (*Fed. Reg.*, 65, 30193, 10 May 2000). Its objective is to provide a companion rule for groundwater sources of supply to the surfacewater treatment rule. Thus, the rule is likely to include MCLGs of zero, disinfection treatment techniques in lieu of MCLs, and so on. It may also include provisions for natural disinfection. The proposed rule provides a treatment that achieves a minimum 99.99% inactivation rate on virus removal. A final regulation was anticipated in November 2000. Currently, only surfacewater systems and systems using groundwater under the direct influence of surface water are required to disinfect water supplies.

4.2.3.5 Total Coliform Rule (TCR)

Promulgated in 1989 (*Fed. Reg.,* 54, 27547, 29 June 1989), the TCR is currently in effect. The rule established approved analytical methods for *Escherichia coli* bacteria. Under the TCR, microbiological samples should be iced during transportation and overviews of sampling points performed. Any coliform-positive sample should be resampled and the test repeated within 24 h of notification. The *MMO-MUG (Colilert) test* should be run on selected selected samples, and another accepted method should be run to check the effectiveness of the MMO-MUG test.

4.2.3.6 Synthetic Organic Chemicals (SOCs) and Inorganic Chemicals (IOCS)

The rule for synthetic organic chemicals (SOCs) and inorganic chemicals (IOCs) was finalized in 1991. Proposed MCLs for *aldicarb, aldicarb sulfoxide*, and *aldicarb sulfon* were scheduled for 1994.

TABLE 4.1
Drinking Water Standards

Parameters	MCl (mg/l)	Analytical Method	Detection Limit (mg/l)
Inorganics		**Primary Standards**	
Arsenic	0.05	EPA 206.2	0.0020
Barium	2.00	EPA 200.7	0.0140
Cadmium	0.005	EPA 200.7	0.0010
Chromium	0.10	EPA 200.7	0.0090
Cyanide	0.20	EPA 335.2	0.0050
Fluoride	4.00	EPA 340.2	0.01
Lead	0.015	EPA 239.2	0.0010
Mercury	0.002	EPA 245.1	0.0002
Nickel	0.100	EPA 200.7	0.0110
Nitrate nitrogen	10.00	EPA 353.2	0.01
Nitrite nitrogen	1.00	EPA 354.2	0.01
Selenium	0.05	EPA 270.2	0.0010
Sodium	160	EPA 200.7	0.226
Antimony	0.006	EPA 204.2	0.0020
Beryllium	0.004	EPA 200.7	0.0020
Thallium	0.002	EPA 279.2	0.0010
Organics			
Trihalomethanes			
Bromoform	—	EPA 502.2	0.00013
Chloroform	—	EPA 502.2	0.00005
Dibromochloromethane	—	EPA 502.2	0.00013
Dichlorobromomethane	—	EPA 502.2	0.00007
Total THMs	0.10	EPA 502.2	
Volatiles			
1,2,4-Trichlorobenzene	70	EPA 502.2	0.310
cis-1,2-Dichloroethylene	70	EPA 502.2	0.0300
Xylenes (Total)	10,000	EPA 502.2	0.170
Dichloromethane	5	EPA 502.2	1.40
o-Dichlorobenzene	600	EPA 502.2	0.140
p-Dichlorobenzene	75	EPA 502.2	0.190
Vinyl chloride	1	EPA 502.2	0.290
1,1-Dichloroethylene	7	EPA 502.2	0.170
trans-1,2-Dichloroethylene	100	EPA 502.2	0.180
1,2-Dichloroethane	3	EPA 502.2	0.0400
1,1,1-Trichloroethane	200	EPA 502.2	0.0300
Carbon tetrachloride	3	EPA 502.2	0.0400
1,2-Dichloropropene	3	EPA 502.2	0.0400
Trichloroethylene	3	EPA 502.2	0.0400
1,1,2-Trichloroethane	5	EPA 502.2	0.0400
Tetrachloroethylene	3	EPA 502.2	0.0800
Monochlorobenzene	100	EPA 502.2	0.0700
Benzene	1	EPA 502.2	0.0500
Toluene	1000	EPA 502.2	0.0800
Ethylene benzene	700	EPA 502.2	0.0600
Styrene	100	EPA 502.2	0.0700
Pesticides and PCBs	*2*	*EPA 508*	*0.01*
Lindane	0.2	EPA 508	0.01
Methoxychlor	40	EPA 508	0.02

TABLE 4.1 (Continued)

Parameters	MCl (mg/l)	Analytical Method	Detection Limit (mg/l)
Toxaphene	3	EPA 508	0.2
Dalapon	200	EPA 515.1	1
Diquat	20	EPA 549	4
Endothal	100	EPA 548	10
Glyphosate	700	EPA 547	10
Di(2-ethylhexyl)adipate	400	EPA 506	1
Oxamyl (Vydate)	200	EPA 531.1	0.5
Simazine	4	EPA 507	0.1
Picloram	500	EPA 515.1	0.2
Dinoseb	7	EPA 515.1	0.2
Hexachlorocyclo-pentadiene	15	EPA 512	0.1
Carbofuran	40	EPA 531.1	0.5
Atrazine	3	EPA 507	0.1
Alachlor	2	EPA 507	0.3
2,3,7,8-TCDD (Dioxin)	0.00003		
Heptachlor	0.4	EPA 508	0.01
Heptachlor epoxide	0.2	EPA 508	0.01
2,4-D	70	EPA 515.1	0.5
2,4,5-T (Silvex)	50	EPA 515.1	0.05
Hexachlorobenzene	1	EPA 508	0.01
Di(2-ethylene hexyl)-phthalate	6	EPA 506	1
Benzo(a)pyrene	0.2	EPA 550	0.01
Pentachlorophenol	1	EPA 515.1	0.05
PCB	0.5	EPA 508	0.05
Dibromochloropropane	0.2	EPA 504	0.005
Ethylene dibromide	0.02	EPA 504	0.005
Chlordane	2	EPA 508	0.05
Radiological analysis			
Gross alpha	5 pCi/l	EPA 900.0	—
Radium-226	15 pCi/l	EPA 900.0	—
Radium-228	50 pCi/l	EPA 900.0	—
Microbiology			
Total coliform	Zero count/100 ml		
Secondary Standards			
Aluminum	0.200	EPA 200.7	0.100
Chloride	250	EPA 300.0	0.5
Copper	1.00	EPA 200.7	0.0040
Iron	0.30	EPA 200.7	0.0500
Manganese	0.05	EPA 200.7	0.0050
Silver	0.10	EPA 200.7	0.0050
Sulfate	250	EPA 300.0	0.0100
Zinc	5.00	EPA 200.7	0.0140
Color	15 C.U.	SM 204A	5.00
Odor	3 TON	SM 207	1.00
pH	6.5–8.5	EPA 150.1	—
Total dissolved solids (TDSs)	500	EPA 160.1	10.0
Foaming agents	0.5	SM 512B	0.0500

Note: SDWA regulations are not health related. They are intended to protect the "public welfare" by offering unenforceable guidelines on the taste, odor, or color of drinking water. Recommended levels are intended mainly to maintain and provide aesthetic and taste characteristics.

MCl = maximum contaminant level; CU = color unit; TON = threshold odor number; pCi/l = picoCurie per liter; µg/l = micrograms per liter; 2,4-D = dichlorophenoxyacetic acid; 2,4,5-T = trichlorophenoxyacetic acid; PCD = polychlorinated biphenyls.

In 1993, the MCL and MCLG for *atrazine* were revised at the request of Ciba-Geigy, the manufacturer of this chemical (*Fed. Reg.*, 56, 3600, 30 January 1991; *Fed. Reg.*, 56, 30266, 1 July 1991).

4.2.3.7 Lead and Copper Rule

This rule defines the action level for lead and copper, establishing monitoring requirements for corrosion control, selecting sampling sites, issuing deadlines for public-education information, requiring monitoring data to be reported to the state, and clarifying which certified laboratories must be used. Monitoring for lead and copper requires the collection of first-draw water samples at taps within consumers' premises. However, most lead and copper content in finished water results from piping, soldering, fixtures, and appliances within consumers' premises over which water utilities have no control. The rule shifts the responsibility for these conditions from consumers to the utility. It imposes on a utility the obligation to proactively control its water through such corrosion control techniques as adjustment to pH, alkalinity, and calcium and additions of phosphates and silicates.

Under the rule, the MCLG for lead is zero, and the action level is 0.015 mg/l. For copper, both the MCLG and action level are 1.3 mg/l, with a nonenforceable MCLG of 1.0 mg/l.

The EPA made what it described as "minor changes" to the lead and copper rule in January 2000. The changes are summarized below:

Clarifications for systems that optimize corrosion control and continue to maintain and operate any corrosion control already in place

Requirement for utilities subject to replacing the lead service-line portions they own to notify residents of lead-level potential in drinking water where the service line is only partially replaced

Revisions of analytical methods and monitoring and reporting requirements

A single national standard for lead is not suitable for every public water system because the conditions of plumbing materials, which are the major source of lead in drinking water, vary across systems and the systems generally do not have control over the sources of lead in their water. In these circumstances, the EPA suggests that requiring public water systems to design and implement customized corrosion control plans for lead will result in optimal treatment of drinking water overall, that is, treatment that deals adequately with lead without causing public water systems to violate drinking water regulations for other contaminants (*Fed. Reg.*, 56, 26487).

4.2.3.8 Sulfate Standard

Sulfates appear to have no adverse chronic health effects. The only impacts are diarrhea and resulting dehydration. The EPA has issued a secondary MCL of 250 mg/l for sulfates. Sulfates are included on the EPA's first list of contaminants for possible regulation. Under the current secondary MCL, the utility should provide public education to protect infants, new residents, and tourists. Bottled water can solve this problem.

4.2.3.9 Arsenic Proposal

One of the EPA's most controversial proposals pertains to arsenic: MCLG of zero and MCL of 0.005 mg/l (*Fed. Reg.*, 65, 38887, 22 June 2000). Arsenic can occur naturally as well as in industrial emissions and effluents. The EPA's proposed minimum levels have been criticized as lacking a scientific basis and being too rigorous upon consideration of compliance costs.

4.2.3.10 Radio Nuclides

The EPA was under court order to promulgate a uranium NPDWR by November 2000. A draft guidance manual pertaining to the anticipated rule on radionuclides was released May 3, 2000. The primary concerns that delayed the issuance of a final rule were the costs and benefits of regulating radon.

4.2.4 National Secondary Drinking Water Regulations (NSDWRs)

The NSDWRs relate to the aesthetics of water, not health effects. These regulations specify maximum levels of a component to ensure a color, taste, or odor that will not cause users to discontinue its use. Secondary maximum contaminant levels (SMCLs) do not cause health risks. At levels above SMCLs, the contaminants may cause users to perceive water to have adverse aesthetic effects, including taste, color, odor, and cosmetic impacts, such as skin or tooth discoloration, staining, and corrosivity. SMCLs are not enforceable as a matter of federal law. However, some states have adopted SMCLs, or regulations above or below SMCLs, as enforceable standards. For example, complaints about iron staining (iron content higher than NSDWRs of 0.3 mg/l constitutes a violation) are common at the state level.

4.3 SURFACEWATER STANDARDS

Freshwater ecosystems fall into two categories — *lakes* and *ponds*, and flowing systems, such as *rivers* and *streams*. Lakes and ponds are more susceptible to pollution because the water is replaced at a slow rate. Complete replacement of a lake's water may take 10 to 100 years or more, and during these years pollutants may build up to toxic levels. In rivers and streams, the water flow easily purges pollutants. If the pollution is continuous and distributed uniformly along river and stream banks, the cleaning effect by purging does not work well.

Rivers, streams, and lakes contain many organic and inorganic nutrients needed by the plants and animals that live in them. These nutrients in higher concentrations may become pollutants. Organic pollutants derive from feedlots, sewage treatment plants, and certain food-processing industries (dairy products, meat packing, etc.). The increased organic matter stimulates the growth of bacteria, which in turn consume the organic matter, and thus help clean up pollution. Unfortunately, bacteria use up oxygen and therefore reduce dissolved oxygen in the water. The lack of dissolved oxygen kills fish and other aquatic organisms, and the aerobic (oxygen-requiring) bacteria population changes to anaerobic (nonoxygen-requiring) bacteria. Anaerobic bacteria produce foul-smelling and toxic gases such as methane and hydrogen sulfide. This process in rivers and streams occurs more readily during the hot summer months. When the organic pollutants are used up, and additional pollutants do not enter the water body, oxygen levels return to normal via oxygen from the air and oxygen released by plants during photosynthesis.

Organic pollutants nourish bacteria and certain inorganic pollutants stimulate the growth of aquatic plants. These pollutants are called *nutrients*, and include nitrogen as ammonia and nitrate, and phosphorus as phosphates. These compounds derive from fertilizers, laundry detergents, and sewage treatment plants. High levels of these nutritional compounds can lead to the dense growth of aquatic plants and thick mats of algae covering lakes and rivers. Excessive plant growth negative affects fishing, swimming, boating, and navigation activities. Aerobic bacteria decompose these plants when they die. The lowered dissolved oxygen content of the water kills aquatic organisms and leads to anaerobic bacteria growth, which in turn produces odorous and toxic gases. Thus, inorganic and organic pollutants cause the same problems in surface waters.

Classification of surface waters is based on water quality and use. The five main groups of surface waters are listed below:

Class I: Potable water supplies
Class II: Shellfish propagation or harvesting
Class III: Recreation — propagation and maintenance of healthy, well-balanced population of fish and wildlife
Class IV: Agricultural water supply
Class V: Navigation, utility, and industrial use

Groundwater contamination via flow from surfacewater is well known. surfacewater flows from open bodies (rivers and lakes) can enter into aquifers where groundwater levels are lower than surfacewater levels. The opposite situation — ground water contaminating surface water — is also possible, and occurs when the water table is high or the surface water is lowered by pumping wells.

Monitoring, maintaining, and regulating the quality of surface waters is the responsibility of state governments.

4.3.1 Clean Water Act (CWA)

The CWA is the primary federal statute that addresses water pollution in the United States. The *Refuse Act* of 1899 was the first federal law affecting water pollution. The Refuse Act, while not a major element of the current federal water pollution control program, is still in effect. The roots of the CWA can be traced to the *Federal Water Pollution Control Act* of 1972. Amendments to the act in 1987 created new programs for controlling toxins, established stormwater regulation, strengthened water-quality-related requirements, and established a loan fund for construction of sewage treatment plants. In 1990, in response to the Exxon *Valdez* oil spill, Congress overhauled the oil spill provisions of the act in the *Oil Pollution Act* of 1990, sometimes referred to as OPA 90.

4.3.1.1 CWA Objectives, Goals, and Policy

The objective of the CWA is to "restore and maintain the chemical, physical, and biological integrity of the nation's waters." To achieve this objective, the act establishes the following goals:

- Elimination of the discharge of pollutants into surfacewaters
- Achievement of a level of water quality that "provides for the protection and propagation of fish, shellfish and wildlife" and "for recreation in and on the water"

The act also establishes a national policy, which states that "the discharge of toxic pollutants in toxic amounts shall be prohibited."

4.3.1.2 Pollutants as Defined by CWA

As defined in the CWA, pollutants include dredged spoil; solid waste; incinerator residue; sewage; garbage; sewage sludge; munitions; chemical wastes; biological materials; heat; wrecked or discarded equipment; rock; sand; cellar dirt; and industrial, municipal, and agricultural waste discharged into water. Despite this specific definition, the term has been broadly interpreted by the courts to include virtually any material, as well as characteristics such as toxicity and acidity.

4.3.1.3 Point Source as Defined by CWA

According to the CWA, a point source is "any discernable, confined and discrete conveyance ... from which pollutants are or may be discharged." This definition has been interpreted to cover almost any natural or manufactured conveyance from which a pollutant may be discharged, including pipes, ditches, erosion channels, and gullies. Vehicles, such as bulldozers or tank trucks, have also been included among point sources. Human beings are not point sources, at least for purposes of criminal enforcement of the act. In other words, a person dumping pollutants into a water body, other than through hose or pipe, for example, would not be in violation of the act's prohibition of discharges from point sources without a permit. The person may, however, be in violation of other laws and regulations.

4.3.1.4 National Pollutant Discharge Elimination System (NPDES) Permit

The NPDES permit program implements the CWA prohibition on unauthorized discharges by requiring a permit for every discharge of pollutants in U.S. waters. Permits, which are issued by the EPA or authorized state government agencies, give the permittee the right to discharge specified pollutants from specified outfalls, normally for a period of 5 years. Currently, 14 states and territories have received permitting authority (40 CFR, 123.24). The implementation and enforcement of the NPDES program depend to a large extent on self-monitoring. Permits require dischargers to monitor their own compliance with permit limitations on a regular basis and to report the results of this monitoring to the permitting authority.

4.3.1.5 Water Quality Standards

Water quality standards are established by the states. The CWA requires all states to classify the waters within the state according to intended use (see Section 4.3).

Water quality criteria quantitatively describe the physical, chemical, and biological characteristics of waters necessary to support designated uses. State criteria are normally based on federal water quality criteria, which have been published for more than 150 pollutants. The EPA published a compilation of its criteria for 157 pollutants in 1998 (*Fed. Reg.*, 63, 67547, 7 December 1998). Normally, a state water quality standard consists of a numeric level of a pollutant that cannot be exceeded in the ambient water in order to protect the designated use. For example, the standard may state that the level of arsenic in a stream designated for trout propagation may not exceed 0.2 mg/l.

4.3.2 EPA Priority Toxic Pollutants

According to the *Federal Pollution Control Act*, the EPA should study particular chemical compounds and classes of compounds for the development of regulations to control discharges into wastewater, using the BAT that is financially viable. Of these 129 priority pollutant compounds, 114 are organic and 15 inorganic. Metals account for 13 of the 15 inorganic pollutants (see Table 4.2).

4.4 AGRICULTURALLY USED WATERS

Water used for irrigation should be free from high salinity and toxic substances. Table 4.3 presents the list of analytical parameters necessary to evaluate *irrigation water quality*. Of particular interest is the ratio of sodium to calcium and magnesium. When sodium-rich water is applied to soil, some of the sodium is taken up by clay and the clay gives up calcium and magnesium in exchange. Clay that takes up sodium becomes sticky and slick when wet and has low permeability. The dry clay shrinks into hard clods that are difficult to cultivate. In other words, when sodium is absorbed into

TABLE 4.2
Priority Toxic Pollutants

Halogenated methanes
Methyl bromide
Methyl chloride
Methylene chloride (dichloromethane)
Bromoform (tribromomethane)
Chloroform (trichloromethene)
Bromodichloromethane
Chlorodibromomethane
Carbon tetrachloride (tetrachloromethane)

Chlorinated hydrocarbons
Chloroethane (ethyl chloride)
Chloroethylene (vinyl chloride)
1,2-Dichloroethane (ethylenedichloride)
1,1-Dichloroethane
1,2-*trans*-Dichloroethylene
1,1-Dichloroethylene (vinylidene chloride)
1,1,2-Trichloroethane
1,1,1-Trichloroethane
Trichloroethylene
Tetrachloroethylene
1,1,2,2-Tetrachloroethane
Hexachloroethane
1,2-Dichloropropane
1,3-Dichloropropylene
Hexachlorobutadiene
Hexachlorocyclopentadiene

Chloroalkyl ethers
bis-(2-Chloroethyl) ether
bis (2-Chloroisopropyl) ether
2-Chloroethylvinyl ether
bis-(2-Chloroethoxy) methane

Haloaryl ethers
4-Chlorophenyl phenyl ether
4-Bromophenyl phenyl ether

Nitrosamines
N-nitrosodimethyl amine
N-nitrosodiphenyl amine
N-nitrosodi-*n*-propyl amine

Nitroaromatics
Nitrobenzene
2,4-Dinitrotoluene
2,6-Dinitrotoluene

Phenols
2,4-Dimeethylphenol

Nitrophenols
2-Nitrophenol
4-Nitrophenol
2,4-Dinitrophenol
4,6-Dinitro-*o*-cresol
Chlorophenols
2-Chlorophenol
4-Chloro-*m*-cresol
2,4-Dichlorophenol
2,4,6-Trichlorophenol
Pentachlorophenol
2,3,7,8-Tetrachlorodibenzol-*p*-dioxin (TCDD)

Benzidines, hydrazine
Benzidine
3,3-Dichlorobenzidine
1,2-Diphenylhydrazine

Phtalate esters
bis-(2-Ethylhexyl) phtalate
Butylbenzyl phtalate
Di-*n*-butyl phtalate
Di-*n*-octyl phtalate
Diethyl phtalate
Dimethyl phtalate

Aromatics
Benzene
Toluene
Ethylbenzene

Polyaromatics
Naphthalene
Acenaphthene
Acenaphthylene
Anthracene
Benzo(*a*)anthracene (1,2-benzanthracene)
Benzo(*a*)pyrene (3,4-benzopyrene)
3,4-Benzofluoranthene (11,12-benzofluoranthene)
Benzo(ghi)perylene (1,12-benzoperylene)
Chrysene
Dibenzo(*a,h*)anthracene (1,2,5,6-dibenzoanthracene)
Fluorene
Fluoranthene
Indenol(1,2,3-*od*)pyrene (2,3-*o*-phenylene pyrene)
Phenanthrene
Pyrene

Chloroaromatics
Chlorobenzene
o-Dichlorobenzene
p-Dichlorobenzene
m-Dichlorobenzene
1,2,4-Trichlorobenzene
Hexachlorobenzene
2-Chloronaphthalene

TABLE 4.2 (Continued)

Polychlorinated Biphenyls (PCBs)	***Miscellaneous***
PCB-1016 (Aroclor 1016)	Acrolein
PCB-1221 (Aroclor 1221)	Acrylonitrile
PCB-1232 (Aroclor 1232)	Isophorone
PCB-1242 (Aroclor 1242)	Asbestos
PCB-1248 (Aroclor 1248)	Cyanide
PCB-1284 (Aroclor 1284)	Metals
PCB-1260 (Aroclor 1260)	Antimony
	Arsenic
Pesticides	Beryllium
Aldrin	Cadmium
Dieldrin	Chromium
Chlordane	Copper
α-Endosulfate	Lead
Endrin	Mercury
Endrin aldehyde	Nickel
Heptachlor	Selenium
Heptachlor epoxide	Silver
α-BHC	Thallium
β-BHC	Zinc
γ-BHC	
δ-BHC	
4,4-DDT	
4,4-DDE (*p,p*-DDX)	
4,4-DDO (*p,p*-TDE)	
Toxaphene	

clay particles, it turns the clay into a cement-like solid that neither water nor roots can penetrate. High concentrations of sodium salts can produce alkali soils in which little or no vegetation can grow. On the other hand, when the same clay carries excess calcium and magnesium ions, it tills easily and has good permeability. If irrigation water contains calcium and magnesium ions sufficient to equal or exceed the sodium ion, enough calcium and magnesium are retained in clay particles to maintain good tilth and permeability.

The sodium effect can be calculated by the *sodium absorption ratio* (SAR) method:

$$SAR = [Na]/([Ca] + [Mg])/2 \tag{4.1}$$

where the [Na], [Ca], and [Mg] values are expressed in milliequivalents per liter.

Waters with SAR values below 10 are acceptable for irrigation, and waters with SAR values of 18 or higher are not recommended for irrigation. Table 4.3 contains the recommended maximum concentrations of trace elements in irrigation water.

4.5 INDUSTRIAL WATERS

Quality requirements for industrial use vary widely according to potential use. *Industrial process waters* must be of much higher quality than *cooling waters* (especially if they are used only once). Municipal supplies are generally good enough to satisfy the quality requirements of most process waters, with the exception of waters used for boilers. *Boiler waters* are specially checked and treated for quality. Silica is an important constituent of the encrusting material or scale formed by many

TABLE 4.3
Recommended Maximum Concentrations of Trace Elements in Irrigation Water

	For Waters Continuously Used on Soils (mg/l)	For Waters Used up to 20 Years on Fine-Textured Soils, pH 6.0–8.5 (mg/l)
Aluminum (Al)	5.00	20.00
Arsenic (As)	0.10	2.00
Beryllium (Be)	0.10	0.50
Boron (B)	[a]	2.00
Cadmium (Cd)	0.01	0.05
Chromium (Cr)	0.10	1.00
Cobalt (Co)	0.05	5.00
Copper (Cu)	0.20	5.00
Fluoride (F)	1.00	15.00
Iron (Fe)	5.00	20.00
Lead (Pb)	5.00	10.00
Lithium (Li)	2.50	2.50
Manganese (Mn)	0.20	10.00
Molybdenum (Mo)	0.01	0.05[b]
Nickel (Ni)	0.20	2.00
Selenium (Se)	0.02	0.02
Vanadium (V)	0.10	1.00
Zinc (Zn)	2.00	10.00

[a] No problem <0.75 mg/l; increasing problem, 0.75–2.00 mg/l; severe problem, >2.00 mg/l.
[b] Only for acidic, fine-textured soils with relatively high iron oxide content.

Source: National Academy of Sciences and National Academy of Engineering, 1972; Driscoll, F.G., *Groundwater and Wells*, 2nd ed., Johnson Division, St. Paul, MN, 1987. With permission.

waters. As a deposit, the scale commonly consists of calcium or magnesium silicate. Silicate scale cannot be dissolved by acids or other chemicals. Therefore, silica-rich water used in boilers must be treated. Sanitary requirements for waters used in processing milk, canned goods, meats, and beverages exceed even those in drinking water.

4.6 WASTE CHARACTERIZATION

The few characteristic properties that qualify waste material under the Resource Conservation and Recovery Act (see Section 4.3.1) are ignitability, corrosivity, reactivity, and toxicity.

Ignitability: This property refers to the characteristics of being able to sustain combustion, including flammability (ability to start fires when heated to temperatures of less than 60°C or 140°F).

Corrosivity: Corrosive wastes may destroy containers, soil, and ground water or react with other materials to cause toxic gas emissions. Corrosive materials provide a very specific hazard to human tissue and aquatic life when pH levels are extreme.

Reactivity: Reactive wastes may be unstable or have a tendency to react, explode, or generate pressure during handling. Pressure-sensitive or water-reactive materials are also included in this category.

Toxicity: Toxicity is an effect of waste materials that may come into contact with water or air and be leached into groundwater or dispersed in the environment. Toxic effects on humans, fish, or wildlife are the principal concerns.

4.7 HAZARDOUS WASTE CHARACTERIZATION

The Resource Conservation and Recovery Act (RCRA) and its amendment, the *Hazardous and Solid Waste Act*, deal with management of solid wastes with an emphasis on hazardous wastes. The goal of the RCRA program is to regulate all aspects of hazardous waste management, from production through treatment and disposal. These wastes include toxic substances, caustics, pesticides, and flammable, corrosive, and explosive materials.

4.7.1 Criteria for Hazardous Waste Evaluation

The criteria for evaluating hazardous waste are as follows:

Ignitability: Flashpoint less than 60°C (less than 140°F)
Corrosivity: pH less than 2.00 or higher than 12.00
Reactivity: Reacts violently or generates pressure; the substance should be free from cyanide (CN) and sulfide (S)
Toxicity: Leaching test — extraction procedure toxicity (EPTOX) and toxicity characteristic leachate procedure (TCLP) — parameters should meet MCLs

TABLE 4.4
Maximum Concentration of Contaminants in Characterization of EP Toxicity

Contaminant	Maximum Concentration (mg/l)
Arsenic (As)	5.0
Barium (Ba)	100.0
Cadmium (Cd)	1.0
Chromium (Cr)	5.0
Lead (Pb)	5.0
Mercury (Hg)	0.2
Selenium (Se)	1.0
Silver (Ag)	5.0
Endrin	0.02
Lindane	0.4
Methoxychlor	10.0
Toxaphene	0.5
2,4-D	10.0
2,4,5-TP Silvex	1.0

Note: The EP toxicity test (EPTOX) was developed to characterize hazardous wastes based on the leaching ability of toxic substances in significant concentrations. 2,4-D = 2,4-Dichlorophenoxyacetic acid; 2,4,5-TP = 2,4,5-trichlorophenoxyacetic acid; EP toxicity = extraction procedure toxicity.

The characterization of hazardous wastes is based on their leaching ability of toxic substances in significant concentrations. In the EPTOX test, the liquid extract or leachate of the material is analyzed for 14 parameters: 8 metals, 4 insecticides, and 2 herbicides. During the migration of the leachate, attenuation and dilution occur with the ratio factor of 100, which is used to establish the maximum concentration level (100 times higher than drinking water standards). Maximum concentrations of contaminants in EPTOX leachate are presented in Table 4.4. The EPA developed the EPTOX test in 1980 (40 CFR, 261.24). (The EPTOX procedure is discussed in Chapter 14.)

In 1986, the EPA expanded the EPTOX characteristic substances by adding 38 organic pollutants. The new procedure is called the *toxicity characteristic leachate procedure* (TCLP). By the application of the TCLP test, the leachate of the waste material containing any of these 52 substances at or above the regulatory level qualifies as hazardous, toxic waste. The TCLP test uses compound-specific dilution/attenuation factors instead of the 100 used in the EPTOX test. The extraction procedure is the same as specified for the EPTOX test. Contaminants and regulatory levels are list in Table 4.5.

4.8 AIR POLLUTION AND CONTROL

4.8.1 Primary and Secondary Air Pollutants

People have known for centuries that air carries "poisons." Coal miners used to take canaries with them into the mine because the death of a bird meant the presence of toxic gases. An important exposure route to hazardous materials is air, and the effects of airborne hazardous materials frequently appear at a great distance from pollution sources. The atmosphere contains hundreds of air pollutants from natural and anthropogenic sources, known as *primary pollutants*. By using the energy from the sun, primary pollutants react with one another or with water vapor in the air and produce dangerous new chemical substances called *secondary pollutants*. These reactions are called *photochemical reactions* because they involve sunlight and chemicals, resulting in a brownish-orange shroud of air pollution called *photochemical smog*. Secondary pollutants include ozone, formaldehyde, peroxyacylnitrate, sulfuric acid, and nitric acid (causes of *acid rain*). Acute health effects include burning or itching eyes and irritated throats, and chronic effects include bronchitis, emphysema, and lung cancer.

4.8.2 Clean Air Act (CAA)

Air pollution control began in 1955. However, the *Clean Air Act* of 1970 (amended in 1975 and 1977) marked the beginning of attempts at effective controls. The two broad regulatory classifications of air pollutants are criteria and noncriteria pollutants.

4.8.2.1 Criteria Pollutants

Federal ambient air quality standards have been established for criteria pollutants, which include gases in the form of nitrogen oxides, ozone, sulfur dioxide, carbon monoxide, and solids in the form of particulate matter and lead (as particulates).

4.8.2.2 Noncriteria Pollutants

Federal ambient air quality standards have not been established for noncriteria pollutants (toxic air contaminants), which include practically every other compound or element that could have an impact on human health or the environment.

TABLE 4.5
Toxic Characteristic Leachate Pollutants (TCLPs) and Regulatory Levels

Contaminant	Regulatory Level (mg/l)
Organics	
Acrylonitrile	5.0
Benzene	0.07
bis-(2-Chloroethyl) ether	0.05
Carbon disulfide	0.07
Carbon tetrachloride	0.03
Chlordane	0.03
Chlorobenzene	1.4
Chloroform	0.07
o-Cresol	10.0
m-Cresol	10.0
p-Cresol	10.0
2,4-D	1.4
1,2-Dichlorobenzene	4.3
1,4-Dichlorobenzene	10.8
1,2-Dichloroethane	0.40
1,3-Dichloroethylene	0.10
2,4-Dinitritoluene	0.13
Endrin	0.003
Heptachlor (and its hydroxide)	0.001
Hexachlorobenzene	0.13
Hexachlorobutadiene	0.72
Hexachloroethane	4.3
Isobutanol	36.0
Lindane	0.06
Methoxychlor	1.4
Methylene chloride	6.6
Methyl ethyl ketone	7.2
Nitrobenzene	0.13
Pentachlorophenol	3.6
Phenol	14.4
Pyridine	5.0
1,1,1,2-Tetrachloroethane	10.0
1,1,2,2-Tetrachloroethane	1.3
Tetrachloroethylene	0.1
2,3,4,6-Tetrachlorophenol	1.5
Toluene	14.4
Toxaphene	0.07
1,1,1-Trichloroethane	30.0
1,1,2-Trichloroethane	1.2
Trichloroethylene	0.07
2,4,5-Trichlorophenol	5.8
2,4,6-Trichlorophenol	0.30
2,4,5-TP (Silvex)	0.14
Vinyl chloride	0.05
Metals	
Arsenic (As)	5.0
Barium (Ba)	100.0
Cadmium (Cd)	1.0
Chromium (Cr)	5.0
Lead (Pb)	5.0
Mercury (Hg)	0.2
Selenium (Se)	1.0
Silver (Ag)	5.0

Note: In 1986, the EPA expanded the EP toxicity characteristic substances (Table 4.3), which included 8 metals, 4 insecticides, and 2 herbicides, to encompass an additional 38 organic substances. The new procedure is called the toxic characteristic leachate procedure (TCLP) test. Through the application of the TCLP test, the extract or leachate of the waste containing any of these 52 substances at or above the regulatory level qualifies as hazardous toxic waste.

Sources: For parameters and regulatory levels, see U.S. Environmental Protection Agency, "Hazardous Waste Management System," 51 CFR, 114, 13 June 1986. For updated TCLP procedure, see 51 CFR, 114, 13 June 1986. For earlier version, see 40 CFR, 261.24, 19 May 1980.

4.8.2.3 Air Quality Regulations

In October 1966, the EPA issued its decision not to set a short-term National Ambient Air Quality Standard (NAAQS) for NO_2 (*Fed. Reg.*, 61, 52852, 1996).

More important, on May 22, 1996, the EPA promulgated a decision not to tighten the NAAQS for SO_2 (*Fed. Reg.*, 61, 25566, 1996). This decision followed an EPA proposal dated November 1994 to revise the SO_2 ambient standard to include a 0.06-ppm, 5-min average standard. Instead of tightening the NAAQS for SO_2, on January 2, 1997, the EPA proposed a program for monitoring and regulation of the 5-min average peak SO_2 concentration in the "emergency powers" provision. On January 30, 1998, in response to a petition from the American Lung Association, the D.C. Circuit Court set aside the EPA's decision on the NAAQS for SO_2 as inadequately justified.

In 1997, the EPA issued proposed rules substantially tightening the NAAQS for particulate matter (PM) and ozone (see *Fed. Reg.*, 62, 38856, 1997, for ozone; *Fed. Reg.*, 62, 38652, 1997, for PM). The EPA's PM rules addressed fine particles of 2.5 microns or less (i.e., PM-2.5) and contain an annual standard of 15 $\mu g/m^3$ (mean) and a 24-h standard of 65 $\mu g/m^3$.

The PM-2.5 standards would result in many new nonattainment areas. Because gaseous emissions react in the atmosphere to form PM-2.5, these new standards established new, more stringent sulfur dioxide (SO_2), nitrogen oxide (NO_x), and volatile organic compound (VOC) emission controls for many industries.

At the same time it promulgated the PM-2.5 standard, the EPA also proposed a new, more stringent NAAQS for ozone of 0.08 ppm, using an 8-h average, with compliance determined on the basis of the third-highest reading. In addition, the EPA issued a new secondary NAAQS for ozone at the same level as the primary NAAQS.

The Clean Air Act gives each state primary responsibility for ensuring that emissions from sources within its borders (including emissions that remain within and travel beyond state borders) are maintained at a level consistent with the NAAQS. This is achieved through the establishment of source-specific requirements in *state implementation plans* that address primary and secondary air quality standards.

4.8.2.4 Specific Noncriteria Standards

Under the 1990 amendments, ozone nonattainment areas are designated as marginal, moderate, serious, severe, or extreme, depending on the severity of the problem. Marginal areas are required to attain the ozone NAAQS within 3 years of enactment of the 1990 amendments, moderate areas within 6 years, serious areas within 9 years, severe areas within 15 years (in some cases, 17 years), and extreme areas within 20 years. CO nonattainment areas are designated as either moderate or serious. Moderate areas had to attain the CO standard by 1995, and serious areas by 2000. Under the 1990 amendments, all PM-10 areas initially were classified as moderate. Serious PM-10 areas were given until 2001 to attain the standard.

4.8.3 Ambient Air Quality Standard (AAQS)

This standard addresses contaminant levels above which adverse health effects occur. Air pollution regulation is focused on pollutant sources. Air pollution sources are classified as follows:

1. *Mobile sources*, including engines, usually associated with transportation (e.g., automobiles, airplanes, trucks, trains, and ships)
2. *Stationary sources*, such as pipelines, factories, boilers, storage vessels, and storage tanks; these sources are classified as *point sources* (e.g., chimneys) and *area sources* (e.g., parking lots and industrial facilities)

The federal government has primary authority to regulate emissions from mobile sources. Regulations for automobile emission controls have become more stringent as increasingly effective technologies emerge. The use of catalytic converters and unleaded gasoline has been a great step forward in the development of better air quality.

To regulate stationary sources, the EPA sets national stationary standards, known as the new source performance standards. The federal government adopts these emission standards on an industry-specific basis for all new sources of air-contaminant-emitting equipment or processes located anywhere in the United States. Local authorities under the jurisdiction of the respective state control these standards. The inspection and maintenance of vehicles for air emissions are also regulated by state laws.

4.9 ISO 14001 AND ENVIRONMENTAL LAW

4.9.1 Environmental Management Systems (EMSs)

Environmental management systems (EMSs) are applications of well-accepted business principles to environmental protection. EMSs identify key issues, establish what to do (policy and objectives), determine how to do it (programs, procedures, and instructions), tell people what to do (communication and training), make sure they do it (implementation, measurement, and auditing), and periodically review the entire process to identify opportunities for improvement. EMSs focus on establishing programs and procedures to integrate environmental performance into everyday operations so that organizations "do it right the first time."

4.9.2 ISO 14001 EMS Standard

ISO 14001, a voluntary, comprehensive EMS standard published by the International Organization on Standards in late 1999, is intended to assist organizations in identifying and meeting their environmental obligations and commitments. The popularity of EMSs is reflected in the rapid and widespread acceptance of ISO 14001. By mid-2000, over 15,000 organizations worldwide had implemented EMSs that were third-party certified as conforming to the ISO 14001 EMS standard, and countless other organizations have been using the standard. Nearly 1000 organizations in the United States have already been certified as conforming to ISO 14001, and this number is expected to increase dramatically.

5 Fundamentals of Spectroscopy

5.1 EARLY HISTORY OF THE NATURE OF LIGHT

For millennia, people have been curious about the nature of light, and particularly of color. From the time of the ancient Greeks to the seventeenth century, scholars and others believed that colors consisted of a mixture of white light and darkness and could be changed by changing the mixture.

This view was radically changed by the work of Isaac Newton (1642–1727). At the age of 24, he began his research on light. In a well-known experiment, a ray of sunlight was passed through a hole into a darkened room and onto a screen. A prism, placed in the beam of the light, dispersed the light into a spectrum of colors in the order red, yellow, green, blue, and violet. Newton concluded that white light is a "confused aggregate of rays imbued with all sort of colors." The function of the prism was merely to separate the light into its component colors.

No more discoveries occurred until 1800, when British astronomer William Herschel discovered the infrared portion of the solar spectrum. Soon after, the ultraviolet part of the spectrum was identified.

In 1802, scientists reported dark lines in the sun's spectrum, but could not provide a satisfactory explanation. In 1817, Joseph Fraunhofer, an optician and instrument maker, noted the same lines. With improved equipment, he proceeded to map the dark lines of the solar spectrum, and calculated the corresponding wavelengths. Still known as *Fraunhofer lines*, this phenomenon is described as "dark lines in the solar spectrum that result from the absorption by elements in the solar chromosphere of some of the wavelengths of the visible radiation emitted by the hot interior of the sun" (*A Concise Dictionary of Chemistry,* 1990, p. 127). During the late eighteenth and early nineteenth centuries, Fraunhofer and others looked at spectra emitted by flames and sparks, and compared them to the spectra emitted by the sun.

During the first half of the nineteenth century, a good deal of experimentation took place with the colored flames produced by injecting various salts into a flame. When light was passed through a slit and prism onto a screen, bright discrete lines were seen against a dark background, the reverse of the solar spectrum. The connection between the two was not made for many years. Robert Bunsen, professor of chemistry at the University of Heidelberg (designer of the Bunsen gas burner), viewed the exhibited colored flames by different salts through a spectroscope. He noted that the colors were linked to the element, not the compound in which it was bound. He realized that the bright lines in the visible region of the spectrum seen with a spectroscope were characteristic of specific elements, and that the method could be used as an extremely sensitive and simple method of element identification. With this new method, Bunsen identified and isolated two new elements, cesium (Cs) and rubidium (Rb).

Gustav Kirchhoff, a professor of physics at Heidelberg University, became interested in Bunsen's work. Kirchhoff examined the dark lines of the spectrum and concluded that the appearance of these lines are due to a process of absorption as the emission rays pass through the cool outer layer of the sun's atmosphere, which causes them to show up dark against the bright background. This phenomenon, called the *absorption spectrum,* is just as characteristic of a specific element as its *emission spectrum.* The effectiveness of Bunsen and Kirchhoff's spectroscopy in chemical analysis was first used as a qualitative method. Modern quantitative methods did not emerge until about the 1920s, when suitable commercial optical equipment began to appear. The method used at this time was emission spectroscopy.

In 1939, Woodson was the first to apply the absorption procedure in the quantitative measurement of elements when he identified characteristics of mercury. *Atomic absorption spectroscopy* was born in 1955, when two independently published papers described the method.

5.2 ELECTROMAGNETIC RADIATION

One of the ways that energy travels through space is electromagnetic radiation. The light from the sun, the energy used for cooking food in a microwave oven, the x-rays used in the medical field, and the radiant heat from a fireplace are all examples of electromagnetic radiation. Although these forms of radiant energy seem quite different, they all exhibit the same type of wavelike behavior and travel at the speed of light in a vacuum. Waves have three primary characteristics: wavelength, frequency, and speed.

Wavelength (symbolized by the Greek letter lambda, λ) is the distance between two consecutive peaks of the wave as shown in Figure 5.1. The *frequency* (symbolized by the Greek letter nu, ν) is defined as the number of waves (cycles) per second that pass a given point in space. Because all types of electromagnetic radiation travel at the speed of light, short-wavelength radiation must have a high frequency. This implies an inverse relationship between wavelength and frequency:

$$\lambda \nu = c \tag{5.1}$$

where

λ = wavelength in meters.
mν = frequency in cycles per second in Hz (1/sec or sec^{-1}).
c = speed of light (2.9979×10^8 m/sec).

Electromagnetic radiation is classified by wavelength range, as illustrated in Figure 5.2. Each portion of the spectrum has a popular name. For example, *radio waves* are electromagnetic radiation with low frequencies and therefore very long wavelengths. *Microwaves* also have low frequencies and are emitted by radar instruments. Microwaves are absorbed by molecules in food, and the energy the molecules take on raises their temperature. This is why foods cook quickly in a

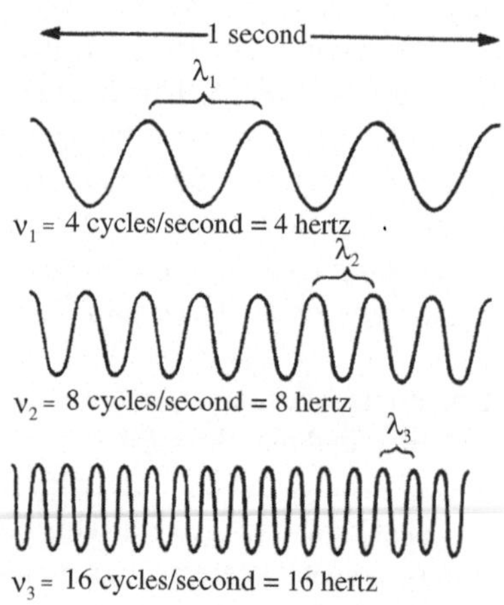

FIGURE 5.1 The nature of waves. Note that the radiation with the shortest wavelength has the highest frequency.

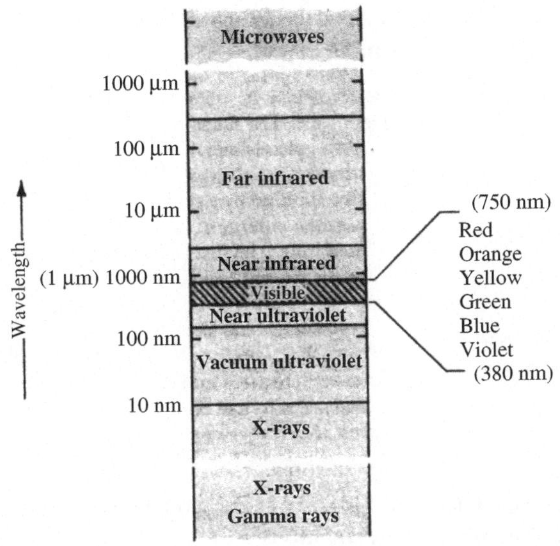

FIGURE 5.2 Electromagnetic radiation.

microwave oven. *Infrared radiation* is emitted by hot objects and consists of the range of frequencies that can make molecules of most substances vibrate internally. Infrared radiation is not visible, but how the body absorbs it can be felt by holding out a hand near a hot radiator; the absorbed radiation warms the hand.

Each substance absorbs a uniquely different set of infrared frequencies. A plot of frequencies absorbed vs. the intensities of absorption is called an *infrared absorption spectrum.* It can be used to identify a compound, because each infrared spectrum is as unique as a fingerprint. *Gamma rays* are at the high-frequency end of the electromagnetic spectrum and are produced by some radioactive elements. *X-rays* are much like gamma rays, but they are usually made by special equipment. Both x-rays and gamma rays easily penetrate living organisms.

Human eyes are able to sense only a narrow band of wavelengths, ranging from about 400 to 700 nm. This band is called the *visible spectrum* and consists of all the colors we can see, from red through orange, yellow, green, blue, and violet. White light is composed of all these colors in roughly equal amounts, and it can be separated into them by focusing a beam of white light through a prism, which spreads the various wavelengths apart. Table 5.1 contains the wavelength region of each color.

5.2.1 The Dual Nature of Light

In 1901, German physicist Max Planck proposed that electromagnetic radiation is emitted only in tiny *packets* or *quanta* of energy, which later became known as *photons*. The energy of one photon is called one quantum of energy.

$$E = h\nu \tag{5.2}$$

where

E = the energy of a photon.
h = *Planck's constant.*
ν = frequency of the electromagnetic radiation absorbed or emitted.

The value of h is 6.626×10^{-34} J (units of energy, Joules) multiplied by time (seconds). Each photon pulses at a frequency and travels at the speed of light. Planck proposed and Albert Einstein

TABLE 5.1
Wavelength Regions by Color

Wavelength Region (nm)	Color
380–450	Violet
450–495	Blue
495–550	Green-yellow
550–570	Green
570–590	Yellow
590–620	Orange
620–750	Red-purple[a]

[a] Purple is visible when equal numbers of photons of blue and red light strike the eye.

(1879–1955) confirmed that the energy of a photon of electromagnetic radiation is proportional to its frequency. According to Einstein's famous equations,

$$E = mc^2 \tag{5.3}$$

$$m = E/c^2 \tag{5.4}$$

where

E = energy.
m = mass.
c = the speed of light.

The main significance of this equation is that energy has mass. We can summarize the conclusions of Planck and Einstein's work as follows:

1. Energy is quantized and can occur in discrete units or quanta.
2. Electromagnetic radiation, which was previously believed to exhibit only wave properties, was found to have certain characteristics of particulate matter. Hence, scientists became aware of the dual nature of light.

5.3 CONTINUOUS AND LINE SPECTRA

5.3.1 Continuous Spectrum

When the light from the sun or from an object heated to a very high temperature (such as a light bulb filament) is split by a prism and displayed on a screen, a *continuous spectrum* forms. The spectrum contains light of all colors, as seen in Figure 5.3. A rainbow seen after a summer shower is a continuous spectrum. In this case, the colors contained in the sunlight are spread out by tiny water droplets in the air. The water droplets act as a prism.

In molecular absorption, both electronic and vibrational transitions are possible, because all wavelengths have a chance of being absorbed to some degree. The result is a continuous spectrum.

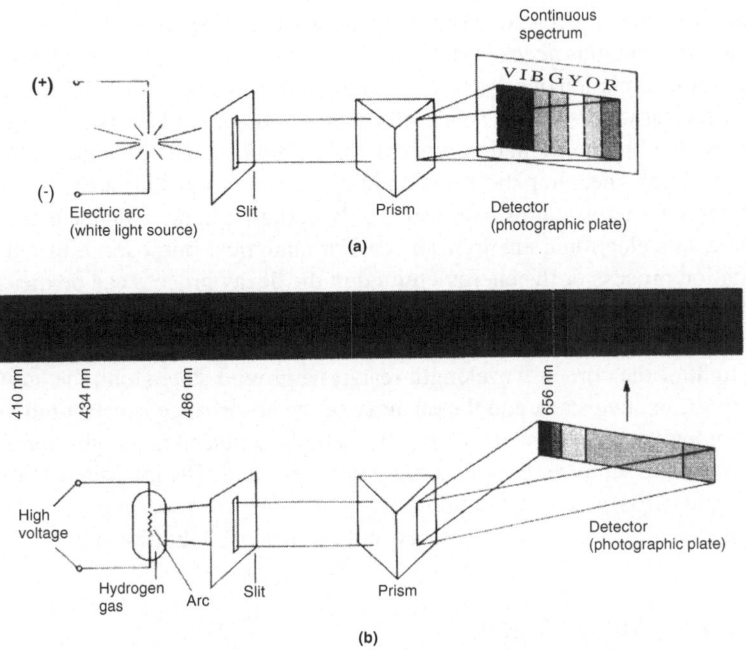

FIGURE 5.3 (a) Continuous spectrum obtaining all wavelengths of visible light (indicated by the initial letters of the colors of the rainbow). (b) The hydrogen line contains only a few discrete wavelengths.

5.3.2 Line Spectrum

When a light given off by an electrical discharge passes through a gas and is separated by a prism, a rather different spectrum can be observed on the screen. The discharge in an electrical current excites or energizes the atoms of the gas. The atoms absorb the energy, electrons are promoted to higher energy levels, and the electrons emit the absorbed energy in the form of light when they return to the lower energy state. When a narrow beam of this light is passed through a prism, only a few colors are observed as a series of discrete lines. This *line spectrum* is illustrated in Figure 5.3.

Atoms have no vibrational energy transition and all energy transfer is electronic. A limited number of wavelengths are absorbed, and only those wavelengths show up in the spectrum. The result is a line spectrum.

5.4 ABSORPTION AND EMISSION

According to the *Bohr model of an atom*, the nucleus is surrounded by electrons that travel around it in discrete orbitals. Every atom has a number of orbitals in which it is possible for electrons to travel. Each of these electron orbitals has an energy level associated with it; in general, the farther away from the nucleus an orbital is, the higher its energy level.

When the electrons of an atom are closest to the nucleus and lowest in energy, the atom is in its most stable state, known as *ground state*. With the addition of sufficient energy to atoms, electrons can be promoted from a lower energy level to a higher, vacant energy level. When light strikes an electron, causing it to be promoted to a higher energy level, the electron now possesses the energy that once was light. This is a less stable configuration, called the *excited state.* An important point concerning the process, however, is that the light coming through must be exactly the same energy as the energy difference between the two energy levels; otherwise, the light will not be absorbed. The atom is less stable in its excited state and will thus *decay* back to a less excited state by losing energy through collision with another particle or by emission of a "particle" of light (electromagnetic radiation), known as a *photon*. The electron will return to its initial, stable orbital position, and radiant energy equivalent to the amount of energy absorbed in the excitation process will be *emitted*. The excitation is forced by supplying energy, but the decay, involving the emission of light, occurs spontaneously.

Because only certain energy jumps can occur, only certain colors can appear in the spectrum. Figure 5.4 illustrates this electronic energy transfer. For analytical purposes, either the energy absorbed in the excitation process or the energy emitted in the decay process can be measured.

Every element has a characteristic set of energy levels and thus a unique set of absorption and emission wavelengths. This property makes *atomic spectrometry* useful in element-specific analytical techniques. If light of the correct wavelength reaches a ground-state atom, the atom absorbs the light and enters into the excited state, and the quantity of the absorbance is measured via *atomic absorption spectrophotometry*. In atomic emissions, the sample is placed in a high-thermal-energy environment, the atoms of the sample are excited, and light is emitted. The intensity of the emitted light is measured via *atomic emission spectrophotometry*.

The other form of interaction between energy and electrons is vibrational energy. Vibration requires less energy.

5.4.1 Molecular vs. Atomic Spectra

The measurement of the absorption and emission of light can be more easily described when the atomic and molecular spectra are understood. The absorption of light by individual, nonbonded atoms must be considered separately from molecular absorption.

5.4.1.1 Atomic Spectrum

To produce an *atomic spectrum*, a compound must first absorb enough energy to vaporize it into a molecular gas and dissociate the molecules into free atoms. In atoms, all energy transitions are electronic; therefore, only individual, discrete, electronic transitions are possible. Each discrete energy increase is due to the absorption of the wavelength corresponding to that energy. Consequently, only those wavelengths are absorbed, and only those wavelengths show up in the atomic spectrum or line spectrum (see Figure 5.3).

Atomic spectra are produced as follows:

Atomic absorption spectra are produced when the free atoms absorb radiant energy at characteristic wavelengths.

Atomic emission spectra are produced when the free atoms are excited by the thermal energy of a flame, arc, spark, or plasma and emit radiant energy at similar wavelengths.

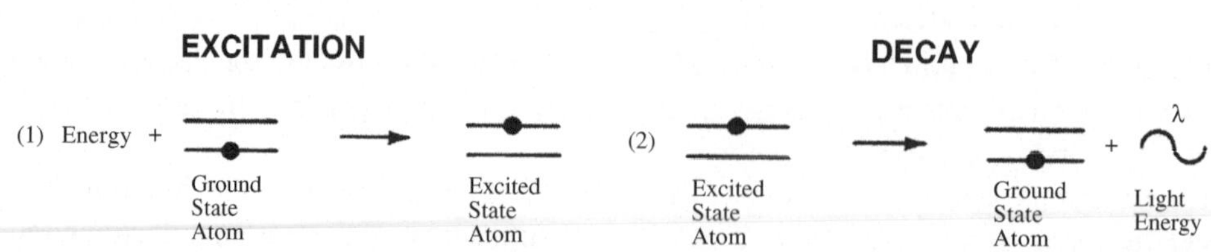

FIGURE 5.4 Electronic energy transition. Step (1), excitation, is forced by supplying energy. The decay process in step (2), involving the emission of light, occurs spontaneously. Because every element has a unique electronic structure, the wavelength of light emitted is a unique property of each individual element.

5.4.1.2 Molecular Absorption

In *molecular absorption*, electronic and vibrational transitions are possible; therefore, a large number of wavelengths are absorbed and produce a continuous spectrum (see Section 5.3.1). Because the amount of light absorbed by a sample is proportional to the concentration of the absorbing species (Beer's law; see Section 5.5), light absorption can be used as an analytical technique in quantitative analytical chemistry. The instrument used for measurement of absorption is the *spectrophotometer*. Molecular spectrophotometry (used in the UV/Vis and IR regions) and its operational techniques are discussed in detail in Chapter 6.

5.5 BEER'S LAW

The amount of light absorbed by a sample is proportional to the concentration of the absorbing species in the sample. There is, then, a linear relationship between absorbance and concentration. This relationship is well defined in the Beer–Lambert law (known simply as Beer's law): The amount of light absorbed or transmitted by a solution is a function of concentration of the substance and the sample path length. The formula follows:

$$A = abc \tag{5.5}$$

where

A = absorbance.
a = absorptivity (sometimes called an extinction coefficient), the ability of the absorbing species to absorb light. Absorptivity depends on the electronic and vibrational transitions in a given species. The numerical value of a depends on the units used for expressing the concentration of the absorbing solution.
b = diameter, or width of the cuvette, called pathlength. A wider cuvette has more of the absorbing species and therefore results in greater absorbance.
c = concentration.

For example, assume that a 2.00 ppm (parts per million = mg/l) standard measured in a 1-cm cuvette shows absorbance of 0.246. What is the concentration of a sample, if the measured absorbance is 0.529 and it is also measured in a 1-cm cuvette?

$$\begin{aligned} c_1 &= 2.00 \text{ ppm} \\ c_2 &= ? \\ b_1 &= 1 \text{ cm} \\ b_2 &= 1 \text{ cm} \\ A_1 &= 0.246 \\ A_2 &= 0.529 \\ a &= ? \end{aligned}$$

Using Beer's law, to calculate the sample concentration, with knowledge of the absorbance and the path length, we need the value of the absorptivity of the species. Based on knowledge of the values of the analyzed standard, we are able to calculate the numerical expression of the absorptivity:

$$\begin{aligned} A_1 &= a \times b_1 \times c_1 \\ 0.246 &= a \times 1 \times 2 \\ a &= 2/0.246 \\ a &= 0.123 \\ A_2 &= a \times b_2 \times c_2 \\ 0.529 &= 0.123 \times 1 \times c_2 \end{aligned} \tag{5.6}$$

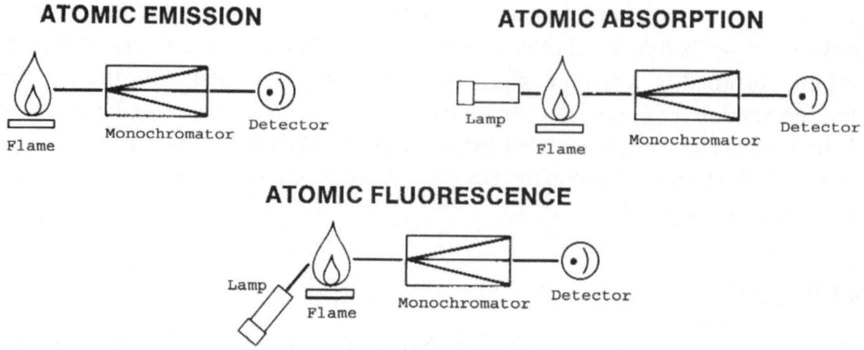

FIGURE 5.5 Atomic spectroscopy systems.

$$c_2 = 0.529/0.123$$
$$c_2 = 4.3 \text{ ppm}$$

5.6 ATOMIC SPECTROSCOPY TECHNIQUES

The most commonly used techniques for identifying trace concentrations of elements in samples are based on atomic spectrometry. These techniques involve electromagnetic radiation (light) that is absorbed by or emitted from atoms of a sample. By using atomic spectroscopy techniques, meaningful quantitative and qualitative information about the sample can be obtained. The qualitative information is related to the wavelengths at which the radiation is absorbed or emitted, and the quantitative information is related to the amount of electromagnetic radiation that is absorbed or emitted.

The sample is decomposed by intense heat into a cloud, or hot gases containing free atoms of the elements of interest. Of the three techniques — atomic absorption, atomic emission, and atomic fluorescence — atomic absorption and atomic emission are the most widely used. Figure 5.5 illustrates the arrangement of instruments in the three techniques. An understanding of atomic structure and of the atomic process involved in each technique is necessary (see Section 5.4).

5.6.1 Atomic Absorption Spectrometry (AAS)

Light of a wavelength characteristic of the element of interest is beamed through the element's atomic vapor. The atoms absorb some of this light. The amount of light absorbed is measured and used to determine the concentration of the element in the sample.

5.6.2 Atomic Emission Spectrometry (AES)

The sample is subjected to temperatures high enough to cause not only dissociation into atoms, but also significant amounts of collisional excitation (or ionization) of the sample atoms. Once the atoms or ions are in their excitation states, they decay to lower states and energy is transmitted (see Section 5.4). The intensity of the light emitted at specific wavelengths is measured and used to determine the concentration of the element of interest.

5.6.3 Atomic Fluorescence Spectrometry (AFS)

This technique incorporates aspects of both atomic absorption and atomic emission. Like atomic absorption, ground-state atoms created in a flame are excited by focusing a beam of light into the atomic vapor. Instead of measuring the amount of light absorbed in the process, the emission resulting from the decay of the atoms excited is measured. The intensity of this "fluorescence" increases with atomic concentration, providing the basis of quantitative determination. The source lamp for the AFS is mounted at an angle to the rest of the optical system, so that the light detector sees only the fluorescence in the flame and not the light from the lamp itself.

5.6.4 Atomization Process and Excitation Sources

Three types of thermal sources are available to dissociate sample molecules into free atoms: flames, furnaces, and electrical discharges. *Flames* and *furnaces* dissociate most types of molecules into free atoms. Because most of the free atoms in typical flames and furnaces are in their ground states, AAS is the preferred method to detect the presence of the element of interest.

Electrical discharges are used as atomization sources in AES. *Arcs* and *sparks* are electrical discharges created by application of electrical currents or potentials across an electrode in an inert gas. These discharges produce temperatures of about 7300°C. More recently, *plasmas* have been used as atomization and excitation sources in AES. Plasma is any form of matter that contains electrons (about 1%) and positive ions in the same quantity. The present state of the art in plasma sources for AES is the argon-supported inductively coupled plasma (ICP). Other plasmas in use are the direct-current plasma and microwave-induced plasma.

5.6.5 Development of Analytical Techniques

In the early twentieth century, the sharp lines that appeared in light emitted from electrical arcs and sparks were used analytically for *qualitative analysis*. During the mid-twentieth century, *quantitative* arc and spark spectroscopy was the best tool that analysts could use for the determination of trace concentrations of elements. While arc/spark emission techniques enjoyed widespread popularity for determination of metals, flame emission spectrometry (also known as flame photometry) was used for determination of alkalis and other easily excited elements. The most widespread use of the technique is in clinical laboratories for determining sodium and potassium levels in blood and other biological materials.

Flame emission spectrometry had the advantage of being simpler than arc/spark emission techniques, but was also limited because flames were not hot enough to cause emission in many elements. In the 1960s and 1970s, both flame and arc/spark atomic emission spectrometry declined in popularity, and *flame atomic absorption* was mostly used to determine trace metals in solutions (solid samples required dissolution prior to analysis). On the other hand, *graphite furnace atomic absorption spectroscopy* (GrAAS) was used when high-sensitivity and low-detection limits were needed. However, the GrAAS technique is not as precise and is subject to more interference. Advances, such as the *stabilized temperature platform furnace* technology and *Zeeman background correction* have reduced or eliminated most interference. Both the flame and graphite-furnace AAS techniques are used today and provide excellent means of trace element analysis. Most atomic absorption instruments are limited to determining only one element at a time.

The first published report on using ICP for elemental analysis was issued in 1973. The obtained law detection limits, freedom from interference, and long linear working ranges proved that it is a superior technique for atomic emission analysis. Besides its ability to determine a large number of elements over a wide range of concentrations, a major advantage of the ICP-AES technique is that many elements can be determined easily in the same analytical run.

Many laboratories are equipped with an ICP-AES instrument to perform moderate-sensitivity, high-sample-throughput, multielement analyses and a graphite-furnace AAS instrument to perform single-element determinations that require high sensitivity.

Inductively coupled plasma mass spectrometry (ICP-MS) is one of the most recently developed techniques for trace element analysis. In this technique, the analyte ions formed in the ICP are sent through a mass spectrometer where they are separated according to mass/charge (m/e) ratios. The number of ions at ratios of interest are then measured and the results used for qualitative and quantitative purposes. See Appendix A for more detail on MS.

5.6.6 Comparison of Techniques Used in Trace Element Analysis

5.6.6.1 Flame Atomic Absorption Spectrophotometry (FAAS)

The two principal advantages of FAAS are low initial cost and simplicity of operation.

5.6.6.2 Graphite Furnace Atomic Absorption Spectrophotometry (GrAAS)

The principal advantage of GrAAS over FAAS and ICP-AES is its greater sensitivity and lower detection limits for most elements. A very small amount of sample can be easily analyzed with GrAAS.

5.6.6.3 Inductively Coupled Plasma Atomic Emission Spectrophotometry (ICP-AES)

The main advantages of ICP-AES over AAS techniques in general are its multielement capabilities, longer linear dynamic ranges, and fewer interferences.

5.6.6.4 Inductively Coupled Plasma Mass Spectrometry (ICP-MS)

A powerful technique for elemental analysis, ICP-MS has the sensitivity and detection limits typical of GrAAS, combined with the multielement capability of ICP-AES. ICP-MS systems are expensive and have severe sample–matrix interferences. Therefore, further development of the technique is limited.

5.6.6.5 Selecting a Technique

Because of the advantages and disadvantages of the various techniques, selecting one for a given circumstance is easy. If an application requires single-element analysis for relatively few samples or if initial cost is an important factor, then FAAS is a good choice. If an application requires very low detection limits for a few elements, use GrAAS. For applications involving multielement analyses from samples in a complicated matrix with moderate sensitivity, ICP-AES is a good choice. When an application requires very low detection limits of many elements per sample, ICP-MS is the best technique.

6 Molecular Spectrophotometry

As discussed in Chapter 5, the absorption properties of atoms and molecules are quite different. Light absorption by individual, nonbonded atoms differs from that of molecules. Consequently, the techniques and instrument designs differ significantly and must be discussed separately.

Absorption of light by molecules or ions causes two types of energy changes: electronic (change in the energy of the electrons of a molecule) and vibrational (change in the internuclear distance of two or more atoms in the molecule). *Electronic transition* requires more energy and thus occurs in the visible–ultraviolet spectral region. *Vibrational changes* in a molecule result from the absorption of low-energy infrared radiation.

6.1 MOLECULAR ABSORPTION AND COLOR

The color of a molecule in solution depends on the wavelengths of light it absorbs. Thus, when a sample solution of a molecule or ion is exposed to white light, certain wavelengths are absorbed, and the remaining wavelengths are transmitted to the eye. The color perceived by the eye is determined only by the wavelengths transmitted. The substance exhibits the color that is complementary to wavelengths absorbed. In simple terms, the color seen is the complementary color of the color absorbed. Table 6.1 shows the general relationship between wavelengths of visible light absorbed and the color observed. For example, if the color of the test solution is yellow, the selected wavelength should be 450 nm, and if the test solution is dark blue the absorbance measured is at the 580-nm wavelength.

6.2 MOLECULAR ABSORPTION SPECTROPHOTOMETRY

The instrument used for measurement of absorbance is known generally as a *spectrophotometer*. In a spectrophotometer, radiant energy of a very narrow wavelength range is selected from a source and passed through the sample solution, which is contained in a glass or quartz cell, called a *cuvette*. The chemicals in the sample absorb some of the radiant energy, and the rest passes on through. Spectrophotometry is a very fast and convenient method for quantitative analysis. The amount of radiation absorbed (absorbance) at a specific wavelength is proportional to the concentration of the light-absorbing chemical in the sample.

6.2.1 Basic Components of Spectrophotometer

The basic spectrophotometer consists of a light source, wavelength selector, sample holder or sample compartment, detector, and readout device, as shown in Figure 6.1.

6.2.1.1 Light Source

The light source provides the light directed at the sample. The selection of the source depends on the region of the electromagnetic radiation needed for the analysis. There are considerable differences in technique and spectrum analysis between methods involving ultraviolet and visible (UV/Vis) light and infrared (IR) light. *UV/Vis spectrophotometry* is used mostly for quantitative

TABLE 6.1
Visible Spectrum and Complementary Colors

Wavelength (nm)	Color	Complementary Color
400–435	Violet	Yellow-green
435–480	Blue	Yellow
480–490	Green-blue	Orange
490–500	Blue-green	Red
500–560	Green	Purple
560–580	Yellow-green	Violet
580–595	Yellow	Blue
595–610	Orange	Green-blue
610–675	Red	Blue-green

analysis; *IR spectrophotometry* is mostly a qualitative technique, although quantitative applications can also be important.

The light sources for visible and UV radiation are presented in Table 6.2. The *tungsten filament lamp* is the only common source for the visible region; it covers part of the UV region, but is generally not used below 320 to 330 nm. At shorter wavelengths, a *deuterium lamp* is used.

6.2.1.2 Wavelength Selector or Monochromator

The function of the monochromator is to select a beam of monochromatic (one-wavelength) radiation. The essential parts of the monochromator follow:

1. The *entrance slit* controls the intensity of the light.
2. The *lens or mirror* causes light to travel as parallel rays.
3. The *dispersion device* selects light of different wavelengths. Dispersion devices include diffraction gratings, prisms, and various optical filters.
    a. A *diffraction grating* is a surface with a large number of parallel grooves. Light striking the grating is diffracted so that different wavelengths come off at different angles. Rotating the grating, by turning the wavelength dial on the instrument, allows radiation of the desired wavelength to be selected.
    b. A *prism* disperses radiation by means of refraction. Radiation of different wavelengths is bent at different angles upon entering and emerging from the prism.
    c. The simplest monochromators rely on *optical filters*. Instruments that use optical filters are inexpensive or portable and can be designed for specific analyses in the field. Filters types include *absorption filters*, which have very wide bandwidths, absorb

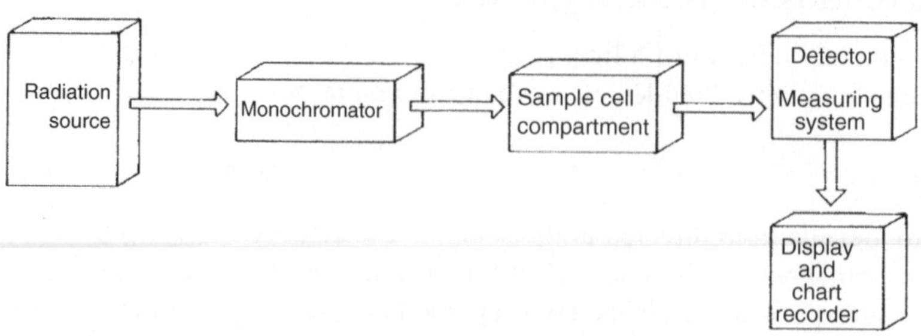

FIGURE 6.1 Basic construction of a simple spectrophotometer.

TABLE 6.2
Ultraviolet and Visible Radiation Sources

Source	Wavelength Range (nm)	Intensity
Tungsten filament lamp	320–2500	Weak below 400 nm; strong above 750 nm
Tungsten halogen lamp (quartz envelope)	250–2500	—
Hydrogen discharge lamp	180–375	Weak at all wavelengths, but best in 200–325 region
Deuterium discharge lamp	180–400	Moderate

certain parts of the spectrum, and are made of colored glass. *Interference filters* reject unwanted wavelengths and transmit a narrow bandwidth.

4. Lenses or mirrors are used to focus the light.
5. The exit slit controls the color (or wavelength) of the light that enters the sample compartment.

6.2.1.3 Sample Holder or Compartment

A sample holder is a tight box where the sample is irradiated by the light emerging from the monochromator. The sample, in the form of solution, is contained in an optically transparent cell, a cuvette, with a known width and optical length. Cells are made of optical glass. Some inexpensive spectrophotometers use circular test-tube cuvettes.

Cells used in the visible region of light are made of optical-quality *borosilicate glass*. At about 320 nm, the glass begins to absorb most of the radiant energy. For lower wavelengths, it is necessary to use more expensive cuvettes made of *quartz* or some other form of *silica*. Of course, these cells can also be used above 320 nm.

Matched cuvettes are identical with respect to path length and reflective and refractive properties in the area where the light beam passes. If the path length is different or if the wall of one cuvette reflects more or less light than another cuvette, then the absorbance measurement could be different, rather than because of the concentration difference. Therefore, the cuvette must be placed in the instrument exactly the same way each time, as path length and refractive properties can change by rotating the cuvette. A vertical line on the cuvette lined up with a similar line on the cuvette holder helps to avoid the abovementioned source of error, as illustrated in Figure 6.2.

Protect cuvettes from scratches. When cleaning cuvettes, use soft cloths, and avoid the use of abrasive cleaning agents. Avoid finger marks, lint, or dirt. When inserting a cuvette into the instrument, grasp it at the top edge. Any liquid or fingerprints adhering to the outside wall of the cuvette must be removed with a soft cloth or soft tissue prior to measurement.

Because of additional reflection from the air to glass surfaces, empty cuvettes transmit less radiation than do cells filled with reference standards, blanks, or deionized water. Do not use an empty cuvette to zero the instrument.

To avoid errors caused by removing a cuvette and then replacing it with another, spectrophotometers are available equipped with a fixed *flow-through cell*. The solution to be measured takes about 30 to 60 sec to flow through the cell. The reading can be taken after the first few seconds, which are needed to wash out the cell and fill it with the new sample.

6.2.1.4 Detector

A photosensitive detector picks up transmitted radiation through the solution. Detectors are *phototubes*, which convert light energy into electrical energy. A typical phototube consists of a half-cylinder

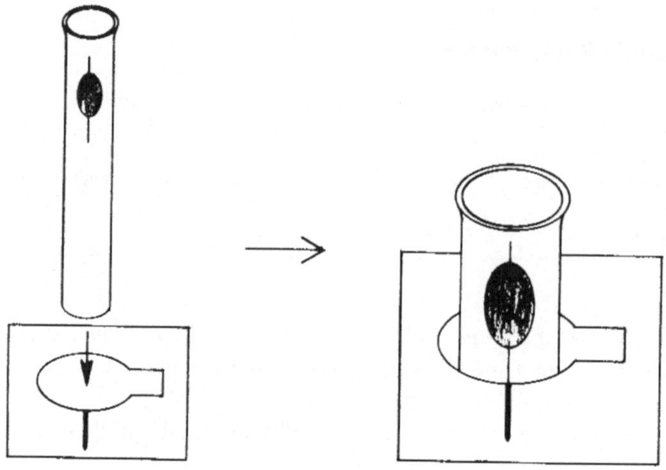

FIGURE 6.2 Lining up a cuvette for insertion into the cuvette holder.

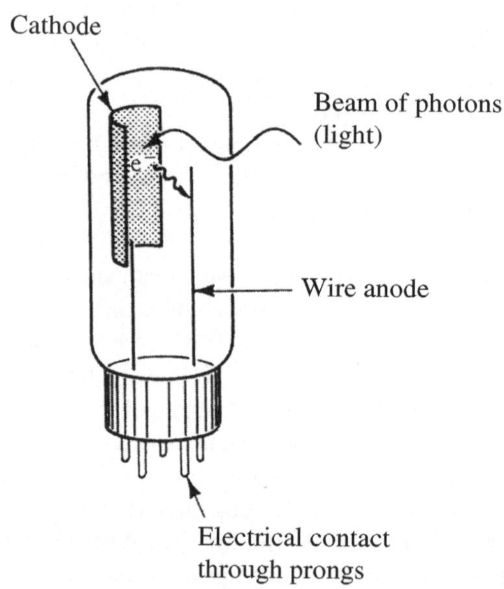

FIGURE 6.3 Schematic diagram of a phototube showing the emission of an electron from the cathode to the anode. A typical phototube consists of a half-cylinder cathode and a wire anode in a sealed evacuated glass tube. A beam of photons passes through the sample and strikes the inner surface of the cathode and ejects electrons from the cathode. The electrons migrate through the vacuum to the positive wire anode and produce a current.

cathode and a wire anode in a sealed, evacuated glass tube. Because the cathode emits electrons when struck by photons, the phototube is called a *photoemissive tube*. The response of the phototube to different wavelengths depends on the composition of the cathode coating. A schematic diagram of a phototube appears in Figure 6.3.

6.2.1.5 Readout Device

Radiation striking the detector generates electric current, which is increased via an amplifier and then transmitted to a recorder or displayed on the spectrophotometer via a digital or scale readout. Digital

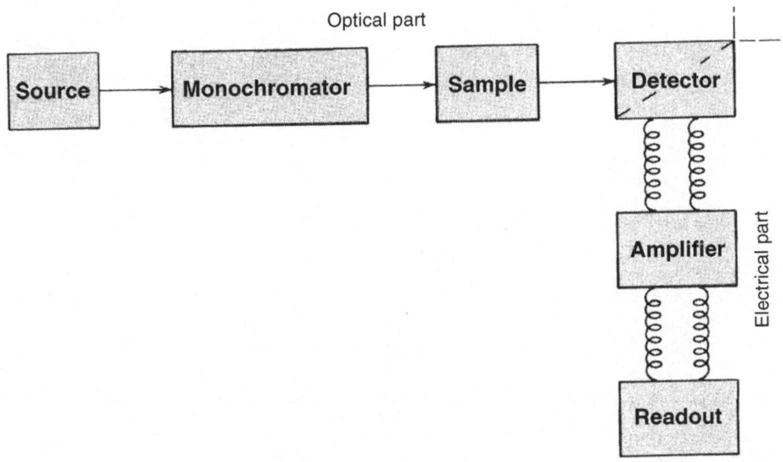

FIGURE 6.4 Block diagram showing components of a single-beam spectrophotometer. The optical and electrical parts of the instrument meet at the detector, which converts radiant energy into electrical energy.

displays are now used except on the most inexpensive instruments. The readout can be either *transmittance* or *absorbance*.

In a conventional spectrophotometer, the measured absorbance is used to calculate the concentration of the measured sample component or to prepare a Beer's law plot. Data manipulation requires more time than the measurements. Modern spectrophotometers with built-in microprocessors or microcomputers can perform rapid computations, store information for later use, and control many meter operations. With such equipment, the operator inserts the cuvette into the instrument and uses the keyboard to perform the measurements. Calibration plots are based on *linear regression* (see Section 6.6.3) and may be graphically displayed.

6.3 SINGLE-BEAM AND DOUBLE-BEAM SPECTROPHOTOMETERS

Two general types of spectrophotometry instruments are available: *single beam* and *double beam*.

6.3.1 Single-Beam Spectrophotometer

In single-beam instruments, all measurements are based on the varying intensity of a single beam of light. All energy from the light source can be directed through the sample cell. Figure 6.4 presents a schematic diagram of a single-beam optical system. The disadvantage of the single-beam instruments is that the light intensity can change due to fluctuations occurring in the line voltage, power source, or light bulb. Thus, an error could result in the sample reading. Single-beam lamp intensity drift has been controlled by designing more stable light sources and lamp power supplies and prewarming of light sources.

6.3.2 Double-Beam Spectrophotometer

The double-beam instrument uses additional optics to divide the light from the lamp into a *sample beam* (directed through the sample cell) and a *reference beam* (directed through the blank). A schematic diagram of a typical double-beam instrument is shown in Figure 6.5. The light coming from the monochromator is directed at one of two paths with a rotating half-mirror, called a *chopper*. At one moment the light passes through the sample, while at the next moment it passes through the

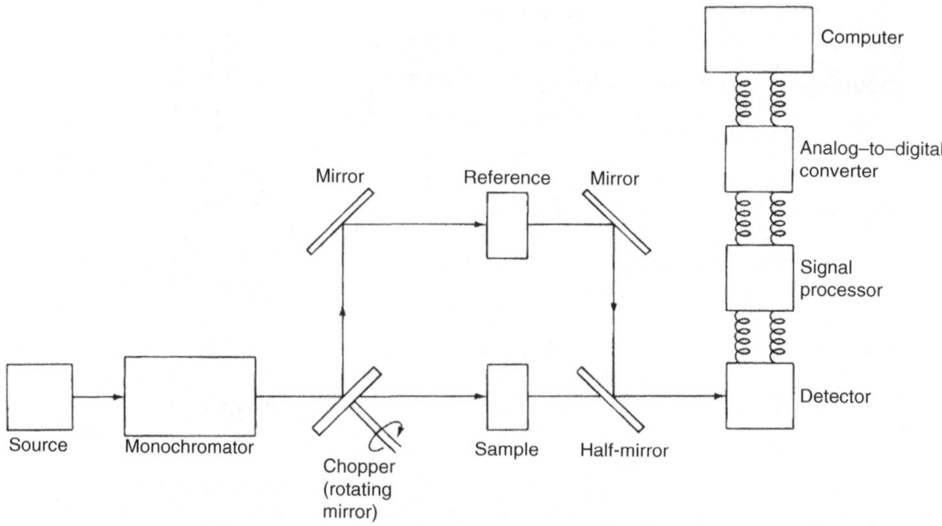

FIGURE 6.5 Schematic diagram of a double-beam spectrophotometer. The light coming through the monochromator is directed along either one of two paths with the use of a "chopper" or rotating half-mirror. At one moment the light passes through the sample, while at the next moment it passes through the blank. Both beams are joined again with a second rotating half-mirror prior to entering the detector.

blank. Both beams are joined again with a second rotating half-mirror prior to entering the detector. The detector sees alternating light intensities and automatically compensates for fluctuations, usually by automatically widening or narrowing the entrance slit to the monochromator. If the beam becomes less intense, the slit is opened; if the beam becomes more intense, the slit is narrowed. Thus, the signal relayed to the readout device is free of effects of intensity fluctuations from the source.

6.4 TYPES OF SPECTROPHOTOMETERS

Spectrophotometric instruments vary greatly in price, performance, and sophistication.

6.4.1 VISIBLE SPECTROPHOTOMETER

These instruments have inexpensive optical glass components and operate in the wavelength range of 325 nm to 900–1000 nm. Older instruments, such as the Spectronic 20, select wavelengths mechanically through a wavelength knob, whereas modern, digital-readout instruments offer electronic wavelength selection via a keyboard. Older instruments rely on blue- and red-sensitive phototubes.

6.4.2 ULTRAVIOLET/VISIBLE (UV/VIS) SPECTROPHOTOMETER

The UV/Vis spectrophotometer is designed for measurements in the ultraviolet and visible regions. Such instruments measure absorption in the 200- to 1000-nm region. For measurements below the 320-nm region, the spectrophotometer must be equipped with an ultraviolet source of radiation. The most common source of radiation in the visible region is the *tungsten filament lamp*, and in the UV region, the *deuterium discharge lamp*. In some spectrophotometers, the *tungsten halogen lamp* can be used for measurements as low as 250 or 220 nm. Light sources for UV/Vis radiation are listed in Table 6.2.

6.4.3 Spectrophotometers with a Built-in Microprocessor or Microcomputer

Modern spectrophotometers with a built-in microprocessor or microcomputer can perform rapid data processing and control many instrument operations. In such instruments, the operator inserts one or more cuvettes into the instrument and uses the keyboard to punch in the necessary operating instructions. Calibration plots are based on linear regression calculations and may be graphically displayed. Linear regression is discussed in Section 6.6.3.

6.4.4 Differences between UV/Vis and IR Spectrophotometric Methods

Methods involving ultraviolet and visible (UV/Vis) light and infrared (IR) light are quite different:

1. UV/Vis spectrophotometry is generally considered a quantitative analysis technique, while IR is considered a qualitative technique. However, both techniques may be utilized in a given analysis.
2. In the UV/Vis technique, absorption spectra are recorded, while transmittance spectra are used in the IR technique.
3. UV/Vis spectra are created from electronic transitions, while IR spectra arise from molecular vibrational transitions. Consequently, the IR technique provides more specific data about molecular structure.
4. UV/Vis and IR instruments are different in design, cuvette materials, and sample preparation techniques.

6.4.5 Infrared (IR) Spectrophotometer

The IR spectrophotometer has the same basic components as UV/Vis instruments, but the radiation source used in the optical system, sample cells, and detectors are different.

6.4.5.1 Light Source

For visible light, the light source is a *tungsten-filament lamp*; for UV light, a *hydrogen discharge lamp* is the most common. For infrared light, a heat source is necessary. The two most important infrared sources are glowing *silicon carbide rods* (Globars) and rods made of the rare earth oxides *zirconium* and *yttrium oxides* (Nernst glowers). Incandescent *nichrome wires* are also common.

6.4.5.2 Monochromator System

The monochromator systems in the IR and UV/Vis spectrophotometers are the same. The only dispersing device used for wavelength selection in the infrared is the *diffraction grating*.

6.4.5.3 Sample Cells

When using visible light, the cuvette may be made of any clear, colorless, transparent material, including glass and plastic. Cuvettes used for measurements in the UV region must be made of quartz glass. Both glass and quartz absorb infrared radiation; therefore, the monochromator optics and cells must be made from ionic materials. Large, polished *sodium chloride* (NaCl) crystals are most often used. Cells made of *lithium fluoride* (LiF) or *calcium fluoride* (CaF_2) provide better resolution at lower wavelengths, and cells made of *potassium bromide* (KBr) or *cesium iodide* (CsI) are more useful at higher wavelengths. The salt crystals are placed in some type of fixture, such as a *demountable cell*. A small amount of liquid sample, introduced through a sample port with a syringe, is held within a gasketed space inside and in the path of the light beam when placed in the instrument. Of course, because salts are highly water soluble, water cannot be used as a solvent for the sample. Usually solvents such as *carbon tetrachloride* (CCl_4) or *methylene chloride* (CH_2Cl_2) are used because their spectra show very little or no absorption in the IR region.

6.4.5.4 Detector

Infrared radiation can be measured by detecting the temperature change of a material in the infrared beam; this type of detector is known as a *thermal detector*. Because the radiant power of infrared radiation is so weak, the response of most thermal detectors is quite low. A preamplifier is usually necessary to obtain a good signal-to-noise ratio in the amplifier. Another problem is heat radiated from objects in the room. To minimize this source of error, the detector must be housed in a vacuum or shielded from direct exposure to heat.

6.4.5.6 Readout

In ultraviolet and visible spectra, absorbance and transmittance are plotted against wavelengths. In infrared spectra, using wavenumbers is preferred over wavelengths. The *wavenumber* is the reciprocal of wavelength expressed in centimeters, and therefore has a unit of cm^{-1}. See Figure 6.6 for wavelength and wavenumber conversion.

The IR region of the spectrum is usually considered to start near the red end of the visible spectrum at the point where the eye no longer responds to dispersed radiation ("infra" means below the red). The fundamental IR region extends from 3600 cm^{-1} (wavenumber) or 2.8 μm (wavelength). The analytically useful IR region extends from 3600 cm^{-1} to somewhere around 300 cm^{-1} or 33 μm.

Infrared spectrophotometers are generally double-beam instruments. The sample cell and reference cell (solvent) are exposed to equivalent beams from the same infrared source. A rotating half-circle mirror is used to direct an equivalent beam alternately through the two cells many times a second. Thus, any condition that affects the sample beam equally affects the reference beam, so that the condition is canceled out in the readout. (Double-beam instruments are discussed in Section 6.3.2 and the schematic diagram of the operation appears in Figure 6.5.)

6.4.5.7 Samples

Samples can be liquids, solids, or gases. They can be organic or inorganic, although inorganic materials sometimes do not give very definitive spectra. The only molecules transparent to IR radiation under ordinary conditions are monatomic and nonpolar molecules, such as Ne, He, O_2, N_2, and H_2.

Liquid samples may be analyzed without dilution or being dissolved in a solvent. Running a liquid sample without a solvent (pure or "neat" sample) is desirable.

Two methods are available for analyzing solids without dissolving them. In the first method, the *potassium bromide (KBr) pellet technique*, a small portion of the dry solid sample is mixed with potassium bromide. A small amount of this mixture is then transferred to a "pellet die," in which the mixture is pressed into a potassium chloride pellet. The pellet is a transparent half-inch disk that can

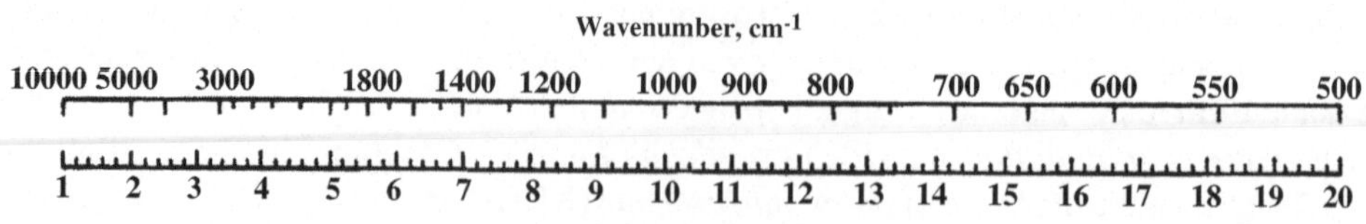

FIGURE 6.6 Conversion of wavelength and wavenumber.

be placed directly in the radiation path. In the other method, known as the *mull method*, the dry solid sample is mixed with mineral oil so that the substance becomes "toothpaste-like." This mixture is then placed between two salt crystals and the spectrum recorded.

6.5 SUMMARY OF MOLECULAR SPECTROPHOTOMETRY

Region	Ultraviolet (UV)	Visible (Vis)	Infrared (IR)
Wavelength	180 to 400 nm	400 to 750 nm	750 to 15,000 nm
Light source	Hydrogen discharge tube	Tungsten-filament lamp	Globar, Nernst glower, incandescent nichrome wire
Cuvette material	Quartz	Glass or clear plastic	Inorganic salt crystal
Detector	Phototube	Phototube	Thermal detector

6.6 SPECTROPHOTOMETER CALIBRATION

Calibrations are performed at the beginning of the analysis to ensure that the instrument is working properly. This *initial calibration* is determined for each parameter tested, based on instrument response for different *calibration standards* against the *calibration blank*. The *optimum concentration range* and the number of these standards are determined by the analytical method. The concentration of calibration standards should be bracketed in the optimum range. The concentration of standards and the measured response (absorbance, transmittance, etc.) of the instruments should plot on the *calibration curve* and be approved by calculating the corresponding *correlation coefficient*. Computerized, modern instruments display the curve and the value of the correlation coefficient. Its value should be greater than 0.9998, which serves as a basis for acceptance or rejection of the calibration curve. In UV/Vis spectrophotometers, the initial calibration is based on a 4- to -6-point standard curve in the optimum linear range as stated in each particular parameter.

After the calibration curve is established, once for each analytical batch (samples that are analyzed together with the same method and the same lot of reagents) or at a 5% frequency, the curve should be approved with a *continuing calibration*. The latter includes the analysis of the *continuing calibration standard* (CCS) and *calibration verification standard* (CVS) and must be analyzed before samples are measured. The CCS value is a midpoint initial calibration standard. Deviation from the original value should be within ±5%. The CVS should be a certified standard or independently prepared from a source other than the calibration standards. Its analyzed value is accepted within ±10% deviation from the 100% recovery.

Sample pretreatments (digestion, distillation, extraction, filtration, etc.) should be verified, and the effects of sample preparations should be monitored. To support these measures, a blank (*preparation blank*) and one standard (*laboratory control standard*, LCS) should be prepared and analyzed together with the samples. The preparation blank or "prep blank" is analyte-free water treated in the same way as the samples. The LCS is a sample taken from the CVS, except that it is carried through the preparation. Accepted values are within ±15% deviation from the 100% recovery.

6.6.1 Frequency of Calibration Curve Preparation

Frequency depends on the instrumentation. For the UV/Vis spectrophotometer, use the available calibration curve until the correct calibration is approved, which is performed every 6 months or on the failure of any continuing calibration standard. Daily calibration is made by zeroing the instrument with a calibration blank and measurement at one continuing calibration standard (CCS) with a ±5%

recovery and with the ±10% recovery of a calibration verification standard (CVS). Once per analytical batch or with 5% frequency, this check should be repeated. If the CCS and CVS fail, the calibration criteria of the analysis must be stopped and a new initial calibration performed. Samples measured before the failed standards must be analyzed again.

6.6.2 General Rules in the Preparation of Calibration Curves

Calibration curves are prepared by taking standards of known concentrations. For the preparation of calibration curves, ordinary rectangular-coordinate paper is generally satisfactory. For some graphs (e.g., measurement of millivolt response by using a pH meter with ion-selective electrodes), semilogarithmic paper is preferable. Plot the independent and dependent variables on the abscissa and ordinate in a manner that can be comprehended easily, and cover as much of the graph paper as possible. Choose the scales so that the slope of the curve approaches unity as nearly as possible and choose the variables so that the plot will be close to a straight line. Graph legends should provide complete information about the conditions under which the data were obtained, including the parameter determined, method and reference, volume of the standards, wavelength used, time between addition of reagents and the reading, data obtained from the linear regression calculation, date of preparation, and name and signature of the preparer. A typical calibration curve appears in Figure 6.7.

6.6.3 Linear Regression Calculation

When a calibration curve is prepared, the sample concentration is obtained by measuring the instrument response under the same conditions used for the standards; sample concentration is read on the horizontal axis of the plot. Although unknown concentrations can be read directly from the graphical plot, better accuracy is possible by using the linear regression calculation, also called the *least squares calculation*. In this calculation, of the possible straight lines that can be drawn through or near the data points, the one chosen minimizes the sum of the squared deviations. The deviation for each point is the difference between the actual data points with the same x-axis value that lies exactly on the straight line. It gives information about the best straight line through the points entered, including the correlation coefficient, intercept, slope, and the predicted x and y values.

The *correlation coefficient* is the correlation between the x and y values in a set of data points. The closer the coefficient is to one, the stronger the direct or positive linear relationship (an increase in one variable is related to an increase in the other). The closer the coefficient is to minus one, the stronger the indirect or negative linear relationship (an increase in one variable is related to a decrease in the other). A value of greater than 0.9998 is accepted.

Intercept b tells whether there is a significant blank measurement even when the concentration of the blank is zero. The intercept value is the y intercept of the best straight line through the points. The calculated value of the *slope m* is the slope of the best straight line. The formulas for calculating these values follow:

$$m = nExy - ExEy/nE_2 - (Ey)_2 \tag{6.1}$$

$$b = nEy^2 - EyExy/nEy^2 - (Ey)^2 \tag{6.2}$$

where

m = slope.
b = intercept.
E = sum.
x = concentration.
y = absorbance (or other response).

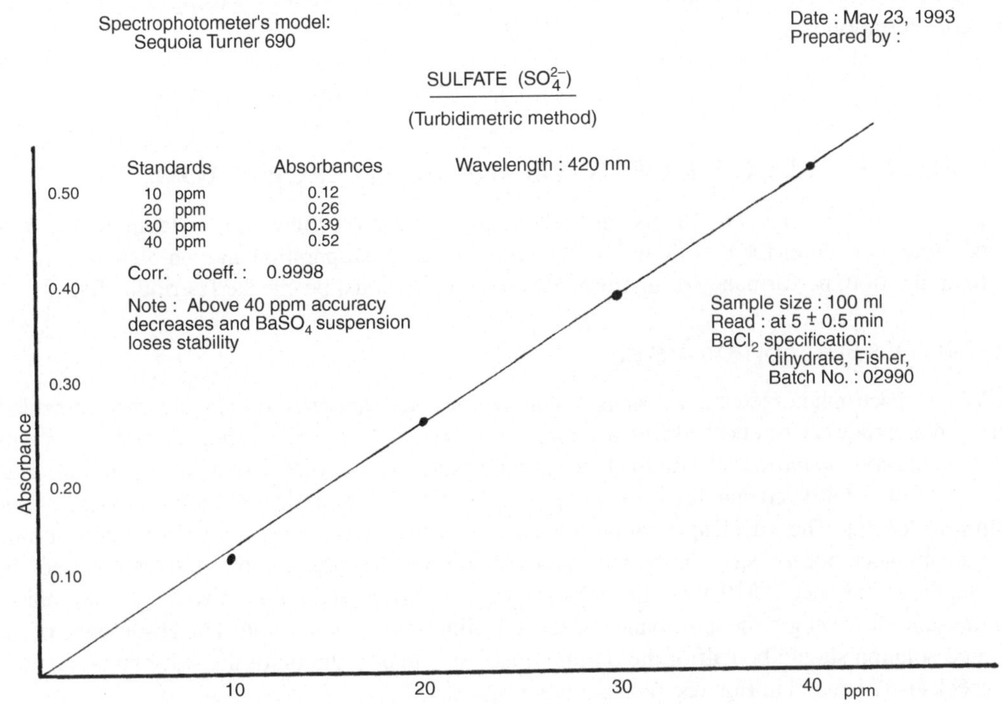

FIGURE 6.7 Typical calibration curve.

When the absorbance is known, the slope and intercept values of the concentration of the sample can be calculated according to the formula,

$$x = my + b \tag{6.3}$$

For example, in a spectrophotometric analysis, the initial calibration provides the following data:

Number of the standards (n)	= 6
Concentration of the standards (x)	= 0.2, 0.4, 0.5, 0.6, 0.8, and 1.0 ppm
Measured absorbances (y)	= 0.206, 0.392, 0.503, 0.598, 0.789, and 0.992
Calculated correlation coefficient	= 0.99985

Using the linear regression calculation, predict the slope and intercept values for the above data with the formulas (6.1), (6.2), and (6.3):

	x	y	y^2	xy
	0.2	0.206	0.042	0.041
	0.4	0.392	0.154	0.157
	0.5	0.503	0.253	0.252
	0.6	0.598	0.358	0.359
	0.8	0.789	0.623	0.631
	1.0	0.991	0.982	0.991
$E =$	3.5	3.479	2.412	2.431

$m = (6 \times 2.431) - (3.5 \times 3.479)/(6 \times 2.412) - (3.479^2) = (14.586 - 12.177)/(14.472 - 12.103) = (2.409/2.369) = 1.107$

$b = (2.412 \times 3.5) - (3.479 \times 2.431)/(6 \times 2.431) - (3.479^2) = (8.442 - 8.457)/(14.472 - 12.103) = -(0.015/2.369) = -0.006$

Absorbance for a sample of unknown concentration measured as 0.246. By using formula 6.3, the sample concentration is:

$$x = 1.017 \times 0.246 + (-0.006) = 0.244 \text{ ppm}$$

6.7 PERFORMANCE CHECK OF UV/VIS AND IR SPECTROPHOTOMETER

Spectrophotometer designs and models vary. The manufacturer's manual and the laboratory's standard operating procedures (SOPs) should be consulted for correct operation and maintenance. In addition to calibration, performance of instruments for accuracy must be checked periodically.

6.7.1 UV/VIS SPECTROPHOTOMETER

For UV/Vis spectrophotometers, *wavelength calibrations* and *linearity checks* are recommended. Wavelength accuracy can be checked with a commercially available *didymium calibration filter*, or with the very simple *cobalt chloride test*. In the latter test, measure the absorbance of a cobalt chloride solution (22 g of $CoCl_2$ dissolved and diluted to 1 liter with 1% HCl solution) on 500-, 505-, 510-, 515-, and 520-nm wavelengths. The wavelength calibration check is satisfactory if maximum absorbance or minimum transmittance occurs between the 505- and 515-nm wavelengths. Perform a linearity check by measuring the absorbance at 510 nm of the cobalt chloride solution used for the wavelength calibration, and at the same wavelength the absorbance of the 1:1 dilution of this solution. The absorbance of the 1:1 diluted solution should be half of the original reading. Documentation of the wavelength and linearity checks is illustrated in Figures 6.8. and 6.9, respectively.

Model							
	Absorbance Read at Wavelength						
Date	500 nm	505 nm	510 nm	515 nm	520 nm	Remarks	Sign.

FIGURE 6.8 Documentation of spectrophotometer wavelength calibration check. nm = nanometer, unit of the wavelength. The calibration check is satisfied when maximum absorbance (or minimum transmittance) occurs between 505 and 515 nm wavelengths.

Spectrophotometer model				
Date	Checking Solution	Absorbance at 510 nm	Remark	Sign.
	Stock Cobalt Soln.			
	1:1 Cobalt Soln.			
	Stock Cobalt Soln.			
	1:1 Cobalt Soln.			
	Stock Cobalt Soln.			
	1:1 Cobalt Soln.			
	Stock Cobalt Soln.			
	1:1 Cobalt Soln.			

FIGURE 6.9 Documentation of spectrophotometer linearity check. The absorbance of the 1:1 diluted cobalt solution should be half reading produced by the stock cobalt solution (22 g $CoCl_2$ in 1 L 1% HCl solution).

6.7.2 IR Spectrophotometer

Satisfactory operation of an IR spectrophotometer is determined with commercially available 0.05-mm-thick *polystyrene film*. Record the spectrum of this film and compare it with the reading supplied by the manufacturer. If the test spectrum is not within the indicated tolerance, adjustment is necessary, probably by a service representative. See Figure 6.10.

6.8 MAINTENANCE OF THE UV/VIS AND IR SPECTROPHOTOMETERS

Proper care and maintenance of the instruments are the basic requirements for accurate and sufficient laboratory results.

6.8.1 UV/Vis Spectrophotometer

Recommended daily, weekly, and quarterly maintenance chores for a UV/Vis spectrophotometer are summarized:

1. On a daily basis, keep the sample compartment and cuvettes sparkling clean.
2. Check lamp alignment on a weekly basis.
3. Under the service contract, an instrumentation specialist must clean the windows.

6.8.2 IR Spectrophotometer

Recommended daily, weekly, and quarterly maintenance chores for a UV/Vis spectrophotometer are summarized below.

1. Clean the sample cell and check for gas leakage every day.
2. Clean windows on a monthly basis.
3. Change the desiccant every quarter.

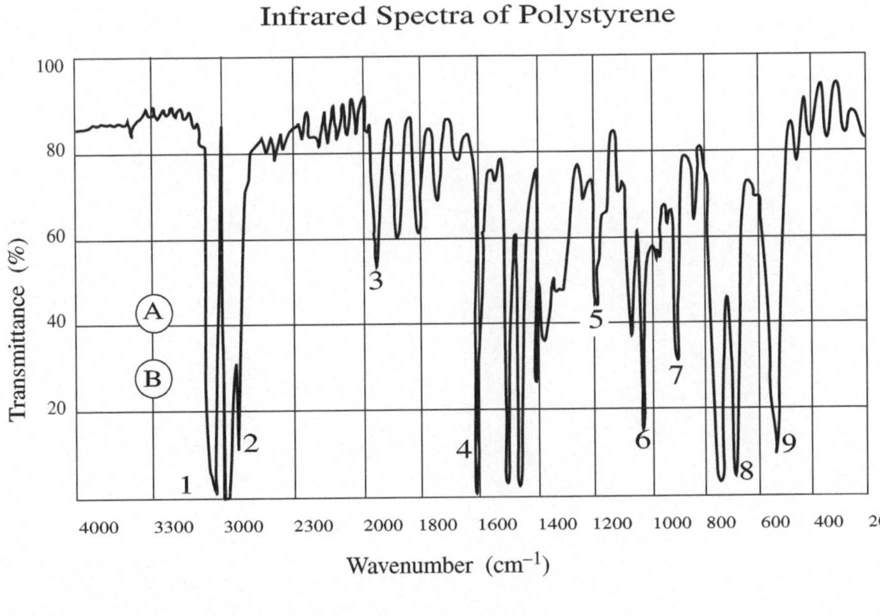

FIGURE 6.10 Performance check of infrared (IR) spectrophotometer.

7 Atomic Absorption Spectrometry

7.1 INTRODUCTION

7.1.1 Atomic Spectrometry (AS)

As discussed previously, AS is a class of elemental analysis techniques that use the interaction of electromagnetic radiation with atoms or ions to detect the presence of elements of interest.

7.1.2 Atomic Absorption (AA)

Atomic absorption occurs when a ground-state atom absorbs energy in the form of light of a specific wavelength and is elevated to an excited state. The amount of light energy absorbed at this wavelength increases as the number of atoms of the selected element in the light path increases. The relationship between the amount of light absorbed and the concentration of analyte present in known standards can be used to determine unknown concentrations by measuring the amount of light absorbed. Instrument readouts can be calibrated to directly display concentrations.

7.1.3 Atomic Absorption Spectrometry (AAS)

Atomic absorption spectrometry is an element analysis technique that uses absorption of electromagnetic radiation to detect the presence of the elements of interest. Molecular spectrophotometry and working techniques were discussed in Chapter 6; this chapter focuses on analytical methods using atomic spectra. This technique has been applied to the determination of numerous elements and is a major tool in studies involving trace metals in the environment and in biological samples. It is also frequently useful in cases where the metal is at a fairly high concentration level in the sample but only a small sample is available for analysis, which sometimes occurs with metalloproteins, for example. The first report of an important biological role for nickel was based on a determination via AA that the urease enzyme, at least in certain organisms, contains two nickel ions per protein molecule.

Light absorption is measured and related to element concentration in both AAS and molecular spectrophotometry (see Chapter 6). The major differences lie in instrument design, especially with respect to the light source, sample cell, and placement of the monochromator. As outlined in previous chapters, the absorption of light by individual, nonbonded atoms must be considered separately from molecular absorption. In atoms, all energy transitions are electronic; therefore, only individual, discrete, electronic transitions are possible. Consequently, atomic spectra are made up of lines, which are much sharper than the bands observed in molecular spectroscopy. Each discrete energy increase is due to the absorption of the wavelength corresponding to an energy transition; therefore, only those wavelengths are absorbed, and only those wavelengths show up in the atomic spectrum, or *line spectrum*. Atomic absorption spectra are produced when the free atoms absorb radiant energy at characteristic wavelengths. To produce an atomic spectrum, a compound must first absorb enough energy to vaporize it to a molecular gas and dissociate the molecules into free atoms. Because the amount of light absorbed by a sample is proportional to the concentration of the absorbing species, light absorption can be used in quantitative analytical chemistry.

Metals in solution can be readily determined by AAS. The method is simple, rapid, and applicable to a large number of metals in different samples. While drinking water that is free of particulate matter can be analyzed directly, samples containing suspended material, sludge, sediment, and other solids are analyzed after proper pretreatment. Sample preparations are discussed in Chapter 15.

7.1.3.1 Atomic Absorption Measurement

The light of a wavelength, which is characteristic of the element of interest, is beamed through an atomic vapor. Some of this light is then absorbed by the atoms of the element. The amount of light that is absorbed by these atoms is then measured and used to determine the concentration of that element in the sample. The use of special light sources and careful selection of wavelengths allow the specific quantitative determination of individual elements in the presence of others. The atom cloud required for atomic absorption measurements is produced by supplying enough thermal energy to the sample to dissociate the chemical compounds into free atoms. Aspirating a solution of the sample into a flame aligned in the light beam serves this purpose. Under the proper flame conditions, most of the atoms will remain in the ground-state form and are capable of absorbing light at the analytical wavelength from a source lamp. The light is then directed onto the detector where the reduced intensity is measured.

7.2 STEPS IN THE ATOMIC ABSORPTION PROCESS

The solvent is evaporated or burned, and the sample compounds are thermally decomposed and converted into a gas of the individual atoms present. The atoms of this element in the *flame* absorb light only from the *hollow-cathode source* that emits the characteristic wavelength of the single element being determined. Some of the light is absorbed and the rest passes through. The amount of light absorbed depends on the number of atoms in the light path. The selected spectral line from the light beam is isolated by a *monochromator.* The wavelength of light selected by the monochromator is directed onto the *detector*. The detector is a photomultiplier tube that produces an electrical current dependent on the light intensity. The electrical current from the photomultiplier is then amplified and processed by the instrument electronics to produce a signal that is a measure of the light attenuation occurring in the sample cell. This signal can be further processed to produce an instrument *readout* directly in concentration units. Steps of the above process are described in the following sections.

7.2.1 Nebulization

Aspirate the sample into the burner chamber. The sample becomes an aerosol and mixes with the fuel and oxidant gases. In this step the metals are still in solution in the fine aerosol.

7.2.2 Evaporation or Desolvation

The aerosol droplets move into the heat of the flame, where the solvent is evaporated and solid particles of the sample remain.

7.2.3 Liquefaction and Vaporization

Heat is applied and the solid particles are liquefied. With additional heat, the particles will vaporize. At this point, the metal of interest (analyte) still contains anions to form molecules.

7.2.4 Atomization

By applying more heat, the molecules are broken down and individual atoms form.

7.2.5 Excitation and Ionization

The ground-state atoms formed during the atomization step will excite and determine the amount of light absorbed. Concentration is determined by comparing the absorbance of the sample to standards with known concentrations.

7.3 ATOMIC ABSORPTION SPECTROPHOTOMETER COMPONENTS

7.3.1 Light Source

As indicated previously, an atom absorbs light at discrete wavelengths. To measure this narrow light absorption with maximum sensitivity, it is necessary to use a light source that emits specific wavelengths which can be absorbed by the atom. In other words, the light emitted from the lamp should be exactly the light required for the particular analysis. To satisfy this criterion, the atoms of the element tested are present in the lamp. When the lamp is on, these atoms are supplied with energy that causes them to enter into *excited states*. When the *promoted atoms* return to their ground state, the light energy will be emitted at the wavelength characteristic to the metal. Thus, each metal analyzed requires a separate source lamp. The most common light sources used in atomic absorption are the hollow cathode lamp and the electrodeless discharge lamp.

The *hollow cathode lamp* (HCL) is an evacuated glass tube filled with either neon or argon gas. The HCL is illustrated in Figure 7.1. The *cathode* (– charged electrode), which is made of the metal to be determined, and the *anode* (+ charged electrode) are sealed in the tube. A window, transparent to the emitted radiation, is at the end of the tube. When the lamp is on, an electrical potential is applied between the anode and cathode, and the gas atoms are ionized. The actively charged gas ions collide with the cathode and liberate metal atoms. These atoms are excited by the energy liberated through the collision. By returning to the ground state, the atoms emit light energy as described above. HCLs have a limited lifetime. Because of the rapid vaporization of the cathode for volatile metals, such as arsenic (As), selenium (Se), and cadmium (Cd), the lifetime of these lamps is especially short.

It is possible to construct a cathode from several metals. This kind of lamp is called a *multi-element lamp*. The intensity of emission for an element in a multielement lamp is not as great as that observed for the element in a single-element lamp. Thus, special consideration is necessary before using multielement lamps in applications where high precision and low detection limits are necessary.

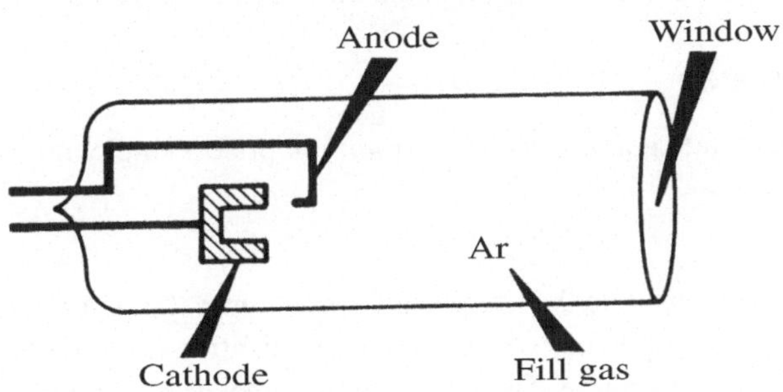

FIGURE 7.1 Hollow cathode lamp.

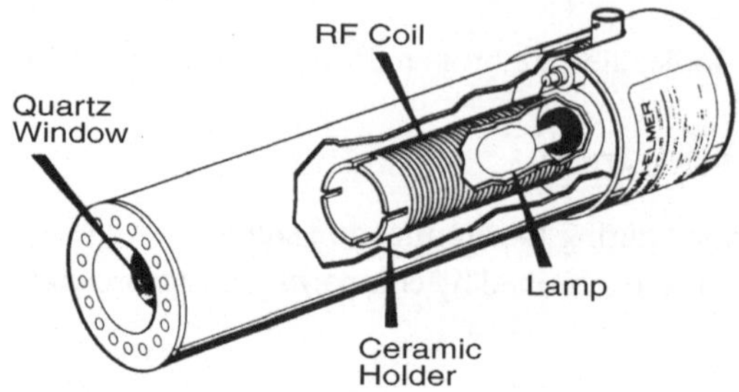

FIGURE 7.2 Electrodeless discharge lamp.

In some applications — primarily in the determination of volatile elements — the resistivity of the HCL is not satisfactory. The analytical performance of these elements by AA can be improved dramatically by using *electrodeless discharge lamps* (EDLs). EDLs offer the analytical advantages of better precision and lower detection limits. In addition to providing superior performance, the useful lifetime of an EDL is much longer than that of a HCL for the same element. EDL design is illustrated in Figure 7.2. A small amount of the metal or its salt is sealed inside a quartz bulb. The bulb is placed inside a ceramic holder on which the antenna from a radio frequency (RF) generator is coiled. When an RF field of sufficient power is applied, the coupled energy will vaporize and excite the atoms inside the bulb, causing them to emit their characteristic spectrum. An accessory power supply is required to operate an EDL.

7.3.2 Flames

In order for the atomic absorption process to occur, individual atoms must be produced from the sample, which starts out as a solution of ions. The function of the flame is to evaporate the solvent, decompose and dissociate molecules, and provide ground-state atoms for absorption of the emitted radiation. All flames require both a fuel and an oxidant.

The two flames used for AA are air–acetylene and nitrous oxide (N_2O)–acetylene. In the case of *air–acetylene flames,* acetylene is the fuel and air is the oxidant. The temperature is 2130 to 2400°C. In the *nitrous oxide–acetylene flame*, acetylene is the fuel but nitrous oxide is used as an oxidant. The temperature of this flame is 2600 to 2800°C.

While the air–acetylene flame is satisfactory for the majority of elements determined by atomic absorption spectrophotometry, the hotter nitrous oxide–acetylene flame is required for many refractory-forming elements. The recommended flame used for any given element is available in reference books or in the application manual issued by the manufacturer of the instrument.

7.3.3 Nebulizer and Burner

Typically, the *nebulizer* (often called *atomizer*) and *burner* comprise a single unit.

7.3.3.1 Nebulizer

The purpose of the nebulizer is to suck up the sample and spray it into the flame at a constant and reproducible rate. In order to provide for the most efficient nebulization for variable sample solution systems, the nebulizer should be adjustable. The most common material of the nebulizer is stainless steel, but this material corrodes in contact with highly acidic samples. A nebulizer made of corrosion-resistant materials, such as plastic or *platinum–rhodium alloy*, is preferable.

7.3.3.2 Burner

Two basic types of burner are used in atomic absorption spectrophotometers: "total consumption burner" and "premix burner."

- In the *total consumption burner*, the channels of the fuel gas, oxidizing gas, and sample meet in a single opening at the base of the flame. The resulting flame is turbulent and non-homogeneous. This type of burner is used in flame photometry.
- The *premix burner* produces a quieter flame that is less turbulent and homogenous; therefore, it is preferable in atomic absorption.

The sample is nebulized and mixed with the fuel and oxidant before introducing it to the flame. Only the finest droplets of the nebulized sample enter the flame; the larger droplets are caught and rejected through a drain. The drain uses a liquid trap to prevent combustion gases from escaping through the drain line.

To deflect larger droplets and remove them from the burner through the drain, an *impact device* is placed in the front of the nebulizer. The impact device can be a *flow spoiler* or a *glass* or *ceramic spoiler*. For routine work, a chemically inert flow spoiler is preferred; glass beads may be used in cases where additional sensitivity is needed. Components of an atomic absorption burner system are shown in Figure 7.3.

Burner heads are constructed of titanium to provide extreme resistance to heat and corrosion. For various types of flames, diverse burner-head geometries are required. For the air–acetylene flame, a *10-cm, single-slit burner head* is used, and, for the nitrous oxide–acetylene flame, a *5-cm slit burner head* is recommended.

7.3.4 Optics and Monochromator System

The function of the monochromator is to isolate a single line of the analyte's spectrum. Light from the source must be focused on the sample cell and directed to the monochromator at the entrance slit and then directed to the *grating* where dispersion takes place. The grating consists of a reflective surface with many fine, parallel lines very close together. Reflection from this surface generates an interference known as *diffraction*, in which different wavelengths of light diverge from the grating at different

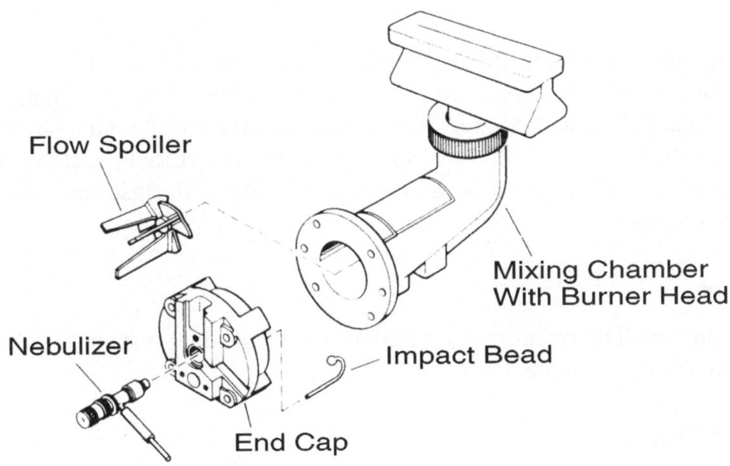

FIGURE 7.3 Premix burner system.

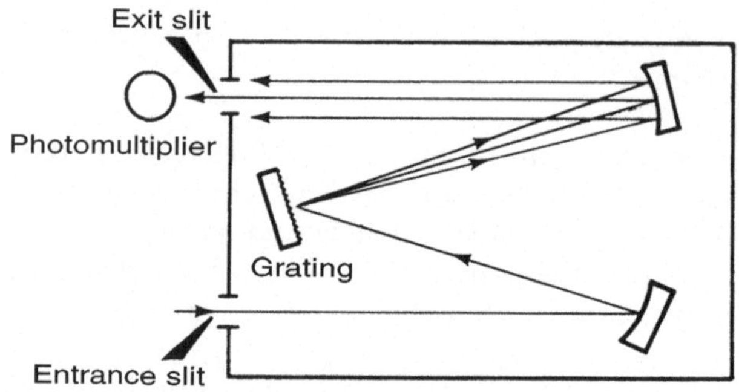

FIGURE 7.4 A monochromator.

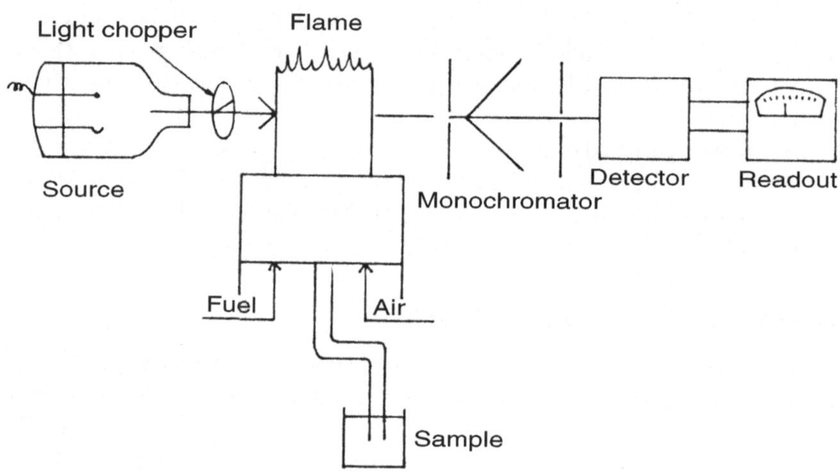

FIGURE 7.5 Basic AA instrument.

angles. By adjusting the angles of the grating, a selected emission light from the source is allowed to pass through the exit slit and focuses on the detector. Curved mirrors within the monochromator comprise the focusing control of the source lamp. A typical monochromator design is shown in Figure 7.4.

The size of the entrance and exit slits should be the same. The size of the slit is variable and adjusted for each element analyzed, according to recommendations by the instrument manufacturer and pertinent reference materials.

7.3.5 Detector

The detector measures the light intensity and transfers it to the readout system. The detector is a multiplier *phototube*, or *photomultiplier* (PM) *tube*.

7.3.6 Readout System

As with molecular spectrophotometry, the readout of the absorbance and transmittance data consists of a meter, recorder, or both. Modern atomic absorption instruments include microcomputer-based electronics. Figure 7.5 shows the basic components of an atomic absorption spectrophotometer.

7.3.7 Automatic Samplers

Automatic samplers offer labor and time savings and thus speed up the analytical process.

7.3.8 Automated Multielement AA Instruments

These instruments set up parameters to preprogrammed values and make it possible to analyze multiple elements in a tray full of samples without operator intervention.

7.3.9 Microcomputer-Based Electronics

Most modern instruments include microcomputer-based electronics. AA instruments are provided with calculation and calibration abilities. Computers can be connected to the instrument output ports to receive, manipulate, and store data and to print reports of calculations.

7.4 SINGLE- AND DOUBLE-BEAM INSTRUMENTS

The differences between single- and double-beam spectrophotometers were discussed in Chapter 6. In the AA technique, the double-beam optical design is generally preferable. Double-beam technology, which automatically compensates for source and common electronics drift, allows these instruments to begin the analysis immediately after the installation of the lamp, with little or no warm-up. This not only reduces analysis time but also prolongs lamp life, as lamp warm-up time is eliminated. Optimized double-beam instruments offer excellent performance, high-speed automation benefits, and operational simplicity. Schematic outlines of the single- and double-beam spectrophotometers are shown in Figures 6.4 and 6.5, respectively.

7.5 ATOMIC ABSORPTION MEASUREMENT TERMS

7.5.1 Calibration

Calibrations are performed at the beginning of the analysis to ensure that the instrument is working properly. Calibrations must be performed according to the analytical methods to be used. *Initial calibration* is determined for each parameter tested and based on the instrument responses for different concentrations of standards, known as *calibration standards.* The number and optimum concentration range of the calibration standards used for each particular method are provided by the approved methodology. A minimum of a blank and three standards must be utilized for calibration. Calibration varies according to the type and model of the equipment. Detailed operation and calibration procedures for each instrument are available in the laboratory's standard operation procedures (SOPs) and the manufacturer's instructions. The instrument response should be linear with the concentration of the introduced standards and plot on a calibration curve, or the instrument software should automatically prepare a curve. Details of calibration curve preparation and the calibration process are provided in Chapter 6.

Calibration accuracy during each analytical run should be ensured via *continuing calibration.* The continuing calibration standard represents the midpoint initial calibration standard. To confirm the calibration curve and to verify the accuracy of the standards and the calibration, run a standard prepared from another source as the calibration standards. Prepare standard solutions of known metal concentrations in water with a matrix similar to the sample.

For samples containing high and variable concentrations of matrix materials, make the major ions in the sample and the standards similar. If the sample matrix is complex and components cannot be matched accurately with standards, use the method of standard addition (see Section 7.7.1). If digestion or another method is used for sample preparation (see Chapter 15), carry standards through the same procedure used for samples.

The range of concentrations over which the calibration curves for an analyte are linear is called the *linear dynamic range*. The highest concentration for an analyte that will result in a linear absorption signal response is the *maximum linear concentration*.

7.5.2 Concentration

When the absorbance of standard solutions containing known concentrations of analyte are measured and the absorbance data plotted against the concentration, a calibration relationship is established. (See calibration details in Section 6.6.) Directly proportional behavior between absorbance and concentration (Beer's law, see Section 5.5) is observed in atomic absorption.

After such calibration, the absorbance of solutions of unknown concentrations may be measured and the concentration determined from the calibration curve. In modern instrumentation, the calibration can be made within the instrument to provide a direct readout of unknown concentrations. Built-in microcomputers make accurate calibration possible, even in the nonlinear region.

7.5.3 Sensitivity

Sensitivity or "characteristic concentration" is a convention for defining the magnitude of the absorbance signal that will be produced by a given concentration of analyte. For flame absorption, this term is expressed in milligrams per liter (mg/l) required to produce a 1% absorption (0.0044 absorbance) signal:

$$\text{Sensitivity (mg/l)} = \text{concentration of standard} \times 0.0044/\text{measured absorbance} \tag{7.1}$$

7.5.4 Detection Limit (DL)

The DL is the smallest measurable concentration at which the analyte can be detected with a specific degree of certainty. The detection limit may be defined as the concentration that will give an absorbance signal of two (sometimes three) times the magnitude of the baseline noise. The baseline noise can be statistically quantitated by making ten or more replicate measurements of the baseline absorbance signal observed for an analytical blank (reagent blank) and determining the standard deviation of the measurements. Therefore, the DL is the concentration that will produce an absorbance signal twice (or three times) the standard deviation of the blank.

Details of the method detection limit, instrument detection limit, and practical detection limit (PDL) are provided in Section 13.8.

7.5.5 Optimum Concentration Ranges

The optimum concentration range usually starts from the concentration of several times the sensitivity and extends to the concentration at which the calibration curve starts to flatten. To achieve best results, use concentrations of samples and standards within the optimum concentration ranges. Sensitivity, detection limits, and optimum ranges vary according to complexity of the matrix, element determined, instrument models, and technique. Table 7.1 shows detection limits obtainable by direct aspiration and furnace techniques for 34 metals.

The concentration range may be extended downward by scale expansion, and extended upward by dilution, using a less sensitive wavelength, rotating the burner head, or utilizing a microprocessor to linearize the calibration curve at high concentrations. Detection limits by direct aspiration may also be extended through concentration of the sample. Lower concentrations may also be detected by

TABLE 7.1
Atomic Absorption Concentration Ranges[a]

	Flame AA		Graphite AA	
Metal	**Detection Limit (mg/l)**	**Optimum Concentration Range (mg/l)**	**Detection Limit (µg/l)**	**Optimum Concentration Range (µ/l)**
Aluminum	0.1	5–50	3	20–200
Antimony	0.2	1–40	3	20–200
Arsenic[b]	0.002	0.002–0.02	1	5–100
Barium (*p*)	0.1	1–20	2	10–200
Beryllium	0.005	0.005–2	0.2	1–30
Cadmium	0.005	0.05–2	0.1	0.5–10
Calcium	0.01	0.2–7	—	—
Chromium	0.05	0.5–10	1	5–100
Cobalt	0.05	0.5–5	1	5–100
Copper	0.02	0.2–5	1	5–100
Gold	0.1	0.5–20	1	5–100
Iridium (*p*)	3	20–500	30	100–1500
Iron	0.03	0.3–5	1	5–100
Lead	0.1	1–20	1	5–100
Magnesium	0.01	0.02–0.5	—	—
Manganese	0.01	0.1–3	0.2	1–30
Mercury[c]	0.0002	0.0002–0.1	—	—
Molybdenum (*p*)	0.1	1–40	1	3–60
Nickel (*p*)	0.04	0.3–5	1	5–100
Osmium	0.3	2–100	20	50–500
Palladium (*p*)	0.1	0.5–15	5	20–400
Platinum (*p*)	0.2	5–75	20	100–2000
Potassium	0.01	0.1–2	—	—
Rhenium (*p*)	5	50–1000	200	500–5000
Rhodium (*p*)	0.05	1–30	5	20–400
Ruthenium	0.2	1–50	20	100–2000
Selenium (2)[b]	0.002	0.002–0.02	2	5–100
Silver	0.01	0.1–4	0.2	1–25
Sodium	0.02	0.03–1	—	—
Thallium	0.1	1–20	1	5–100
Tin	0.8	10–300	5	20–300
Titanium (*p*)	0.4 5–100	10	50–500	
Vanadium (*p*)	0.2	2–100	4	10–200
Zinc	0.005	0.05–1	0.05	0.2–4

Note: The listed furnace values are expected when using a 20-µl injection and normal gas flow except in the cases of As and Se where gas interrupt is used. The *p* indicates use of pyrolytic graphite with the furnace procedure.

[a] The concentrations shown should be obtainable with any good-quality AAS.

[b] Gaseous hydride method.

[c] Cold-vapor technique.

using furnace techniques. In cases where flame AAS does not provide adequate sensitivity, specialized furnace procedures are used, such as the gaseous hydride method (see Section 7.6.3 and Chapter 11) for arsenic and selenium, the cold vapor technique (see Section 7.6.4 and Chapter 10) for mercury, and the chelation-extraction procedure (see Section 7.6.2). Table 7.1 contains the detection limits and optimum concentration ranges of atomic absorption spectrophotometers.

7.6 TECHNIQUES IN AAS MEASUREMENT

Atomic absorption is a mature analytical technique. Interferences are well documented and, for the most part, easy to control. Various atomizer alternatives make atomic absorption one of the most versatile analytical techniques, capable of determining a great number of elements over wide concentration ranges.

7.6.1 Direct-Aspiration or Flame Atomic Absorption Spectrophotometry (FAAS)

In direct-aspiration atomic absorption or flame atomic absorption spectrophotometry (FAAS), a sample is aspirated and atomized in a flame. A light beam from a hollow cathode lamp (HCL) or an electrodeless discharge lamp (EDL) is directed through the flame into a monochromator and onto a detector that measures the amount of absorbed light. Absorption depends on the presence of free excited ground-state atoms in the flame. Because the wavelength of the light beam is characteristic of only the metal being determined, the light energy absorbed by the flame is a measure of the concentration of that metal in the sample. This principle is the basis of AAS. Flames used in the FAAS technique are discussed in Section 7.3.2, and details of the technique appear in Chapter 8.

7.6.2 Chelation-Extraction Method

Many metals at low concentrations — including Cd, Cr, Co, Cu, Fe, Pb, Mn, Ni, Ag, and Zn — can be determined by the chelation-extraction technique. A chelating agent, such as *ammonium pyrrolidine dithiocarbamate* (APDC), reacts with the metal, forming the metal chelate that is then extracted with methyl isobutyl ketone (MIBK). An aqueous sample of 100 ml is acidified to a pH 2 to 3 with 1 ml of 4% APDC solution. The chelate is extracted with MIBK by shaking the solution vigorously for 1 min. If an emulsion formation occurs at the interface of the water and MIBK, use anhydrous sodium sulfate (Na_2SO_4). The extract is aspirated directly into the air–acetylene flame. APDC chelates of certain metals such as Mn are not very stable at room temperature. Therefore, analysis should commence immediately after extraction.

The chelation-extraction method determines Cr in the hexavalent state. In order to determine total Cr, the metal must be oxidized with potassium permanganate ($KMnO_4$) at boiling temperature and the excess $KMnO_4$ is destroyed by hydroxylamine hydrochloride prior to chelation and extraction.

Low concentrations of Al and Be can be determined by chelating with 8-hydroxyquinoline and extracting the chelates into MIBK and aspirating into an N_2O–acetylene flame.

Calibration standards of the metal are similarly chelated and extracted in the same manner, and the absorbances are plotted against concentrations.

7.6.3 Hydride Generation Method

Samples are reacted in an external vessel with a reducing agent, usually sodium borohydride. Gaseous reaction products are then carried to the sampling cell in the light path of the AA spectrophotometer. The gaseous reaction products are not free analyte atoms, but rather volatile hydrides.

To dissociate the hydride gas into free atoms, the sample cell must be heated. The cell is heated by an air–acetylene flame or with another electricity-driven system. The maximum absorption reading or peak height is understood as the analytical signal. This technique is discussed in Chapter 11.

7.6.4 Cold Vapor Atomic Absorption Spectrophotometer

Because atoms cannot exist in the free ground state at room temperature, heat must be applied to the sample to break the bonds connecting atoms into molecules. The only notable exception to this general rule is mercury. Free mercury atoms can exist at room temperature; therefore, mercury can be measured by atomic absorption without a heated sample cell. The cold vapor method, which is applicable to the determination of mercury, is described in Chapter 10.

7.6.5 Electrothermal or Graphite Furnace Atomic Absorption Spectrophotometry (GrAAS)

When using the furnace technique in conjunction with an atomic absorption spectrophotometer, a representative aliquot of a sample is placed in the *graphite tube* in the furnace, evaporated to dryness, charred, and atomized. As a greater percentage of available analyte atoms is vaporized and dissociated for absorption in the tube rather than the flame, the use of smaller sample volumes or detection of lower concentrations of elements is possible.

The principle underlying GrAAS and FAAS is essentially the same, except that a furnace instead of a flame, respectively, is used to atomize the sample. Radiation from a given excited element is passed through the vapor containing ground-state atoms of that element. The intensity of the transmitted radiation decreases in proportion to the amount of ground-state element in the vapor. The metal atoms to be measured are placed in the radiation by increasing the temperature of the furnace, thereby causing the injected specimen to be volatilized. A monochromator isolates the characteristic radiation from the hollow cathode lamp or electrodeless discharge lamp, and a photosensitive device measures the attenuated transmitted radiation. Electrothermal methods generally increase sensitivity. This technique is described in Chapter 9.

7.7 INTERFERENCE IN AAS TECHNIQUES

When the sample alters one or more steps of the above process (Section 7.3) from the performance of the standard, interference exists. If the interference is not recognized and corrected, or compensated, the measured concentration will be inaccurate. Interferences in AAS can be divided into two general categories: nonspectral and spectral.

7.7.1 Nonspectral Interferences

Nonspectral interferences affect the formation of analyte atoms.

7.7.1.1 Matrix Interference

If the sample is more viscous or has different surface tension characteristics than the standards, the sample nebulization may be different from the standards. Consequently, the number of the atoms and thus the absorbance of the standards and samples will not correlate. This situation is known as matrix interference.

For example, *negative interference* is caused by increased acids or dissolved solids in the sample. *Positive error* is caused by the presence of organic solvent in a sample, resulting in increased absorption. To eliminate this interference, any acid or other reagent added to the sample during preparation should also be added to the standard and blank in a similar concentration.

7.7.1.1.1 Correction by Standard Addition Method

In this technique, the accurate concentration of the analyte is obtained without the elimination of the interfering substance. Aliquots of the standard are added to portions of the sample, which allow the interfering substance in the sample to affect the standard as well.

1. Take equal volumes of aliquot from the sample.
2. Add nothing to the first aliquot.
3. Add a measured amount of standard to the second aliquot. The volume of the standard is selected to give the approximate concentration of the analyte in the sample.
4. Add twice the volume of the same standard to the third aliquot of the sample.
5. Add three times the first addition of the standard volume to the next aliquot of the sample.
6. Finally, all portions are diluted to the same volume so that the final concentrations of the original sample constituents are the same in each case. Only the added analyte differs by a known amount.

The absorbance for all of the solutions must fall within the linear portion of the working curve. If there is no interference in the sample, a plot of the measured absorbance vs. concentration of the added standard would be parallel to the aqueous standard calibration. If no interfering substance is present, the absorbance also increases in the added standards and will be proportional to the analyte in the sample. Therefore, the result is also a straight line, but because of the interference substance, its slope will be different from the aqueous standards. Continue the concentration calibration on the abscissa backward from zero and extrapolating the calibration line backward until it intercepts the concentration axis. This will be the concentration corresponding to the absorbance of the unspiked sample. Thus, the presence of interference in the sample can be determined easily by the standard addition method. If the calibration curve of the spiked sample is not parallel with the calibration line of the aqueous standards, interference is present. The standard addition technique is illustrated in Figure 7.6.

7.7.1.2 Chemical Interferences

During the atomization process, sufficient energy should be available to dissociate the molecular form of the analyte and create free atoms. If the sample contains a component that forms a thermally

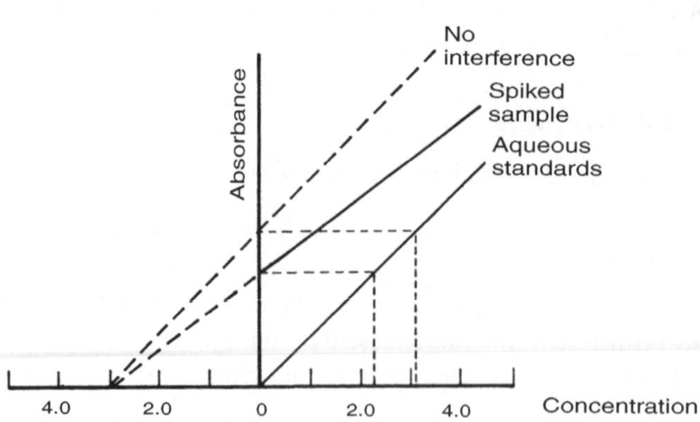

FIGURE 7.6 Standard additions method.

stable compound with the analyte, complete decomposition is not possible. For example, phosphate causes this effect in calcium, because calcium phosphate does not totally dissociate in an air–acetylene flame.

7.7.1.2.1 Correction by Addition of Excess of One Chemical Element or Compound

The first solution to chemical interferences is the addition of an excess of one chemical element or compound that also forms a thermally stable compound with the interfering substance. In the case of calcium, the addition of lanthanum is the solution. Lanthanum combines with phosphate and allows the calcium to be completely atomized.

7.7.1.2.2 Correction by the Use of Nitrous Oxide–Acetylene Flame

The second solution to this kind of interference problem is to use a hotter flame. The nitrous oxide–acetylene flame is considerably hotter than the air–acetylene type and can therefore be used in interference elimination.

7.7.1.3 Ionization Interference

Most of this type of interference occurs in a hot nitrous oxide–acetylene flame. During the dissociation process of the molecules, the excess energy causes atoms to easily lose electrons and become ions. In this case, the excited number of atoms decreases and atomic absorption is reduced, and ionization interference exists.

7.7.1.3.1 Correction by Addition of Ionization Suppressant

Ionization interference can be eliminated with the addition of one element that is very easily ionized, creating a large number of free electrons in the flame and suppressing ionization of the analyte. Potassium (K), rubidium (Rb), and cesium (Cs) salts are commonly used as ionization suppressants. For example, in barium (Ba) determination, adding 1000 to 5000 mg/l potassium (K) to all standards and samples can eliminate the ionization effect. Recommended additions of ionization suppressants are listed below.

1. In *aluminum* (Al), *barium* (Ba), and *chromium* (Cr) *determination*, the addition of 1000 µg/ml (1000 µg/l = 1 mg/l) of potassium (K) is recommended.
   a. Preparation of the stock K solution: Dissolve 95 g of potassium chloride (KCl) in analyte free water and dilute to 1 liter.
   b. Add 2 ml of this stock solution into each 100-ml standard and each 100 ml of sample prior to analysis.
2. In *calcium* (Ca) and *magnesium* (Mg) *determination*, the addition of 1000 µg/ml of lanthanum (La) is advised.
   a. Preparation of the stock La solution: Dissolve 29 g of lanthanum oxide (La_2O_3) in 250 ml of HCl concentrate (be careful, reaction is violent!), and dilute to 500 ml with analyte-free water.
   b. Add 2 ml of this stock solution into each 100-ml standard and sample prior to analysis.
3. In *molybdenum* (Mo) and *vanadium* (V) determination, the addition of 1000 mg/ml of aluminum (Al) is helpful.
   a. Preparation of stock solution: Dissolve 139 g of aluminum nitrate nonahydrate ($Al(NO_3)_3.9H_2O$) in 150 ml of analyte-free water by heating. After the solution is completely cool, dilute to 200 ml.
   b. Add 2 ml of this stock solution to each 100-ml standard and sample prior to analysis.

7.7.2 Spectral Interferences

Spectral interferences are present when the measured light absorption is higher than the absorption of the analyte. This type of interference is the result of light absorption by a nonanalyte element in the sample.

7.7.2.1 Background Absorption

In reality, not all of the matrix materials in the sample are 100% atomized. Undissociated matrix molecules may have broadband absorption spectra, and tiny particles in the flame may scatter light over a wide wavelength region. This type of absorption covers the atomic absorption wavelength of the analyte, and background absorption occurs.

7.7.2.1.1 Continuous Source Background Correction

To eliminate this type of interference, the background absorption must be measured and subtracted from the total absorption to verify the analyte's absorption. The lack of accuracy of this method led to the development of a more convenient technique called *continuous source background correction*, which automatically measures and compensates for any background absorption. This method incorporates a continuum light source in the optical system. Light from both the primary and continuum lamps are combined and pass through the flame and monochromator and reach the detector. The instrument itself separates the detector's response. The continuum source signal can be subtracted electronically from the primary signal source, which contains the sum of the background and atomic absorption signals. The displayed absorbance and concentration will correspond to the difference of the absorbance of the lamps. Figure 7.7 shows how background absorption is eliminated by using this technique.

7.7.2.1.2 Zeeman Background Correction

The other method for correcting background absorbance is the Zeeman background correction. The principle of this technique is that the energy levels of an atom change when the atom is placed in a strong magnetic field. The spectrum of the atom, which is a measure of energy levels, also changes, but the background absorption is usually unaffected by the magnetic field. When an atom is placed in a magnetic field and its atomic absorption is observed with a *polarized light* (polarized light vibrates in only one plane, in contrast to ordinary light, which vibrates in all planes, as can seen in Appendix G), the normal single-line atomic absorption is split into two components symmetrically displaced about the normal position, as seen in Figure 7.8. In the Zeeman background correction, a

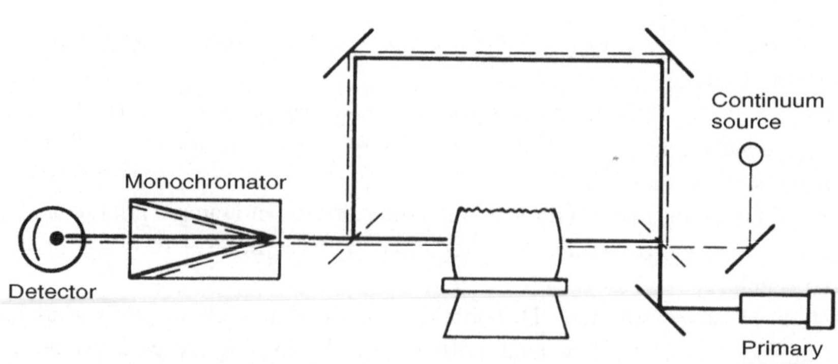

FIGURE 7.7 Continuum source background corrector.

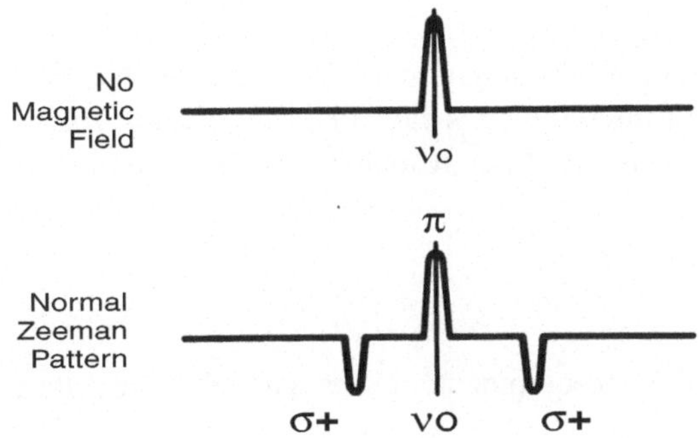

FIGURE 7.8 Zeeman effect.

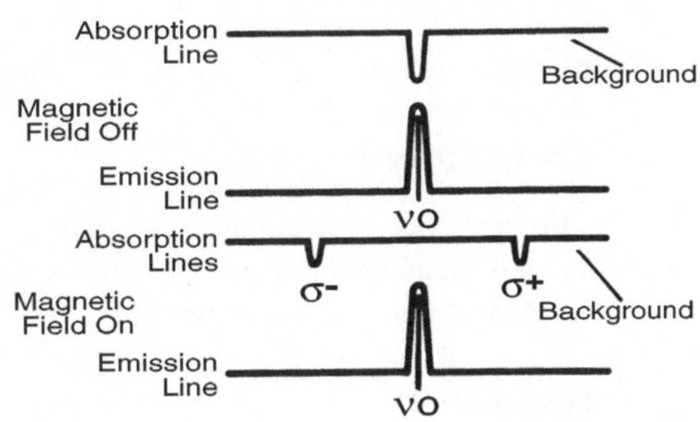

FIGURE 7.9 Zeeman effect background correction.

magnetic field is placed around the atomizer and makes alternating absorption measurements with the magnet off and on. At the "magnet off" position, the uncorrected total absorbance can be measured, and in the "magnet on" position, the background absorbance reading is made, as seen in Figure 7.9. The comparison automatically made by the instrument to compensate for background correction is similar to the continuum source technique. In the Zeeman background correction, the emission profile of the light source is identical in both AA and background measurements. As a result, most complex structured background situations can be accurately corrected with the Zeeman background correction.

7.7.3 Summary of Interferences

7.7.3.1 Nonspectral Interference

1. *Matrix interference*: Sample nebulization is different from standards, so the absorbance of samples is not correlated with standards. This type of interference can be eliminated by special care and selection of the sample preparation technique, or by using the standard addition method.
2. *Chemical interference*: If the sample contains a component that forms a thermally stable component with the analyte, it is not able to complete decomposition. It can be clarified by the addition of an excess of an appropriate chemical element or compound, or by changing to a hotter nitrous oxide-acetylene flame.
3. *Ionization interference*: In the hot nitrous oxide–acetylene flame during the dissociation process, atoms lose electrons and become ions. Consequently, the number of atoms and thus the atomic absorptions are reduced. The interference may be reduced by adding an excess of an element (called a suppressant) that is very easily ionized and suppresses the ionization of the analyte.

7.7.3.2 Spectral Interference

The measured absorption is higher than the absorption of the analyte, caused by the light absorption by a nonanalyte element in the sample. This type of interference is called *background absorption*, and is compensated by instrument correction, such as the continuum source background correction and Zeeman background correction techniques.

7.8 SAFETY IN AAS WORK

Important safety considerations regarding the use of AAS equipment are summarized in the following sections.

7.8.1 FLAMMABILITY OF ACETYLENE

Acetylene, a common fuel in AAS work, is flammable. Its flammability poses a safety problem.

7.8.2 COMBUSTION PRODUCTS

During equipment operation, combustion products can easily pollute the laboratory atmosphere. For this reason, an independently vented fume hood is placed above the burner to remove burned and unburned fuel from the area as illustrated in Figure 7.10.

7.8.3 FLASHBACKS

Flashbacks are minor explosions due to improperly mixed fuel and air. Some specific causes of flashbacks are:

1. Air being drawn back through the drain hole in the mixing chamber of the premix burner. This problem can be avoided by connecting a 6-ft-long tube to the drain hole and forming the tube into a loop, which is then filled with water. The other end is then placed into a container of water (see Figure 7.10).
2. Shutting off all air to the burner before the fuel has been shut off. The solution here is proper operation and proper instruction of operators.
3. Improper proportioning of the fuel air while adjusting the fuel or airflow rate. The solution to this problem is the same as in (2).

Because of the danger of flashbacks, safety glasses must be worn at all times while operating the instrument. General laboratory safety considerations are discussed in Chapter 19.

7.9 QUALITY CONTROL

See Chapter 13 for detailed quality control procedures to be followed during analysis.

7.10 MAINTENANCE OF AA SPECTROPHOTOMETERS

Constant care and routine maintenance are the secret for maintaining proper working conditions of laboratory instruments. Maintenance activities for each instrument are found in the manufacturer's manual. A written maintenance schedule for each instrument must be available. The laboratory must have a maintenance expert on staff or contract with the vendor to provide a specialist for maintenance

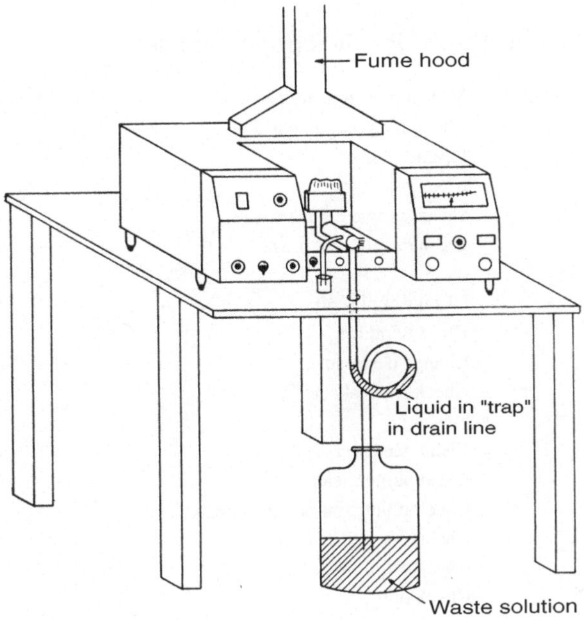

FIGURE 7.10 AA instrument showing fume hood and drain line.

activities. Routine maintenance activities are based on recommendations of the manufacturer of each type and model. Maintenance frequency may also change according to the workload and the type of samples analyzed. Scheduling of maintenance activities by AAS type is presented in Table 7.2.

7.11 AAS PERFORMANCE CHECKS

Performance checks should occur every time a different metal is analyzed as part of the analytical procedure. The performance check is an indicator of deterioration of the lamps or the spectrophotometer and reveals the instrument's optimal operating condition. Performance is measured via a "sensitivity check standard" based on a concentration specific to the method for each metal. The absorbance of this standard should be 0.200. If the absorbance differs by more than ±10%, the instrument is not performing correctly and has to be corrected. Metal concentrations used in sensitivity check data for flame and graphite techniques are presented in Tables 8.1 and Table 9.4, respectively.

7.12 SAMPLE COLLECTION AND SAMPLE PREPARATION

Collection and preparation of environmental samples for analysis using atomic absorption and atomic emission spectrophotometers are discussed in Chapters 14 and 15.

TABLE 7.2
Maintenance of Atomic Absorption Spectrophotometer

Instrument Type	Maintenance Activity	Frequency
UV/Vis	Check lamp alignment	W
	Replace lamp	AN
	Clean windows	Q (I) (C)
	Clean sample compartment	D
	Clean cuvettes after use	D
IR	Clean sample cell	D
	Clean windows	M
	Change dessicant	Q
	Check gas leakage	D
AA flame	Clean nebulizer	D
	Clean burner head	D
	Check tubing, pump, and lamps	D
	Clean quartz windows	W
	Check electronics	SA (I) (C)
	Check optics	A (I) (C)
AA graphite	Check graphite tube	D
	Flush autosampler tubing	D
	Clean furnace housing and injector tip	W
	Check electronics	SA (I) (C)

Note: D = daily; W = weekly; M = monthly; Q = quarterly; SA = semiannually; A = annually; AN = as needed; I = instrumentation specialist; C = on contract.

8 Direct Aspiration or Flame Atomic Absorption Spectrometry (FAAS)

8.1 PRINCIPLE

Flame atomic absorption spectrometry is a rapid and precise method of analysis. In this atomic absorption spectrometry technique (see Section 7.6.1), the sample is vaporized and atomized in a high-temperature flame. Atoms of the analyte element absorb light of a specific wavelength from a hollow cathode lamp (HCL), passing through the flame. The amount of energy absorbed by these atoms is measured and is proportional to the number of atoms in the light path. A light beam is directed through the flame into a monochromator and onto a detector that measures the amount of light absorbed by the atomized element in the flame. The amount of energy at the characteristic wavelength absorbed in the flame is proportional to the concentration of the element in the sample over a limited concentration range. Table 8.1 shows the FAAS concentration ranges.

Determinations of analyte concentrations in a milligram-per-liter concentration region are routine for most elements. However, trace metal analyses at microgram-per-liter and nanogram-per-liter levels are also needed. Because the thermal energy from the flame is responsible for producing the absorbing species, flame temperature is an important parameter governing the flame process. The two premixed flames used almost exclusively for atomic absorption are *air–acetylene* (2125–2400°C) and *nitrous oxide–acetylene* (2000–2800°C).

While the air–acetylene flame is satisfactory for most elements determined by atomic absorption, the hotter nitrous oxide–acetylene flame is required for many refractory-forming elements. The nitrous oxide–acetylene flame is also effective in the control of some types of interferences.

8.2 DIRECT AIR–ACETYLENE FLAME METHOD

8.2.1 General Discussion

Because the thermal energy from the flame is responsible for producing the absorbing species, flame temperature is an important parameter governing the flame process. Flames used in the AAS technique are discussed in Section 7.3.2. The air–acetylene flame is satisfactory for most elements determined by atomic absorption, including Sb, Bi, Cd, Cs, Cr, Co, Cu, Au, Ir, Fe, Pb, Li, Mg, Mn, NI, Pa, Pt, K, Rh, Ru, Ag, Na, Sr, Ta, Sn, and Zn.

8.2.2 Instrumentation

See Section 7.3.

TABLE 8.1
Atomic Absorption Concentration Ranges, FAAS Technique

Element	Wavelength (nm)	Flame Gas	Instrument Detection Limit (mg/l)	Sensitivity (mg/l)	Optimum Concentration Range (mg/l)
Ag	328.1	A–Ac	0.01	0.06	0.1–4
Al	309.3	N–Ac	0.1	1	5–50
Au	242.8	A–Ac	0.01	0.25	0.5–20
Ba	553.6	N–Ac	0.03	0.4	1–20
Be	234.9	N–Ac	0.005	0.03	0.05–2
Bi	223.1	A–Ac	0.06	0.4	1–50
Ca	422.8	A–Ac	0.03	0.08	0.2–20
Cd	228.8	A–Ac	0.02	0.025	0.05–2
Co	240.7	A–Ac	0.03	0.2	0.5–10
Cr	357.9	A–Ac	0.02	0.1	0.2–10
Cs	852.1	A–Ac	0.02	0.3	0.5–15
Cu	324.8	A–Ac	0.01	0.1	0.20–10
Fe	248.3	A–Ac	0.02	0.12	0.3–10
Ir	264.0	A–Ac	0.6	8	—
K	766.5	A–Ac	0.005	0.04	0.1–2
Li	670.8	A–Ac	0.002	0.04	0.1–2
Mg	285.2	A–Ac	0.0005	0.007	0.2–2
Mn	279,5	A–Ac	0.01	0.05	0.1–10
Mo	313.3	N–Ac	0.1	0.5	1–20
Na	589.0	A–Ac	0.02	0.015	0.03–1
Ni	232.0	A–Ac	0.02	0.15	0.3–10
Os	290.9	N–Ac	0.08	1	—
Pb[a]	283.3	A–Ac	0.05	0.5	1–20
Pt	265.9	A–Ac	0.1	2	5–75
Rh	343.5	A–Ac	0.5	0.3	—
Ru	349.9	A–Ac	0.07	0.5	—
Sb	217.6	A–Ac	0.07	0.5	1–40
Si	251.6	N–Ac	0.3	2	5–150
Sn	224.6	A–Ac	0.8	4	10–200
Sr	460.7	A–Ac	0.03	0.15	0.3–5
Ti	365.3	N–Ac	0.3	2	5–100
V	318.4	N–Ac	0.2	1.5	2–100
Zn	213.9	A–Ac	0.05	0.2	0.05–2

Note: A–Ac = air–acetylene; N–Ac = nitrous oxide–acetylene.
[a] The more sensitive 217.0-nm wavelength is recommended for instruments with background correction capabilities.

8.2.3 Reagents

8.2.3.1 Air

The air used should be cleaned and dried through a suitable filter to remove oil, water, and other foreign substances. Sources include a compressor or commercially bottled gas.

8.2.3.2 Acetylene

The acetylene used should be a standard commercial grade. Acetone, which is always present in acetylene cylinders, can be prevented from entering and damaging the burner head by replacing a cylinder when its pressure has fallen to 689 kPa (100 psi) acetylene. Because of differences among makes and models of AASs, it is not possible to formulate measurements applicable to every instrument.

8.2.3.3 Metal-Free Water

Use metal-free water for preparing all reagents and calibration standards and as dilution water. Prepare metal-free water by deionizing and distilling, redistilling, or sub-boiling tap water. Always check laboratory pure water to determine whether the element of interest is present in trace amounts.

Note: If the source water contains Hg or other volatile metals, single or redistilled water may not be suitable for trace analysis, because these metals remain in distilled water. In such cases, use sub-boiling to prepare metal-free water.

8.2.3.4 Stock Metal Solutions

Stock metal solutions are commercially available or can be prepared according to recipes in Appendix H.

8.2.4 Operation

Because of differences among makes and models of AASs, it is not possible to formulate instructions applicable to every instrument. Follow the manufacturer's instructions, but in general proceed as follows:

1. Install hollow cathode lamp for the desired metal.
2. Set wavelength dial as specified by the analytical methodology (see Table 8.1).
3. Set slit width according to manufacturer's suggested setting.
4. Turn on instrument, apply the hollow cathode lamp current suggested by the manufacturer, and let the instrument warm up until energy source stabilizes, usually about 10 to 20 min.
5. Readjust current if necessary after warm-up. Adjust wavelength dial until optimum energy gain is obtained.
6. Align lamp in accordance with manufacturer's instructions.
7. Install suitable burner head, and adjust its position. A 10-cm, single-slot burner head is recommended for air–acetylene flames.
8. Turn on air, and adjust flow rate according to manufacturer's instructions to give maximum sensitivity for the metal being measured.
9. Turn on acetylene and adjust flow rate to value specified.
10. Ignite flame and let it stabilize for a few minutes.
11. Aspirate blank and zero instrument.
12. Aspirate a standard solution and adjust aspiration rate of nebulizer to obtain maximum sensitivity.
13. Adjust burner both vertically and horizontally to obtain maximum response.
14. Aspirate blank again and re-zero instrument.
15. Aspirate a standard with a concentration near the middle of the linear range and record absorbance.
16. The instrument is now ready to operate.
17. When analyses are finished, extinguish flame by turning off acetylene first and then air.

8.2.5 Standardization

1. Select at least three concentrations of each standard metal solution to bracket expected metal concentration of sample. (See Table 8.1 for the optimum concentration ranges of metals.)
2. Aspirate blank and zero instrument.
3. Aspirate each standard and record absorbances. Prepare the calibration curve (Section 6.6). For instruments equipped with direct concentration readout, this step is unnecessary. Preparation of standards and complete calibration processes are discussed in Section 6.6.

Calibrate Ca and Mg based on concentration of standards before dilution with *lanthanum solution*; Fe and Mn based on original concentration of standards before dilution with *calcium solution*; and Cr based on original standards before addition of H_2O_2. See methodologies for determining these metals in Chapter 18.

8.2.6 Sample Analysis

1. Rinse the nebulizer by aspirating water with 1.5 ml of concentrated HNO3/l.
2. Aspirate blank and zero instrument. Analyze samples.

8.2.7 Calculations

1. Calculate from calibration curve, or read directly from instrument the concentration in milligrams or micrograms per liter according to calibration.
2. If the sample has been diluted, multiply concentration readout by the appropriate dilution factor.
3. If the sample has been concentrated, divide concentration readout by appropriate concentration factor.

8.3 DIRECT NITROUS OXIDE–ACETYLENE FLAME METHOD

8.3.1 General Discussion

The hotter nitrous oxide—-acetylene flame (2600–2800°C) is required for many refractory-forming elements. The nitrous oxide–acetylene flame is also effective in the control of some types of interference. This method is applicable to the determination of Al, Ba, Be, Mo, Os, Re, Si, Th, Ti, and V.

8.3.2 Apparatus

Use atomic absorption spectrophotometer and associated equipment. See Section 7.3.

8.3.2.1 Nitrous Oxide Burner Head

Use special burner head as suggested in the manufacturer's manual. At roughly 20-min intervals of operation, it may be necessary to dislodge the carbon crust that forms along the slit surface with a carbon rod or appropriate alternative. Usually a special 5-cm head is required when using a nitrous oxide–acetylene flame. Burner heads are discussed in Section 7.3.3.

8.3.2.2 T-Junction Valve

Use a T-junction valve or other switching valve for rapidly changing from nitrous oxide to air, so that the flame can be turned on or off with air as the oxidant to prevent flashbacks.

8.3.3 Reagents

8.3.3.1 Air

See Section 8.2.3.

8.3.3.2 Acetylene

See Section 8.2.3.

8.3.3.3 Nitrous Oxide

Nitrous oxide is commercially available in cylinders. Fit the nitrous oxide cylinder with a special nonfreezable regulator, or wrap a heating coil around an ordinary regulator to prevent flashback at the burner caused by reduction in nitrous oxide flow through a frozen regulator. Some AASs have automatic gas control systems that will shut down a nitrous oxide flame safely in the event of a reduction in nitrous oxide flow rate.

8.3.3.4 Metal-Free Water

See Section 8.2.3.

8.3.3.5 Potassium Chloride Solution

Dissolve 250 g of KCl and dilute to 1000 ml with water.

8.3.3.6 Aluminum Nitrate Solution

Dissolve 139 g of $Al(NO_3)_3.9H_2O$ in 150 ml of water. Acidify slightly with concentrated HNO_3 to prevent hydrolysis and precipitation. Warm to dissolve completely. Cool and dilute to 200 ml.

8.3.3.7 Stock Metal Solutions

These solutions are commercially available or can be prepared according to recipes in Appendix H.

8.3.4 Operation

Instrument operation follows the air–acetylene flame method, as discussed in Section 8.2.4. After steps 1 to 6, proceed as follows:

7. After adjusting wavelength, install a nitrous oxide burner head.
8. Turn on acetylene (without igniting flame) and adjust flow rate according to value specified by manufacturer for a nitrous oxide–acetylene flame.
9. Turn off acetylene.
10. With both air and nitrous oxide supplies turned on, set T-junction valve to nitrous oxide and adjust flow rate according to manufacturer's specifications.
11. Turn switching valve to the air position and verify that the flow rate is the same.
12. Turn acetylene on and ignite to a bright yellow flame.
13. With a rapid motion, turn switching valve to nitrous oxide. The flame should have a red cone above the burner. If it does not, adjust fuel flow to obtain red cone. After nitrous oxide flame has been ignited, let burner come to thermal equilibrium before starting analysis.
14. Atomize water containing 1.5 ml of concentrated HNO_3 per liter and check aspiration rate. Adjust if necessary to a rate between 3 and 5 ml/min.

15. Aspirate standard of desired metal with concentration near the midpoint of the optimum concentration range and adjust burner (both vertically and horizontally) in light path to obtain maximum response. The instrument is ready for analysis.
16. To extinguish flame, turn switching valve from nitrous oxide to air and turn off acetylene. This procedure eliminates the danger of flashback that may occur on direct ignition or shutdown of nitrous oxide and acetylene.

8.3.5 Standardization

1. Select at least three concentrations of standard metal solution to bracket expected metal concentration of sample. See Table 8.1 for optimum concentration ranges of metals.
2. Aspirate blank and zero instrument.
3. Aspirate each standard and record absorbances. Prepare calibration curve (Section 6.6). For instruments equipped with direct concentration readout, this step is unnecessary. Preparation of standards and complete calibration processes are discussed in Section 6.6.

For 100 ml of Al, Ba, and Ti standards, add 2 ml of KCl solution (dissolve 250 g of KCl and dilute to 1000 ml). For 100 ml of Mo and V standards, add 2 ml of $Al(NO_3)_3.9H_2O$ solution. (Dissolve 139 g of $Al(NO_3)_3.9H_2O$ in 150 ml of water, acidify slightly with concentrated HNO_3, and warm to dissolve completely. Cool and dilute to 200 ml.)

Most modern instruments are equipped with microprocessors and digital readouts that permit calibration in the direct concentration range.

8.3.6 Analysis of Samples

1. After standardization, rinse atomizer by aspirating water with 1.5 ml of concentrated HNO_3 per liter and zero instrument.
2. Analyze samples. When analyzing Al, Ba, and Ti, add 2 ml of KCl solution to a 100-ml sample before analysis. When analyzing Mo and V, add 2 ml of $Al(NO_3)_3$ solution to a 100-ml sample.

8.3.7 Calculations

See Section 8.2.7.

8.4 INTERFERENCES, SAFETY, AND QUALITY CONTROL REQUIREMENTS IN FAAS

See Sections 7.7, 7.8, and 7.9, respectively.

8.5 MAINTENANCE OF FAA SPECTROPHOTOMETER

Maintenance activities for the FAAS are listed in Table 8.2.

8.6 PERFORMANCE CHECK OF FAA SPECTROPHOTOMETER

Performance of the AAS is checked every time a different metal is analyzed. The performance check is an indicator of deterioration of the lamps or the spectrophotometer and reveals the optimal operating condition of the instrument. Performance is measured by a "sensitivity check standard" based

TABLE 8.2
Maintenance of FAAS

Maintenance Activity	Frequency
Clean bebulizer	D
Clean burner head	D
Check tubing, pump, and lamps	D
Clean quartz windows	W
Check electronics	SA (I) (C)
Checks optics	A (I) (C)

Note: D = daily; W = weekly; SA = semiannually; A = annually; I = instrumentation specialist; C = on contract.

on a concentration specific to the method for each metal. The absorbance of this standard should be 0.200. If it differs by more than 10%, the instrument is not performing correctly and has to be corrected. Metal concentrations used in performance checks for the FAAS technique are presented in Table 8.3. Standard conditions for the FAAS technique are summarized in Table 8.4.

TABLE 8.3
FAAS Performance Check

Element	Concentration of Sensitivity Standard (mg/l)
Aluminum (Al)	50
Antimony (Sb)	25
Barium (Ba) 20	
Beryllium (Be)	1.5
Calcium (Ca) at 422.7 nm	4.0
Calcium (Ca) at 287.4 nm	60
Cadmium (Cd) at 228.8 nm	1.5
Cadmium (Cd) at 368.4 nm	850
Cobalt (Co) 7.0	
Chromium (Cr)	4.0
Copper (Cu) 4.0	
Iron (Fe) 5.0	
Lead (Pb) 20	
Potassium (K)	2.0
Magnesium (Mg)	0.3
Manganese (Mn)	2.5
Molybdenum (Mo)	30
Nickel (Ni) 7.0	
Silicon (Si) 100	
Sodium (Na) 0.5	
Strontium (Sr)	5.0
Tin (Sn) 150	
Titanium (Ti)	80
Thallium (Ta)	30
Tungsten (W)	450
Zinc (Zn) 1.0	
Zirconium (Zr)	300

Note: Performance of the FAA should be checked every time a metal is analyzed by using a sensitivity check standard. The sensitivity check data here pertain to the metal concentration (mg/l) in aqueous solution, which will give a reading of approximately 0.20 absorbance units.

TABLE 8.4
Standard Conditions for Flame AAS

Metal	Fuel Oxidant	SBW (nm)	Wavelength (nm)	Optimum Range (mg/l)	Sensitivity Check (mg/l)	Detection Limit (mg/l)	Addition of Suppressant
Ag	Ac–air	0.7	328.1	0.1–4	2.5	0.01	[a]
Al	Ac–N_2O	0.7	324.7	5–50	50.0	0.1	—
Ba	Ac–N_2O	0.4	553.6	1.0–20	20.0	0.1	2 ml KCL per 100-ml sample
Be	Ac–N_2O	0.7	234.9	0.05–2	1.5	0.005	—
Cd	Ac–air	0.7	228.8	0.05–2	1.5	0.005	—
Ca	Ac–N_2O	0.7	422.7	0.2–7	4.0	0.01	1 ml $LaCl_3$ per 10-ml sample
Cr	Ac–N_2O	0.7	357.9	0.5–10	4.0	0.05	2 ml KCL per 100-ml sample
Co	Ac–air	0.2	240.7	0.5–5	7.0	0.05	—
Cu	Ac–air	0.7	324.7	0.2–5	4.0	0.02	—
Fe	Ac–air	0.2	248.3	0.3–5	5.0	0.03	—
K	Ac–air	1.4	766.5	0.1–2	2.0	0.01	[a]
Mg	Ac–air	0.7	285.2	0.02–0.05	0.3	0.001	1 ml $LaCl_3$ per 10-ml sample
Mn	Ac–air	0.2	279.5	0.1–3	2.5	0.01	[a]
Mo	Ac–N2O	0.2	313.3	1–40	30	0.1	—
Ma	Ac–air	0.4	589.0	0.03–1	0.5	0.002	[a]
Ni	Ac–air	0.2	232.0	0.3–5	7.0	0.04	—
Pb	Ac–air	0.7	283.3	1–20	20	0.1	[a]
Sb	Ac–air	0.2	217.6	1–40	25	0.2	—
Si	Ac–N_2O	0.2	251.6	—	100	—	—
Sn	Ac–N_2O	0,7	286.3	10–300	150	0.8	—
Sr	Ac–N_2O	0.4	460.7	—	5.0	—	—
Ti	Ac–Air	0.7	276.8	1–20	30	0.1	[a]
V	Ac–N_2O	0.7	318.4	2–200	90	0.2	2 ml $Al(NO_3)_2$ per 100-ml sample
Zn	Ac–air	0.7	213.9	0.05–1	1.0	0.005	a

Note: Potassium chloride (KCl) solution: Dissolve 95 g of KCl in analyte-free water and dilute to 1 liter. Lanthanum chloride ($LaCl_3$) solution: Dissolve 29 g of La_2O_3 in 250 ml of concentrated HCl (be careful, because reaction is violent!), and dilute to 500 ml with analyte-free water. Aluminum nitrate ($Al(NO_3)_2$) solution: Dissolve 139 g of aluminum nitrate nonahydrate ($Al(NO_3)_2.9H_2O$) in 150 ml of analyte-free water. Heat. After dissolution, cool to room temperature, and dilute to 200 ml. Suppressant should be added to the blanks, standards, and samples. Addition of alkali salt is recommended to control ionization. Aluminum is added to improve sensitivity and linearity. Lanthanum is added to improve sensitivity.

[a] The use of an impact bead will improve sensitivity by about 2%.

9 Graphite Furnace Atomic Absorption Spectrometry

9.1 GENERAL DISCUSSION

9.1.1 APPLICATION

An atomic absorption spectrophotometer equipped with a graphite furnace or an electrically heated atomizer, instead of a standard burner head, offers better sensitivity and a much lower detection limit compared to the flame method (see Chapter 8). The sensitivity of electrothermal AAS or graphite furnace AAS (GrAAS) makes it ideal for trace metals analysis. The GrAAS determination of most metallic element sensitivities and detection limits is 20 to 1000 times better than that of conventional flame techniques. Many elements can be determined at concentrations as low as 1.0 µg/l.

The GrAAS technique is subject to more interference than the FAA procedure and requires more analysis time. However, extensive studies of the furnace technique combined with the development of improved instrumentation have changed the GrAAS into a highly reliable, routine technique for trace metals analysis. It is also much more automated than the other techniques in this field. The entire process is automated once the sample has been introduced and the furnace program initiated. With the use of automatic samplers, a completely unattended operation is possible.

An additional benefit of the GrAA technique is the use of microliter sample sizes. Routine determination at the microgram-per-liter (ppb) level for most elements makes it ideal for environmental applications, but it is also suitable for biological and geological samples, and many clinical analyses. The graphite furnace method can determine most elements, measured by atomic absorption, in a wide variety of matrices.

9.1.2 PRINCIPLE

Electrothermal atomic absorption spectroscopy (GrAAS) is based on the same principle as direct-flame atomic absorption spectroscopy (FLAAS). In the GrAAS technique, a tube of graphite is located in the sample compartment of the AAS, with the light path passing through it. A small volume of sample solution is quantitatively placed into the tube, normally through a sample injection hole located in the center of the tube wall. The tube is heated through a programmed temperature sequence until finally the analyte present in the sample is dissociated into atoms. The resultant ground-state atomic vapor absorbs monochromatic radiation from the source. As atoms are created and diffuse out of the tube, the absorbance rises and falls in a peak-shaped signal. The peak height or integrated peak area is used as the analytical signal for quantitation. The detection limits, optimum concentration ranges, and wavelengths used are presented in Table 9.1. For a comparison with the flame AA technique, see Table 8.1. (Sensitivity, detection limits, and optimum concentration range are discussed in Sections 7.5.3, 7.5.4, and 7.5.5, respectively).

Use of the *stabilized temperature platform furnace* (STPF) *technique* also offers significant interference reduction with improved sensitivity. See Section 9.5 for a detailed discussion of the STPF.

Sensitivity changes with sample tube age. Discard graphite tubes when significant variations in sensitivity or poor reproducibility are observed. The use of high acid concentrations, brine samples, and matrix modifiers (see Sections 9.3.2 and 9.4.1) often drastically reduce tube life. Use of the L'vov platform (Section 9.4.2) in such situations is preferable.

9.2 APPARATUS

9.2.1 Atomic Absorption Spectrophotometer

A single- or dual-channel or single- or double-beam instrument has a grating monochromator, photomultiplier detector, adjustable slits, and a wavelength range of 190 to 800 nm, and is equipped with a strip-chart recorder (see Section 9.2.5). The instrument must have background correction capability.

9.2.2 Burner

The burner recommended by the instrument manufacturer should be used. For certain elements, the nitrous oxide burner is required.

9.2.3 Hollow Cathode Lamps

Single-element lamps are preferred but multielement lamps are also used. Electrodeless discharge lamps (EDLs) may also be used when available (Section 7.3.1).

9.2.4 Graphite Furnace

Any furnace device capable of reaching the specified temperatures is satisfactory. Use an electrically heated device with electronic control circuitry designed to carry a graphite tube through a heating program that provides sufficient thermal energy to atomize the element of interest. Fit the furnace into the sample compartment of the spectrometer in place of the conventional burner assembly. Use argon as a purge gas to minimize oxidation of the furnace tube and to prevent the formation of metallic oxides. The graphite furnace is made up of three major components: atomizer, power supply, and programmer.

9.2.4.1 Atomizer

The atomizer is located in the sampling compartment of the AAS. The basic graphite furnace atomizer is composed of the following components: graphite tube, electrical connection, water-cooled housing, and inert gas purge control. Figure 9.1 illustrates the graphite furnace atomizer.

9.2.4.1.1 Graphite Tube

Normally the graphite tube is the furnace's heating element. This cylindrical tube is aligned horizontally in the optical path of the spectrophotometer and serves as the sampling cell. A few microliters (5–50 μl) of sample are measured and dispensed through a hole in the center of the tube wall or graphite platform. Use graphite tubes with L'vov platforms to minimize interference and to improve sensitivity.

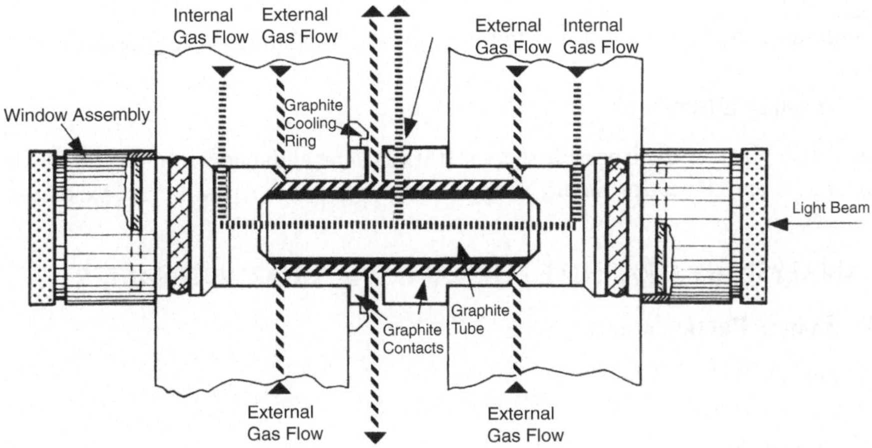

FIGURE 9.1 Graphic furnace atomizer.

9.2.4.1.2 Electrical Connection

The tube is held in place between two graphite contact cylinders that provide the electrical connection. An electrical potential applied to the contacts causes current to flow through the tube, which heats the tube and the sample.

9.2.4.1.3 Water-Cooled Housing

The entire assembly is mounted within an enclosed water-cooled housing. Quartz windows at each end of the housing allow light to pass through the tube.

9.2.4.1.4 External and Internal Gas Flows

The heated graphite is protected from air oxidation by the end windows and two streams of argon. An external gas flow surrounds the outside of the tube, and a separately controlled internal gas flow purges the inside of the tube. During atomization, the internal gas flow is reduced or, preferably, completely interrupted. This maximizes sample residence time in the tube and increases the measurement signal.

9.2.4.2 Power Supply

The power supply controls the *electrical power* of the graphite tube.

9.2.4.3 Programmer

The temperature of the tube is controlled by a user-specified *temperature program*. Through the programmer, the operator selects the sequence of the temperature vs. time during atomization. The programmer also controls the *internal inert gas flow rate* and certain *spectrometer functions*.

9.2.5 Strip-Chart Recorder

A recorder is recommended for furnace work to generate a permanent record and as a means to easily recognize problems, such as drift, incomplete atomization, losses during charring, changes in sensitivity, peak shape, and so on.

9.2.6 Water Supply for Cooling

Cool with tap water flowing at 1 to 4 l/min or use a recirculating cooling device.

9.2.7 Sample Dispensers

Use microliter pipets with disposable tips (5–100 ml) or an automatic sampling device designed for the specific instrument. Pipet tips should be checked prior to use as a possible source of contamination.

9.3 ANALYSIS BY GRAPHITE FURNACE SPECTROPHOTOMETER

9.3.1 Sample Pretreatment

See Chapter 15.

9.3.2 Reagents

9.3.2.1 Metal-Free, Reagent-Grade Water

Use metal-free water for preparing all reagents and calibration standards and as a dilution water. Always check deionized or distilled water to determine whether the element of interest is present in trace amounts. If the source water contains Hg or other volatile metals, distilled or redistilled water may not be suitable for trace analysis because these metals remain in distilled water. In such cases, use sub-boiling to prepare metal-free water.

9.3.2.2 Hydrochloric Acid (HCl)

Both concentrate and 1:1 dilution are used.

9.3.2.3 Nitric Acid (HNO_3)

Both concentrate and 1:1 dilution are used.

9.3.2.4 Stock Metal Solutions

These solutions are commercially available or can be prepared according to recipes in Appendix H.

9.3.2.5 Matrix Modifiers

9.3.2.5.1 Ammonium Nitrate, 10% (w/v)
Dissolve 100 g of NH_4NO_3 and dilute to 1000 ml with reagent-grade water.

9.3.2.5.2 Ammonium Phosphate, 40% (w/v)
Dissolve 40 g of $(NH_4)_2HPO_4$ and dilute to 100 ml with reagent-grade water.

9.3.2.5.3 Calcium Nitrate, 20,000 mg Ca per Liter
Dissolve 11.8 g of $Ca(NO_3)_2.4H_2O$ and dilute to 100 ml with reagent-grade water.

9.3.2.5.4 Nickel Nitrate, 10,000 mg Ni per Liter
Dissolve 49.56 g of $Ni(NO_3)_2.6H_2O$ and dilute to 1000 ml with reagent-grade water.

9.3.2.5.5 Phosphoric Acid, 10% (v/v)
Dilute 10 ml of H_3PO_4 concentrate to 100 ml with reagent-grade water.

9.3.2.6 Metal-Free Seawater or Brine

1. Fill a 1.4-cm × 20-cm borosilicate glass column to within 2 cm of the top with purified chelating resin (see Section 9.3.2.7 below).
2. Elute resin with successive 50-ml portions of 1N HCl, metal-free water, 1N NaOH, and metal-free water at the rate of 5 ml/min just before use.
3. Pass salt water or brine through the column at a rate of 5 ml/min to extract trace metals present. Discard the first 300 ml of eluate.

9.3.2.7 Chelating Resin

1. Purify 100- to 200-mesh (Chelex 100 or equivalent) by heating at 60°C in 19N NaOH for 24 h.
2. Cool resin and rinse ten times each with alternating portions of 10N HCl, metal-free water, 1N NaOH, and metal-free water.

9.3.3 Instrument Operation

1. Mount and align the furnace device according to the manufacturer's instructions.
2. Turn on the instrument and strip-chart recorder.
3. Select the appropriate light source and adjust the recommended electrical setting.
4. Select the appropriate wavelength and set all conditions according to the manufacturer's instructions, including background correction. Background correction is important when elements are determined at short wavelengths or when the sample has a high level of dissolved solids. In general, background correction is usually not necessary at wavelengths longer than 350 nm. Above 350 nm, deuterium arc background correction is not useful and other types of correction must be used.
5. Through the programmer, enter a sequence of selected temperature vs. time to carefully dry, pyrolyze, and finally, atomize the sample. The program may also include settings for the internal inert gas flow rate and, in some cases, the selection of an alternate gas.

9.3.4 Multi-Step Temperature Program

1. *Drying step*: After the sample is placed in the furnace, it must be dried at a sufficiently low temperature to avoid sample splattering. Temperatures of 100 to 120°C are common for aqueous samples. During the drying step, the internal gas flow is left at a 300-ml/min value to purge out the vaporized solvent.
   a. Ramp time is a variable time over which the temperature is increased. A longer ramp time provides a slower, more gentle increase in heat. Ramp time is usually 1 sec with the platform, and longer when the sample is atomized from the tube wall.
   b. Hold time is the time that the sample is held at the selected temperature until dry. Because only a few microliters of sample are used, it is usually less than a minute.
2. *Pyrolysis step*: The temperature is increased as high as possible to volatilize the components, but low enough to prevent loss of the analyte. Temperature selection depends on the nature of the analyte and the matrix.
3. *Cool-down or preatomization step*: Cool the furnace prior to atomization.
4. *Atomization step*: The temperature is increased until the volatilized molecules are dissociated and produce an atomic vapor of the analyte. Atomization temperature is a property specific to each element and recommended in the analytical procedure. Excessively high temperatures cause decreased sensitivity and shorten the lifetime of the graphite tube.

Interrupting the internal gas flow during atomization is desirable. This increases the residence time of the atomic vapor in the furnace, maximizing sensitivity and minimizing interference. At the beginning of this step, the spectrometer "read" function is triggered to begin the measurement of light absorption.

5. *Clean-out step*: After atomization, increase the temperature to burn out all remaining sample residue in the graphite tube.
6. *Cool-down step*: Allow the furnace to cool down to near-ambient temperature before introducing the next sample. In some systems, this step is automatically presented and does not need to be programmed.

9.3.5 Measuring the Graphite Furnace AA Signal

In FAA work, a constant absorbance is observed, but in a GrAA the signal is transient. As atomization begins, analyte atoms are formed, and the signal increases. The signal will continue to increase until the rate of atom generation becomes less than the rate of atom diffusion out of the furnace. The falling atom population results in a signal that decreases until all atoms are lost and the signal has fallen to zero. To determine the analyte content of the sample, the resulting peak-shaped signal must be quantitated.

Modern instrumentation provides the capability of integrating absorbance during the entire atomization period, yielding a signal equal to the integrated peak area, which is the area under the peak signal. The peak area represents a count of all atoms present in the sample aliquot, regardless of whether the atoms were generated early or late in the atomization process. Integrated peak-area measurements are independent of the atomization rate and are therefore much less subject to matrix effects. The peak area is preferred for graphite furnace analysis.

9.3.6 Instrument Calibration

1. Prepare standards for instrument calibration by dilution of the metal stock solutions. Prepare standards fresh daily.
2. Prepare a blank and at least three standards in the appropriate concentration range (see Table 9.1). Match the acid background similar to the sample. Add the same matrix modifier (if required for sample analysis) to the standard solutions.
3. Inject a suitable portion of each standard solution in order of increasing concentration. Analyze each standard solution in triplicate to verify method precision.
4. Construct an analytical curve by plotting absorbance of the standard solutions vs. concentration (see Section 6.6). Alternatively, use electronic instrument calibration if the instrument has this capability.

9.3.7 Sample Analysis

1. Measure and dispense a known volume of sample into the furnace. The analytical range of the furnace analysis can be controlled by varying the sample volume.

   For very low concentrations, the maximum sample volume can be used.

   For higher concentrations, the sample volume can be reduced. Smaller sample volumes can also be used when sample availability is limited.

   Sample *maximum volume* is up to 100 μl when the graphite platform is not used, and less than 50 μl when the platform is in place. A *convenient sample volume* for analysis is 20 μl. The use of an *autosampler* is recommended because it provides excellent reproducibility and superior results.
2. Subject the sample to a multistep temperature program (see Section 9.3.4). When the temperature is increased to the point where sample atomization occurs, the atomic absorption measurement is taken.

TABLE 9.1
Detection Limits and Concentration Ranges for GrAAS

Element	Wavelength (nm)	Detection Limit (µg/l)	Optimum Concentration Range (µg/l)
Aluminum (Al)	300.3	3	20–200
Antimony (Sb)	217.6	3	20–300
Arsenic (As)	193.7	1	5–100
Barium (Ba)	553.6	2	10–200
Beryllium (Be)	234.9	0.2	1–30
Cadmium (Cd)	228.8	0.1	0.5–10
Chromium (Cr)	357.9	2	5–100
Cobalt (Co)	240.7	1	5–100
Copper (Cu)	324.7	1	5–100
Iron (Fe)	248.3	1	5–100
Lead (Pb)	283.3	1	5–100
Manganese (Mn)	279.5	0.2	1–30
Molybdenum (Mo)	313.3	1	3–60
Nickel (Ni)	232.0	1	5–100
Selenium (Se)	196.0	2	5–100
Silver (Ag)	328.1	0.2	1–25
Selenium (Se)	224.6	5	20–300

Note: For Pb determination, the most sensitive 217.0 wavelength is recommended for instruments with background correction capabilities.

3. Repeat until reproducible results are obtained.
4. Compare the absorbance value to the calibration curve to determine the concentration of the element of interest. Alternatively, read results directly if the instrument is equipped with this capability.
5. If the absorbance (or concentration) of the most concentrated sample is greater than the absorbance (or concentration) of the standard, dilute that sample and reanalyze. If very large dilutions are required, another technique (e.g., ICP) may be more suitable for the sample. If the sample is diluted with water, add acid and matrix modifier to restore the concentration of both to the original.

9.3.8 Calculation

See Section 8.2.7.

9.4 INTERFERENCE AND THE GRAPHITE FURNACE

Determinations by GrAA are subject to significant interference. At the beginning of GrAA work, interference was accepted as an unavoidable part of the technique. Using today's instrumentation with new corrected methodologies makes the current graphite furnace technology a potential and excellent tool in metals analysis. Interference is divided into two categories: spectral and nonspectral.

9.4.1 Spectral Interference

Spectral interference occurs when the measured light absorption is erroneously high due to absorption by a species other than the analyte element. This type of interference results from light absorption by molecules or by atoms other than those of the analyte element.

9.4.1.1 Emission Interference

The intense light emitted by the hot graphite tube or platform, called *black body radiation,* blinds the photomultiplier tube and interferes with readings during atomization. The maximum intensity of this interference is in the near-infrared wavelength region. Elements determined at the UV wavelength range — for example, zinc at 213.9 nm — are free from this type of disturbance. However, chromium at 357.9 nm and calcium at 422.7 nm are greatly affected, and barium at 553.6 nm is especially vulnerable.

Black body radiation can be controlled by reducing the monochromator slit height. Instruments usually have two sets of monochromator slits; the "low" (sometimes called "alternate") slits are used with the graphite furnace to eliminate the emission interference problem. Attention to furnace alignment and maintenance is still required. Cleanliness of the graphite furnace and sample compartment windows must also be maintained to prevent light scattering. Atomization temperature should be not higher than that required for efficient analyte atomization.

9.4.1.2 Background Absorption

Background absorption or molecular absorption may occur when components of the sample matrix volatilize during atomization, resulting in broadband absorption (sometimes covering tens or hundreds of nanometers).

Several background correction techniques are available commercially to compensate for this type of interference, including sample treatment, furnace control procedures, and optical background controls.

9.4.1.2.1 Addition of Matrix Modifier or Matrix Modification

This technique helps to control matrix and analyte volatilities. It is desirable that the matrix be more volatile than the analyte, so that during the pyrolysis step all matrix components from the sample are volatilized and analyte atoms are not lost. The matrix modifier is selected to increase matrix volatility or decrease analyte volatility. For example, add *ammonium nitrate* (NH_4NO_3) to samples with a high sodium chloride (NaCl) matrix according to the following reaction:

$$NaCl + NH_4NO_3 \rightarrow NaNO_3 + NH_4Cl \qquad (9.1)$$

NaCl is a relatively nonvolatile compound, and requires pretreatment temperatures that would result in the loss of many analytes. By adding ammonium nitrate, however, the sample matrix is converted into more volatile components that can be driven off with high efficiency at lower pyrolysis temperatures. Decomposition temperatures are:

NaCl — 1400°C
NH_4NO_3 — 210°C
$NaNO_3$ — 380°C
NH_4Cl — 330°C

The other type of matrix modification is adding matrix modifier to make the analyte less volatile. An example is the addition of *nickel nitrate* to selenium determination. Selenium is highly volatile, but in the presence of nickel it can be heated to 900°C or more without loss. This process allows the removal of the sample matrix, which otherwise could not be driven off without loss of the selenium. A mixed modifier, such as *palladium plus magnesium nitrate*, can be used with various elements with excellent results. Table 9.2 lists substances added to the sample to remove interference in the GrAA method.

TABLE 9.2
Matrix Modifiers Added to Sample To Eliminate Interference, GrAAS Technique

Element	Matrix Modifiers
Aluminum (Al)	$Mg(NO_3)_2$
Antimony (Sb)	$Mg(NO_3)_2$ $Ni(NO_3)_2$
Arsenic (As)	$Mg(NO_3)_2$, $Ni(NO_3)_2$
Beryllium (Be)	$Mg(NO_3)_2$, $Al(NO_3)_2$
Cadmium (Cd)	$Mg(NO_3)_2$, $NH_4H_2PO_4$, $(NH_4)_2SO_4$, $(NH_4)_2S_2O_8$
Chromium (Cr)	$Mg(NO_3)_2$
Cobalt (Co)	$Mg(NO_3)_2$, $NH_4H_2PO_4$, ascorbic acid
Copper(Cu)	NH_4NO_3, ascorbic acid
Iron (Fe)	NH_4NO_3
Lead (Pb)	$Mg(NO_3)_2$, NH_4NO_3, $NH_4H_2PO_4$, $LaCl_3$, HNO_3, H_3PO_4, ascorbic acid, oxalic acid
Manganese (Mn)	$Mg(NO_3)_2$, NH_4NO_3, ascorbic acid
Nickel (Ni)	$Mg(NO_3)_2$, $NH_4H_2PO_4$
Seleium (Se)	$Ni(NO_3)_2$, $AgNO_3$, $Fe(NO_3)_3$, $(NH_4)_6MO_7O_{24}$
Silver (Ag)	$(NH_4)_2HPO_4$, $NH_4H_2PO_4$
Tin (Sn)	$Ni(NO_3)_2$, NH_4NO_3, $(NH_4)_2HPO_4$, $Mg(NO_3)_{22}$, ascorbic acid

9.4.1.2.2 Varying the Sample Volume

This technique is also an effective way to control background absorption. Larger sample volumes improve the ability to detect low analyte concentrations. Smaller sample size reduces the mass of the background-producing sample matrix and reduces with it background absorption.

9.4.1.2.3 Using Different Wavelengths

This method can also reduce background absorption.

9.4.1.2.4 Continuum Source Background Correction

This technique employs the use of a continuum source to measure the background contribution to the total measured signal. Instrument electronics then automatically remove the unwanted background contribution, and provide the corrected result.

9.4.1.2.5 Zeeman Effect Background Correction

This correction provides the best precision and accuracy for the elimination of background absorption. See Section 7.7.2.

9.4.2 Nonspectral Interference

Nonspectral interference results when diverse components in the sample matrix inhibit the formation of free analyte atoms. Accurate compensation of this interference is more difficult than correcting background absorption.

9.4.2.1 Standard Additions Method

In this method, a known amount of analyte is added to an aliquot of the sample. The absorbance values of the unspiked and spiked samples are calculated or measured and compared to the added analyte. For detailed information, see Section 7.7.1.

1. Inject a measured volume of the sample into the furnace device.
2. Dry, char or ash, and atomize samples according to the preset program, as described in Sections 9.3.3 and 9.3.4.
3. Record the instrument response in absorbance or concentration as appropriate.
4. Add a known concentration of the element of interest to a separate portion of the sample so as not to change significantly the volume of the sample. Mix well and determine instrument response.
5. Add a known concentration, preferably twice as much as was used in the first addition (as in step 4) to a separate sample portion. Mix well and determine the concentration.
6. Plot the absorbance, or the instrument response for the sample, and the two portions with known additions on the vertical axis with the concentration of the element added on the horizontal axis of linear graph paper.
7. Draw a straight line connecting the three points and extrapolate to zero absorbance. The intercept of the horizontal axis is the concentration of the sample. The concentration axis to the left of the origin should be a mirror image of the axis to the right.

Figure 9.2 illustrates a typical standard addition plot.

9.4.2.2 Graphite Tube Surface

A number of elements tend to form nonvolatile carbide by their interaction with the surface of the graphite tube. The wall of the tube is porous, allowing the sample solution to soak into the structure of the graphite tube during drying, and during atomization the atomic vapor interacts with the porous surface of the graphite and forms analyte carbides. These actions decrease free atom populations. A pyrolytically coated graphite tube offers a denser surface and prevents nonspectral interference. Pyrolytic coating can also increase the useful lifetime of the graphite tube.

9.4.2.3 L'vov Platform

B.V. L'vov, a pioneer in GrAAS, developed the use a small platform made from a flat piece of solid pyrolytic graphite, which is placed in the bottom of the graphite tube (Figure 9.2). The sample is pipeted into a shallow depression on the platform. Because the platform is prepared from pyrolytic graphite, this dense surface prevents the sample from soaking into the surface, as well as carbide formation, and also tolerates the high acid content of the sample. The other benefit of the platform is that the sample experiences the temperature of the platform, not the temperature of the tube wall.

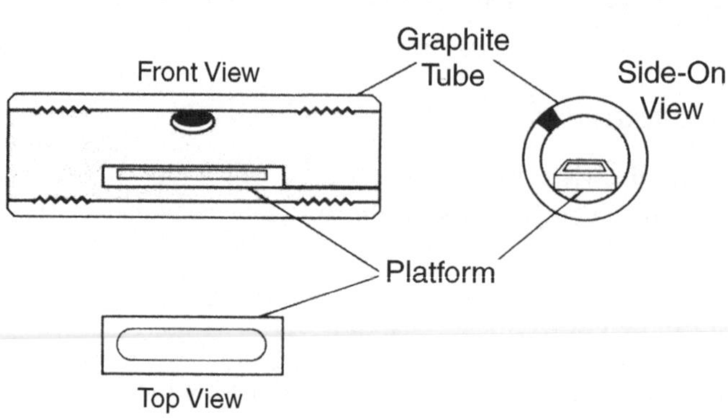

FIGURE 9.2 L'vov platform.

Because the platform is heated by radiation from the tube wall, temperature changes in the sample on the platform — and therefore in the vapor within the tube — are delayed compared with the tube wall. Instead of volatilizing the analyte as the temperature is changing, the appropriate conditions can be attained to volatilize the analyte after the tube wall and the gas phase have reached a more stable or steady-state condition.

9.4.2.4 Matrix Modification

Temperature is increased by adding a matrix modifier (see Section 9.4.1). This process delays the release of the analyte into the furnace, allowing additional time to establish a constant furnace temperature before atomization. When recommended, a matrix modifier is always used to improve resistance to nonspectral interference. The addition of a matrix modifier delays the atomization until a constant furnace temperature is reached.

9.4.2.5 Maximum Power Atomization

The release of analyte atoms into the same furnace environment for all analyte forms is the prerequisite for the elimination of nonspectral interference. *Rapid heating,* or *maximum power atomization,* increases the temperature of the tube atmosphere more rapidly and the analyte is volatilized into a hotter environment. By increasing the temperature of the furnace during atomization, the tube wall and atmosphere are heated much faster than the platform, thus ensuring a stabilized tube atmosphere temperature during atomization. Some furnace systems employ an optical temperature sensor, or the time for maximum power heating is programmed based on the desired final atomization temperature.

9.4.2.6 Fast Electronics

In order to provide accurate analytical results, a graphite furnace AA system must be capable of accurately quantitating the peak absorbance signal. One potential limitation is the speed of the instrument's electronics. Fast electronics provide accurate measurement of the rapidly changing signal.

9.4.2.7 Baseline Offset Correction (BOC)

One complication of peak-area measurement has been the exaggerated effect of baseline drift. Even a slight baseline change becomes noticeable when it accumulates over several seconds. To eliminate this potential problem, baseline offset correction was developed. BOC measures the baseline reading immediately prior to atomization. Each reading during the peak-area integration is then automatically corrected for baseline offset. This eliminates all drift effects and improves the accuracy of peak-area measurement. It also maintains the correction for sample blanks implemented with the automatic zero adjustment.

9.5 STABILIZED TEMPERATURE PLATFORM FURNACE (STPF)

The goals of an analytical technique should be able to control all interference and deliver accurate measurements. In the stabilized temperature platform furnace (STPF) system, all previously discussed techniques for eliminating interferences are simultaneously applied, resulting in an interference-free analysis. Every part of the system is crucial to the effectiveness of the system in providing accurate results.

The functions of each element comprising the STPF system are as follows:

High-quality pyrolytic graphite tubes provide an impervious, nonreactive surface.
L'vov platform delays atomization until stable temperature conditions are achieved.
Maximum power atomization hastens establishment of a stable atomization temperature and enhances the temperature lag between heating of the tube wall (and atmosphere) and the platform.
Internal gas stop maximizes residence time of atoms in the furnace.
Fast spectrometer electronics provide accurate measurement of the rapidly changing signal.
Peak area measurement quantitates all analyte atoms passing through the furnace independent of the matrix-dependent analyte volatilization rate.
Baseline offset correction improves accuracy of peak-area measurement by compensating for small changes in the baseline.
Matrix modification improves matrix removal during pretreatment.
Zeeman effect background correction corrects for high sample background, structured background absorption, and spectral interference.

9.6 QUALITY CONTROL REQUIREMENTS

See Chapter 14.

9.7 MAINTENANCE OF GRAPHITE ATOMIC ABSORPTION SPECTROPHOTOMETER

See Table 9.3.

9.8 PERFORMANCE CHECK OF GRAPHITE ATOMIC ABSORPTION SPECTROPHOTOMETER

The criteria of the performance check of the GrAAS is the same as for FAAS (see Section 8.6) with the exception of the concentration of the sensitivity check standards. These performance check standards are listed in Table 9.4.

TABLE 9.3
Maintenance of GrAAS

Maintenance Activity	Frequency
Check graphite tube	D
Flush autosampler tubing	D
Clean furnace housing and injector tip	W
Check electronics	SA (I) (C)

Note: D = daily; W = weekly; SA = semiannually; I = instrumentation specialist; C = on contract.

TABLE 9.4
Performance Check for GrAAS

Element	Concentration of Sensitivity Check Standard (μg/l)
Aluminum (Al)	0.05
Antimony (Sb)	0.04
Arsenic (As)	0.04
Barium (Ba)	0.03
Cadmium (Cd)	0.003
Cobalt (Co)	0.03
Chromium (Cr)	0.007
Copper(Cu)	0.01
Iron (Fe)	0.01
Manganese (Mn)	0.005
Molybdenum (Mo)	0.025
Nickel (Ni)	0.04
Lead (Pb) at 283 nm	0.025
Lead (Pb) at 217 nm	0.016
Selenium (Se)	0.06
Silicon (Si)	0.14
Tin (Sn) at 286 nm	0.14
Tin (Sn) at 224.6 nm	0.08
Titanium (Ti)	0.21
Vanadium (V)	0.15
Zinc (Zn)	0.0007

Note: The sensitivity check data here pertain to the metal concentrations in milligrams per liter in aqueous solution, which will give a reading of approximately 0.200 absorbance units when 20 μl (microliters) are used.

10 Cold-Vapor Atomic Absorption Spectrometry

10.1 GENERAL DISCUSSION

Because the atoms of most atomic absorption elements cannot exist in a free ground state at room temperature, heat must be applied to break the bonds and release free atoms. The atoms are produced by supplying enough heat energy to the sample to dissociate the compounds into free atoms.

Only one element, mercury (Hg), is an exception. Because free Hg atoms can exist at room temperature, Hg can be measured by atomic absorption without a heated device. This process is called the *cold-vapor technique.* In the cold-vapor technique, the Hg is reduced by the addition of a stannous (Sn^{2+}) compound, which is a strong reducing agent. The volatile-free Hg is taken from the reaction vessel by bubbling air through the solution; Hg atoms go through the tubing to the absorption cell, which is placed in the light path of the AA spectrometer. As the Hg atoms pass into the sampling cell, measured absorbance rises, indicating the increasing concentration of Hg atoms in the light path. The highest absorbance observed during the measurement is taken as the analytical signal. The procedure is based on the absorption of radiation at 253.7 nm by mercury vapor. The absorbance is measured as a function of Hg concentration and recorded in the usual manner.

10.1.1 ADVANTAGES

One of the advantages of the cold-vapor technique is high sensitivity, which is achieved through a 100% sampling efficiency. Because all mercury contained in the sample is released for measurement, increasing the sample volume means that more mercury atoms are available to be transported to the sample cell and measured.

10.1.2 LIMITATIONS

The cold-vapor technique is limited to Hg determination, because no other element offers the possibility of chemical reduction to a volatile-free atomic state at room temperature. In addition to inorganic forms of Hg, organic mercurials may also be present. These organo-mercury compounds will not respond to the cold-vapor technique unless they are first broken down and converted to mercuric ions. Potassium permanganate ($KMnO_4$) oxidizes many of these compounds, and the subsequent addition of potassium persulfate ($K_2S_2O_8$) completely solves this problem.

10.1.3 DETECTION LIMIT

When using the cold-vapor technique and a 100-ml sample size, the detection limit for Hg is 0.0002 mg/l or 0.2 μg/l for liquid samples and 5 μg/g for solid samples.

10.2 APPARATUS

When possible, dedicate glassware for use in Hg analysis. Avoid glassware previously exposed to a high level of Hg, such as glassware used in chemical oxygen demand (COD), total Kjeldahl nitrogen (TKN), and chloride (Cl) determination.

10.2.1 Atomic Absorption Spectrophotometer

Any AAS unit with an open sample presentation area in which to mount the absorption cell is suitable. Instrument settings recommended by the manufacturer should be followed. Instruments designed specifically for the measurement of mercury using the cold-vapor technique are commercially available and may be substituted for the AAS.

10.2.2 Mercury Hollow Cathode Lamp (HCL) or Electrodeless Discharge Lamp (EDL)

Both HCL and EDL lamps are discussed in Section 7.3.1.

10.2.3 Recorder

Any multirange, variable-speed recorder that is compatible with the UV detection system is suitable.

10.2.4 Absorption Cell

Typically, the absorption cell is a glass or plastic tube approximately 2.5 cm in diameter. An 11.4-cm-long tube has been found to be satisfactory but a 15-cm-long tube is preferable. Grind the tube ends perpendicular to the longitudinal axis and cement the quartz windows in place. Attach the gas inlet and outlet ports (6.4-mm diameter) 1.3 cm from each end.

10.2.5 Cell Support

Strap the cell to the flat nitrous oxide burner head or on another suitable support and align it in the light beam to provide maximum transmittance.

10.2.6 Air Pump

Any peristaltic pump capable of delivering 2 liters of air per minute can be used. (Some references recommend 1 liter of air per minute.) Any other regulated compressed air system or air cylinder is also satisfactory.

10.2.7 Flowmeter

Flowmeters capable of measuring airflow of 2 l/min are recommended. (Some references accept 1 l/min.)

10.2.8 Aeration Tubing

Use a straight glass frit with coarse porosity. Tygon tubing is used for the passage of the mercury vapor from the sample bottle to the absorption cell and back.

10.2.9 Reaction Flask

The reaction flask is typically a 250-ml Erlenmeyer flask or 300-ml BOD bottle fitted with a rubber stopper to hold the aeration tube.

10.2.10 Drying Tube

Typically, the drying tube is 150 mm × 18 mm in diameter and contains 20 g of magnesium perchlorate ($Mg(ClO_4)_2$). A small reading lamp with a 60-W bulb with a suitable shade may be substituted to prevent condensation of moisture inside the cell. The lamp is positioned to shine on the absorption cell, maintaining the air temperature in the cell at about 10°C above ambient temperature.

10.2.11 Connecting Tubing

Glass tubing is used to pass the Hg vapor from the reaction flask to the absorption cell and to interconnect all the other components. Clear vinyl plastic (tygon or equivalent) tubing may be substituted for glass. The apparatus for Hg determination by the cold-vapor technique is shown in Figure 10.1.

10.3 PROCEDURE

10.3.1 Sample Collection, Preservation, and Handling

All samples must be collected using a sampling plan that addresses the considerations discussed in Chapter 14. Due to the extreme sensitivity of the analytical procedure and the omnipresence of mercury, care must be taken to avoid extraneous contamination. The sampling devices and sample containers should be free of mercury and the sample should not be exposed to conditions in the laboratory that may result in contact with airborne mercury contamination. Plastic or glass containers are suitable for sample collection. All containers must be prewashed with detergent, acid, and reagent grade water, as discussed in Section 14.3.

10.3.1.1 Aqueous Samples

Aqueous samples must be acidified to a pH of less than 2 with HNO_3. The maximum holding times for these samples are 38 days in glass containers and 13 days in plastic containers.

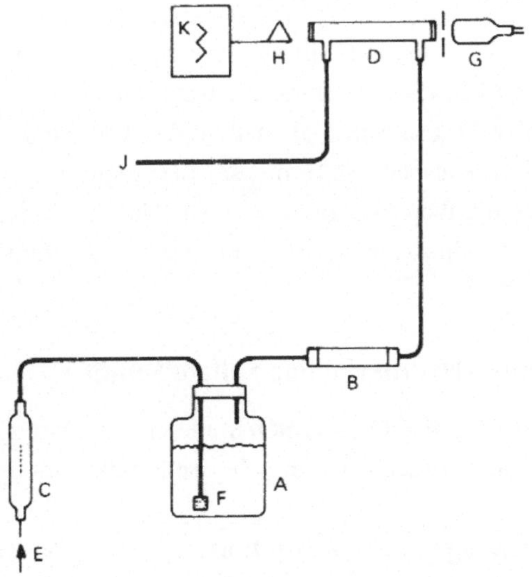

FIGURE 10.1 Schematic arrangement of equipment for measurement of mercury by the cold-vapor atomic absorption technique.

10.3.1.2 Nonaqueous Samples

Nonaqueous samples should be refrigerated, when possible, and analyzed as soon as possible.

10.3.1.3 Solid or Semisolid Samples

These samples may be analyzed directly and their moisture content determined separately. Determination of moisture content of a sample is described in Appendix I. Analysis of dry samples is more convenient. Moisture can be removed in a drying oven at a temperature of 60°C; no mercury losses have been observed when using this drying step. The dry sample must be pulverized and thoroughly mixed before the aliquot is weighed for analysis.

10.3.2 Reagents

All chemicals and the reagent-grade water should be mercury free!

10.3.2.1 Aqua Regia

Prepare immediately before use by carefully adding three volumes of HCl concentrate to one volume of HNO_3 concentrate.

10.3.2.2 Sulfuric Acid (H_2SO_4) Concentrate

10.3.2.3 H_2SO_4, 0.5N

Dilute 14 ml of concentrated H_2SO_4 to 1 liter.

10.3.2.4 Nitric Acid (HNO_3) Concentrate

10.3.2.5 Stannous Ion (Sn^{2+}) Solution

Use either stannous chloride or stannous sulfate to prepare this solution containing about 7 g Sn^{2+} per 100 ml. Dissolve 10 g of $SnCl_2$ in analyte-free water containing 20 ml of HCl concentrate and dilute to 100 ml, *or* dissolve 11 g of $SnSO_4$ in analyte-free water containing 7 ml of H_2SO_4 concentrate and dilute to 100 ml. If a suspension forms, stir the reagent continuously during use. Both solutions decompose with aging; therefore, prepare fresh solutions daily. A reagent volume of 100 ml is sufficient for 20 samples; adjust the volumes prepared to accommodate the number of samples processed.

10.3.2.6 Sodium Chloride–Hydroxylamine Sulfate Solution

Dissolve 12 g of NaCl and 12 g $(NH_2OH)_2.H_2SO_4$ in mercury-free, reagent-grade water and dilute to 1 liter A 10% hydroxylamine hydrochloride may be used in place of hydroxylamine sulfate.

10.3.2.7 Potassium Permanganate 5% Solution

Dissolve 50 g of $KMnO_4$ in reagent-grade water and dilute to 1 liter. Store the solution in a glass-stoppered, amber-colored glass bottle.

10.3.2.8 Potassium Permanganate 0.1N Solution

Dissolve 3.2 g of $KMnO_4$ in about 100 ml of reagent-grade water and dilute to 1000 ml. Allow the solution to stand in the dark for a few days and then filter through a fine-porosity sintered glass crucible. Do not use filter paper! Store the solution in a glass-stoppered, amber-colored glass bottle.

10.3.2.9 Potassium Persulfate 5% Solution

Dissolve 50 g of $K_2S_2O_8$ in reagent-grade water and dilute to 1 liter.

10.3.2.10 Stock Mercury Solution

Dissolve 0.1354 g of $HgCl_2$ in 75 ml reagent-grade water. Add 10 ml of HNO_3 concentrate and adjust the volume to 100 ml.

10.3.2.10.1 Working Standard Mercury Solution, 1 ml = 0.1 μg Hg
Dilute from the stock mercury solution in two steps:

1. Dilute 1 ml of stock solution to 100 ml: 1 ml = 0.01 mg Hg = 10 μg Hg.
2. Dilute 1 ml of solution in number 1 to 100 ml: 1 ml = 0.0001 mg Hg = 0.1 μg Hg.

This working standard and the dilutions from the stock solution should be prepared fresh daily. The acidity of the working standard should be maintained at 0.15% HNO_3. This acid should be added to the flask before the aliquot is added.

10.3.3 Instrument Operation

Because of differences among makes and models of atomic absorption spectrophotometers, it is not possible to formulate instructions applicable to every instrument. See the manufacturer's operation manual. In general, proceed according to the following steps:

1. Install the hollow cathode lamp (HCL) for Hg in the instrument.
2. Set the slit width according to the manufacturer's suggestion.
3. Turn on the instrument.
4. Apply the current suggested by the manufacturer to the HCL and let the instrument warm up until the energy source stabilizes (generally 10–20 min).
5. Readjust the current as necessary after warm-up.
6. Set the wavelength to 253.7 nm.
7. Install the absorption cell and align it in the light path to provide maximum transmission.
8. Connect the associated equipment to the absorption cell with glass or vinyl plastic tubing as indicated in Figure 10.1.
9. Turn on the air and adjust the flow rate to 2 l/min. Allow the air to flow continuously.

10.3.4 Standardization

1. Set out correct number and type of reaction flasks (Section 10.2.9). Start with a blank and follow with standards (calibration standards, continuing calibration standard, and calibration verification standard, or quality control sample). For detailed discussion of these standards, see Section 13.6.2.
2. Transfer 0, 0.5, 1.0, 2.0, 5.0, and 10.0 ml of aliquot of the working standard solution containing 0.1 μg Hg per ml (Section 10.3.2) to the bottles marked as calibration standards.
3. Add enough analyte-free water to each bottle to make a total volume of 100 ml.

4. Add 5 ml of H_2SO_4 concentrate followed by 2.5 ml of HNO_3 concentrate to each flask.
5. Add 15 ml of potassium permanganate solution (Section 10.3.2) to each bottle and allow to stand for at least 15 min.
6. Add 8 ml of potassium persulfate solution (Section 10.3.2) to each bottle and heat for 2 h in a water bath maintained at 95°C. Cool to room temperature.
7. Treating each flask individually, add 6 ml of sodium chloride–hydroxylamine sulfate (Section 10.3.2) solution (or more if necessary) to reduce the excess permanganate.
8. When the solution has been decolorized, wait 30 sec, add 5 ml of Sn^{2+} solution (Section 10.3.2), and immediately attach the bottle to the aeration apparatus. At this point the sample is allowed to stand quietly without manual agitation. The circulating pump (Section 10.2.6), which has previously been adjusted to a rate of 2 l/min, is allowed to run continuously. As the Hg is volatilized and carried into the absorption cell, absorbance will increase and reach a maximum within 30 sec.
9. As soon as the recorder returns approximately to the baseline, remove the stopper holding the frit from the reaction flask and replace it in a flask containing reagent-grade water.
10. Flush system for a few seconds and run the next standard in the same manner.
11. Close the bypass valve, remove the stopper and frit from the BOD bottle, and continue aeration.
12. Construct a calibration curve by plotting the absorbances of the standards vs. the microgram of mercury. Preparation and checking of the calibration curve are discussed in Section 6.6.

10.3.5 Sample Analysis

10.3.5.1 Liquid Samples

The cold-vapor method is applicable to liquid samples, such as ground waters, drinking waters, aqueous wastes, and mobility procedure extracts.

1. Transfer a 100-ml sample or portion diluted to 100 ml, containing not more than 1.0 μg of Hg, to a 300-ml BOD bottle (Section 10.2.9).
2. Add 5 ml of H_2SO_4 concentrate and 2.5 ml of HNO_3 concentrate, mixing after each addition.
3. Add 15 ml of potassium permanganate solution (Section 10.3.2.7) to each sample flask. Sewage samples may require additional $KMnO_4$ solution; add until the purple color persists for at least 15 min.
4. Add 8 ml of potassium persulfate solution (Section 10.3.2) to each bottle and heat for 2 h in a water bath maintained at 95°C.
5. Cool and add 6 ml of sodium chloride–hydroxylamine sulfate solution (Section 10.3.2.6) to reduce the excess permanganate.
6. Follow the procedure described in Section 10.3.4, steps 8 through 10.

Note: Seawaters, brines, and effluents high in chloride require up to 25 ml of additional potassium permanganate solution. During the oxidation step, chlorides are converted to free chlorine, which absorbs at 253 nm. Remove the free chlorine before the Hg is reduced and swept into the cell by using an excess (25 ml) of hydroxylamine sulfate (Section 10.3.2.6) reagent. In addition, the dead air space in the BOD bottle must be purged before adding the Sn^{2+} solution (Section 10.3.2.5). All samples that suffer from matrix interference should be analyzed by the standard addition method (Section 7.7.1.1.1).

10.3.5.2 Solids and Semisolids

This method is applied to determine the total mercury content in soils, sediments, bottom deposits, and sludge-type materials. All samples must be subjected to an appropriate dissolution step prior to analysis.

1. Weigh three 0.2-g portions of untreated sample and place them in the bottom of a BOD bottle.
2. Add 5 ml of analyte-free, reagent-grade water.
3. Add 5 ml of aqua regia (Section 10.3.2.1).
4. Heat for 2 min in a water bath at 95°C.
5. Cool and then add 50 ml of analyte-free, reagent-grade water and 15 ml potassium permanganate solution (Section 10.3.2.7).
6. Mix thoroughly and place in a water bath for 30 min at 95°C.
7. Cool and add 6 ml of sodium chloride–hydroxylamine sulfate (Section 10.3.2.6) to reduce the excess potassium permanganate. Add this material under a hood, as chlorine (Cl_2) could evolve.
8. Add 55 ml of analyte-free water.
9. Treating each bottle individually, add 5 ml of stannous sulfate reagent (Section 10.3.2.5) and immediately attach the bottle to the aeration apparatus.
10. Follow the procedure described in Section 10.3.4, steps 8 through 10.

Note: An alternate digestion procedure employing an autoclave may also be used. In this case use the following steps:

1. Weigh three 0.2-g portions of untreated sample and place them in the bottom of a BOD bottle.
2. Add 5 ml of analyte-free, reagent-grade water.
3. Add 5 ml of H_2SO_4 concentrate and 2 ml of HNO_3 concentrate to the 0.2-g sample.
4. Add 5 ml of saturated potassium permanganate solution.
5. Cover the bottle with a piece of aluminum foil.
6. Autoclave the samples at 121°C and 15 lb for 15 min.
7. Cool and dilute to a volume of 100 ml with reagent-grade water.
8. Add 6 ml of sodium chloride–hydroxylamine sulfate solution (Section 10.3.2.6) to reduce the excess permanganate.
9. Follow the procedure described in Section 10.3.4, steps 8 through 10.

10.4 INTERFERENCE

10.4.1 SULFIDES

Sulfide interference is eliminated by the addition of $KMnO_4$. Concentrations as high as 20 mg/l of sulfide as sodium sulfide do not interfere with the recovery of added inorganic mercury from distilled water.

10.4.2 COPPER

Copper has also been reported to interfere; however, copper concentrations as high as 10 mg/l have no effect on the recovery of Hg from spiked samples.

10.4.3 Seawaters, Brines, and Industrial Effluents High in Chlorides

These materials require additional permanganate (up to 25 ml). During the oxidation step, the chlorides are converted to free chlorine, which also absorbs radiation at 253.7 nm. Care must be taken to ensure that free chlorine is absent before the mercury is reduced and swept into the cell. This may be accomplished by using an excess of hydroxylamine sulfate reagent (Section 10.3.2.6) (25 ml). In addition, the dead air space in the BOD bottle must be purged before adding stannous sulfate (Section 10.3.2.5). Both inorganic and organic mercury spikes have been recovered from seawater by using this technique.

10.4.4 Certain Volatile Organic Materials

These materials also absorb radiation at 253.7 nm, which may cause interference. A preliminary run without reagents should determine if this type of interference is present. In order to remove any volatile materials, the dead air space in the BOD bottle should be purged before adding the stannous reagent.

10.5 QUALITY CONTROL REQUIREMENTS

1. All quality control data should be maintained and available for easy reference or inspection.
2. Calibration curves must be composed of at least one blank and three standards. A calibration curve should be made for every hour of continuous sample analysis.
3. Dilute samples if they are more concentrated than the highest standard or if they fall on the plateau of a calibration curve.
4. Employ a minimum of one blank per sample batch to determine if contamination or memory effects are occurring.
5. Verify calibration with an independently prepared calibration verification standard (CVS) every 15 samples.
6. Run one spiked duplicate sample for every ten samples. A duplicate sample must be brought through the entire sample preparation and analytical process.
7. The standard addition method (Section 7.7.1) should be used in all extraction procedure toxicity (EPTOX) tests, analyses submitted as part of a de-listing petition, and analysis of a new sample matrix.

Detailed quality control requirements are discussed in Chapter 13.

10.6 CALCULATIONS AND REPORTING

Calculate metal concentrations from the calibration curve. All dilution and concentration factors must be taken into account. Report the results for liquid samples as micrograms or nanograms per liter; for solid samples, use micrograms or nanograms per gram. All results must be appropriately qualified for dry weight (see Appendix I).

Report Hg concentration for liquid samples as milligrams or micrograms per liter; for solid samples, report as micrograms per gram on the dry-weight basis. (See Appendix I for calculation.) Report Hg concentrations as follows: below 0.1 μg/g; between 0.1 and 1.0 μg/g to the nearest 0.01 μg; between 1.0 and 10 μg/g to the nearest 0.1 μg; and above 10 μg/g to the nearest microgram.

10.7 SAFETY

Because Hg vapor is toxic, precautions must be taken to avoid inhalation. Therefore, a bypass has been included in the system either to vent the mercury vapor into an exhaust hood or to pass the vapor through some absorbing medium, such as equal volumes of 0.1N $KMnO_4$ and 10% H_2SO_4, or 0.25% iodine in a 3% potassium iodide (KI) solution. A specially treated charcoal that will absorb mercury vapor is also available from Barnebey and Cheney (East 8th Ave. and North Cassidy St., Columbus, OH 43219; Cat. No. 580–13 or 580–22).

11 Hydride-Generation Atomic Absorption Technique

11.1 PRINCIPLE

Hydride-generation sampling systems for atomic absorption bear some resemblance to cold-vapor mercury systems. Samples are reacted in an external vessel with a reducing agent, usually *sodium borohydride*. Gaseous reaction products are then carried to a sampling cell in the light path of the AA spectrometer. Unlike the mercury technique, the gaseous reaction products are not free-analyte atoms but *volatile hydrides*. These molecular species are not capable of causing atomic absorption. To dissociate the hydride gas into free atoms, the sample cell must be heated.

In some hydride systems, the absorption cell is mounted over the burner head of the AA spectrometer, and the cell is heated by an air–acetylene flame. In other systems, the cell is heated electrically. In either case, the hydride gas is dissociated in the heated cell into free atoms, and the atomic absorption rises and falls as the atoms are created and then escape from the absorption cell. The maximum absorption reading, or peak height, is taken as the analytical signal. Recommended wavelengths are 193.7 nm for As and 196.0 nm for Se.

11.1.1 ADVANTAGE

The advantage of the technique is the easily achievable detection limits below micrograms per liter.

11.1.2 DISADVANTAGE

The disadvantage of the technique is that its results depend heavily on a variety of parameters, including the valence state of the analyte, reaction time, gas pressures, concentration, and cell temperature. Therefore, the success of the hydride generation technique will vary with the care taken by the operator in attending to the required detail. The formation of analyte hydrides is also suppressed by a number of common matrix components, leaving the technique subject to chemical interference.

11.2 APPLICATION

The method is applicable to the determination of *arsenic* (As) and *selenium* (Se) via conversion to their hydrides with *sodium borohydride* reagent and aspiration into an atomic absorption atomizer.

Arsenous acid and *selenous acid*, the As(III) and Se(IV) oxidation states of As and Se, respectively, are instantaneously converted by sodium borohydride reagent in acid solution to their volatile hydrides. The hydrides are purged continuously by argon or nitrogen into an appropriate atomizer of an AA spectrometer and converted to the gas-phase atoms. The sodium-borohydride reducing agent, by rapid generation of the elemental hydrides in an appropriate reaction cell, minimizes dilution of the hydrides by the carrier gas and provides rapid, sensitive determinations of As and Se.

Caution: As and Se and their hydrides are toxic. Handle with care!

At room temperature and solution pH values of 1 or less, *arsenic acid*, the As(V) oxidation state of As, is reduced relatively slowly by sodium borohydride to As(III), which is then instantaneously converted to *arsine*. The arsine atomic-absorption peaks are commonly lower for As(V) than for As(III). Determination of total As requires that all inorganic arsenic compounds be in the As(III) state. Organic and inorganic forms of As are first oxidized to As(V) by acid digestion. The As(V) is then reduced to As(III) with sodium or potassium iodide before reaction with sodium borohydride.

Selenic acid, the Se(VI) oxidation state of Se, is not measurably reduced by sodium borohydride. To determine total Se, first reduce the Se(VI) formed during the acid digestion procedure to Se(IV), being careful to prevent reoxidation by chlorine. Reduction efficiency depends on temperature reduction time and HCl concentration. For 4N HCl concentration, heat 1 h at 100°C; for 6N HCl, boiling for 10 min is sufficient. Recommended wavelengths are 193.7 nm for As and 196.0 nm for Se.

11.2.1 Detection Limit and Concentration Range

For both As and Se, the method detection limit is 0.002 mg/l and the optimum concentration range is 0.002 to 0.02 mg/l.

11.3 APPARATUSES AND MATERIALS

11.3.1 Atomic Absorption Spectrometer

Use an AA spectrometer equipped with gas-flow meters for argon (or nitrogen) and hydrogen.

11.3.2 Arsenic and Selenium Hollow Cathode Lamp or Electrodeless Discharge Lamp

Use an As and Se hollow cathode lamp (HCL) or electrodeless discharge lamp (EDL) with power supply.

11.3.3 Background Correction at Measurement of Wavelength

11.3.4 Strip-Chart Recorder

A high-quality 10-mV recorder with high sensitivity and fast response time is required.

11.3.5 Atomizer

Certain atomic absorption atomizers and hydride reaction cells are available commercially for use with the sodium borohydride reagent. Three types of atomic absorption atomizers are commonly used in the measurement of As and Se:

- *Boling-type burner*: For argon (or nitrogen) air-entrained hydrogen flame
- *Cylindrical quartz cell externally heated*: 10 to 20 cm long, electrically heated by external nichrome wire to 800–900°C.
- *Cylindrical quartz cell with internal fuel, oxygen-hydrogen or air-hydrogen flame*: The sensitivity of quartz cells deteriorates over several months of use. Sensitivity may be restored by treatment with 40% HF.

Caution: HF is extremely corrosive. Avoid all contact with skin. Handle with care!

11.3.6 Reaction Cell For Producing As and Se Hydride

A commercially available system is acceptable. (See Figure 11.1.)

11.3.7 Eye Dropper or Syringe

The eye dropper or syringe should be capable of delivering 0.5 to 3.0 ml of sodium borohydride reagent (Section 11.3.9.1).

11.3.8 Vent

Place a vent about 15 to 30 cm above the burner to remove fumes and vapors from the flame. This precaution protects laboratory personnel from toxic vapors, protects the instrument from corrosive vapors, and prevents flame stability from being affected by room drafts.

Commercially available continuous hydride generator units make the operation simpler than the manual method.

11.3.9 Reagents

11.3.9.1 Sodium Borohydride Reagent

Dissolve 8 g of $NaBH_4$ in 200 ml of 0.1N NaOH. Prepare fresh daily.

11.3.9.2 Sodium Hydroxide (NaOH), 0.1N

Dissolve 4 g of NaOH and dilute to 1000 ml.

11.3.9.3 Sodium Iodide Prereductant

Dissolve 50 g of NaI in 500 ml of reagent water. Prepare fresh daily.

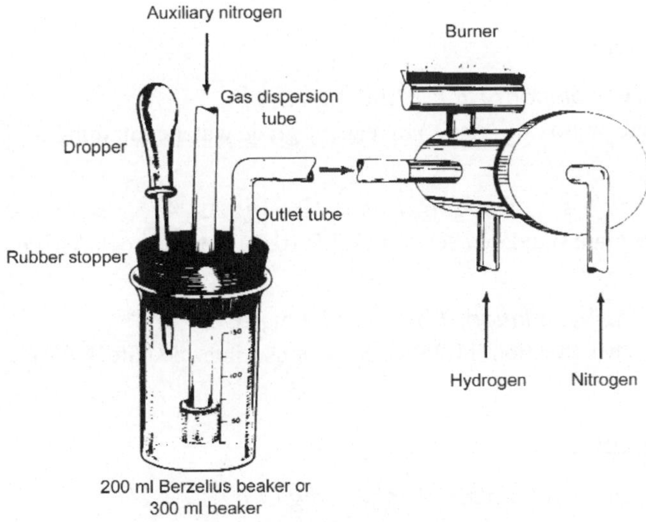

FIGURE 11.1 Manual reaction cell for producing As and Se hydrides.

11.3.9.4 Sulfuric Acid (H_2SO_4), 18N

Very carefully make a 50% dilution (1:1) of H_2SO_4 concentrate (H_2SO_4 concentrate is 36N).

11.3.9.5 Sulfuric Acid (H_2SO_4), 2.5N

Cautiously add 35 ml of H_2SO_4 concentrate to about 400 ml of reagent-grade water, let cool, and adjust volume to 500 ml.

11.3.9.6 Potassium Persulfate, 5%

Dissolve 25 g of $K_2S_2O_8$ in reagent-grade water and dilute to 500 ml. Store in glass bottle and refrigerate. Prepare fresh weekly.

11.3.9.7 Nitric Acid (HNO_3) Concentrate

11.3.9.8 Hydrochloric Acid (HCl) Concentrate

11.3.9.9 Perchloric Acid ($HClO_4$) Concentrate

11.3.9.10 As(III) Solutions

11.3.9.10.1 Stock As(III) Solution, 1 ml = 1.0 mg As(III)
Either procure a certified aqueous standard from a supplier or dissolve 1.32 g of arsenic trioxide (As_2O_3) in reagent-grade water containing 4 g of NaOH, and dilute to 1000 ml.

11.3.9.10.2 Intermediate As(III) Solution, 1 ml = 10 mg As(III)
Dilute 10 ml of stock As(III) solution (Section 11.3.9.10.1) to 1000 ml with reagent-grade water containing the same concentration of acid used for sample preservation (2–5 ml of HNO_3 concentrate).

11.3.9.10.3 Standard As(III) Solution, 1 ml = 0.100 mg As(III)
Dilute 10 ml of intermediate As(III) solution (Section 11.3.9.10.2) to 1000 ml with reagent-grade water containing acid at the same concentration as used for sample preservation (2 to 5 ml HNO_3 concentrate). Prepare the solution fresh daily.

11.3.9.11 As(V) Solutions

11.3.9.11.1 Stock As(V) Solution, 1 ml = 1.00 mg As(V)
Dissolve 1.534 g of arsenic pentoxide (As_2O_5) in reagent-grade water containing 4 g of NaOH. Dilute to 1 liter.

11.3.9.11.2 Intermediate As(V) Solution, 1 ml = 10.0 mg As(V)
Prepare as As(III) intermediate standard (Section 11.3.9.10.2), but use stock As(V) solution (Section 11.3.9.11.1).

11.3.9.11.3 Standard As(V) Solution, 1 ml = 0.110 mg As(V)
Prepare as for As(III) above (Section 11.3.9.10.3), but use intermediate As(V) solution (Section 11.3.9.11.2).

11.3.9.12 Organic As Solutions

11.3.9.12.1 Stock Organic As Solution, 1 ml = 1 mg Org As
Dissolve 1.842 g dimethyl-arsinic (cacodylic acid, $(CH_3)_2AsOOH$) in reagent-grade water containing 4 g of NaOH. Dilute to 1 liter.

11.3.9.12.2 Intermediate Organic As Solution, 1 ml = 10.0 mg Org As
Prepare as for As(III) (Section 11.3.9.10.2) but use stock organic As solution.

11.3.9.12.3 Standard Organic As Solution, 1 ml = 0.100 mg Org As
Prepare as for As(III) above (Section 11.3.9.10.3), but use intermediate organic As solution (Section 11.3.9.12.2).

11.3.9.13 Se(IV) Solutions

11.3.9.13.1 Stock Se(IV) Solution, 1 ml = 1.00 mg Se(IV)
Use a commercially available 1000 mg/l Se standard solution or prepare by dissolving 2.190 g of sodium selenite (Na_2SeO_3) in reagent-grade water containing 10 ml of HCl concentrate and dilute to 1000 ml. (Alternatively, you can use 0.3453 g of selenious acid (H_2SeO_3) and dilute to 200 ml.)

11.3.9.13.2 Intermediate Se(IV) Solution, 1 ml = 10.0 mg Se(IV)
Dilute 10 ml of stock Se(IV) solution to 1000 ml with reagent-grade water containing 10 ml of HCl concentrate.

11.3.9.13.3 Standard Se(IV) Solution, 1 ml 0.100 mg As(IV)
Dilute 10 ml of intermediate Se(IV) solution to 1000 ml with water containing the same acid concentration used for sample preservation (2–5 ml of HNO_3 concentrate). Prepare the solution daily.

11.3.9.14 Se(VI) Solutions

11.3.9.14.1 Stock Se(VI) Solution, 1 ml = 1.00 mg As(IV)
Dissolve 2.393 g of sodium selenate (Na_2SeO_4) in reagent-grade water containing 10 ml of HNO_3 concentrate and dilute to 1000 ml.

11.3.9.14.2 Intermediate Se(VI) Solution, 1 ml = 10 mg As(VI)
Dilute 10 ml of stock Se(VI) (Section 11.3.9.14.1) to 1000 ml with reagent-grade water.

11.3.9.14.3 Standard Se(VI) Solution, 1 ml = 0.100 mg As(VI)
Prepare as Se(IV) standard solution (Section 11.3.9.13.3), but use intermediate Se(VI) solution (Section 11.3.9.14.2).

11.4 INTERFERENCES

Interference is minimized because the As and Se hydrides are removed from the solution containing interfering substances. Interferences depend on system design. Certain waters and wastewaters contain interferences in sufficient concentration to suppress absorption responses. If average analytical recoveries of the sample are less than 90%, use an alternative analytical procedure.

11.4.1 POSSIBLE INTERFERENCES

- Low concentrations of noble gases (100 mg/l)
- Concentrations of Cu, Pb, and Ni at or greater than 1 mg/l
- Concentrations between 0.1 and 1 mg/l of hydride-forming elements such as Bi, Sb, Sn, and Te
- Interference by transition metals that depends strongly on HCl concentration; 4N HCl or 6N HCl (see Section 11.2) is recommended
- Reduced nitrogen oxide resulting from HNO_3 digestion, suppressing instrumental response
- Large concentrations of iodide interfering with Se determination by reducing Se to its elemental form
- Chlorine gas produced in the reduction of Se(VI) to Se(IV) preventing generation of the hydride within a few hours of the reduction steps

11.5 SAMPLE COLLECTION, PRESERVATION, AND HANDLING

- All samples must have been collected using a sampling plan.
- All sample containers must be prewashed with detergents, acids, and reagent-grade water. Glass and plastic containers are both suitable as discussed in Section 14.3.
- Aqueous samples must be acidified to a pH of less than 2 with HNO_3 (see Section 14.4).
- Nonaqueous samples should be refrigerated, when possible, and analyzed as soon as possible.

11.6 PREPARATION OF SAMPLES AND STANDARDS FOR TOTAL ARSENIC AND SELENIUM

1. Add 50 ml of sample or standard to a 200-ml Berzelius beaker or 300-ml beaker.
2. Add 1 ml of 2.5N H_2SO_4 and 5 ml of 5% $K_2S_2O_8$. Boil gently on a preheated hot plate for approximately 30 to 40 min or until a final volume of 10 ml is reached. Do not let sample evaporate to dryness. Alternatively, heat in an autoclave at 121°C for 1 h in capped containers.
3. After manual digestion, dilute to 50 ml for arsenic measurement and 30 ml for selenium measurement.

11.7 PROCEDURE

11.7.1 Apparatus Setup

See Figure 11.1 or follow manufacturer's instructions.

1. Connect inlet of reaction cell with auxiliary, purging gas controlled by flow meter.
2. If a dryimg cell between the reaction cell and atomizer is necessary, use only anhydrous $CaCl_2$ (but not $CaSO_4$ because it may retain SeH_2).
3. Optimize operating parameters. Aspirate diluted aqueous solutions of As and Se directly into the flame to facilitate atomizer alignment. Align quartz atomizers for maximum absorbance.
4. Aspirate a blank until memory effects are removed.
5. Establish purging gas flow, concentration and rate of addition of sodium borohydride reagent, solution volume, and stirring rate for optimum instrument response for the species analyzed.
6. If a quartz atomizer is used, optimize cell temperature.
7. If sodium borohydride reagent is added too quickly, rapid evolution hydrogen will unbalance the system.
8. If the volume to be analyzed is too large, the absorption signal will be decreased.

11.7.2 Instrument Calibration Standards

1. Transfer 0, 1, 2, 5, 10, 15, and 20 ml of standard solution of As(III) (Section 11.3.9.10.3) or Se(IV) (Section 11.3.9.13.3) into 100-ml volumetric flasks.
2. Bring to volume with reagent-grade water containing the same acid concentration used for sample preservation (commonly 2–5 ml/l of HNO_3 concentrate).

These steps yield blank and standard solutions of 0, 1, 2, 5, 10, 15, and 20 μg/l of As or Se. Prepare fresh daily.

11.7.3 Determination of As and Se with Sodium Borohydride

11.7.3.1 Determination of As

1. Measure 50 ml of standard or sample in a 200-ml Berzelius beaker or 300-ml beaker.
2. Add 5 ml of HCl concentrate and mix.
3. Add 5 ml of NaI solution (Section 11.3.9.3), mix, and wait at least 30 min.
4. Attach one Berzelius beaker at a time to the rubber stopper containing the gas dispersion tube for the purging gas, the sodium borohydride reagent inlet, and the outlet to the atomizer.
5. Turn on the strip-chart recorder and wait until the baseline is established by purging gas and expelling all air from the reaction cell.
6. Add 0.5 ml of sodium borohydride reagent (Section 11.3.9.1).
7. After the instrument absorbance has reached a maximum and returned to the baseline, remove beaker, rinse dispersion tube with water, and proceed to the next sample or standard.
8. Check for presence of chemical interferences by treating a digested sample with 10 mg/l As(III) or As(V) as appropriate. Average recoveries should be no less than 90%.

11.7.3.2 Determination of Se

1. Measure 30 ml of standard or sample into a 200-ml Berzelius beaker or 300-ml beaker.
2. Add 15 ml of HCl concentrate and mix.
3. Heat for a predetermined period at 90 to 100°C. Alternatively, autoclave at 121°C in capped containers for 60 min or heat for a predetermined period in open test tubes at 90 to 100°C in hot water bath or an aluminum block digester. Effective heat exposure for converting Se(VI) to Se(IV) ranges from 5 to 60 min when open beakers or test tubes are used. Check effectiveness of selected heating by demonstrating equal instrument responses for calibration curves prepared either from Se(IV) or Se(VI) solutions. Do not digest these solutions used for this check!
4. Attach Berzelius beakers one at a time to the purge apparatus, turn on the strip-chart recorder, and wait until the baseline is established.
5. Add 0.5 ml of sodium borohydride reagent (Section 11.3.9.1).
6. After the instrument absorbance has reached a maximum and returned to the baseline, remove beaker, rinse dispersion tube, and proceed to the next sample or standard.
7. Check for the presence of chemical interferences by treating a digested sample with 10 mg/l Se(IV). Average recoveries should be not less than 90%.

11.7.4 Calculations

Construct a calibration curve by plotting peak heights vs. concentration of standards, and read concentration from curve. On instruments so equipped, read concentration directly after calibration. If sample was diluted or concentrated before digestion, apply the appropriate factor.

11.8 QUALITY CONTROL REQUIREMENTS

See Chapter 13.

12 Inductively Coupled Plasma Atomic Emission Spectroscopy

12.1 ATOMIC EMISSION SPECTROSCOPY (AES)

In AES, the sample is subjected to temperatures high enough to cause not only dissociation into atoms, but also significant amounts of collisional excitation (and ionization) of the sample atoms. Once the atoms and ions are in their excitation states, they can decay to lower states through thermal or radioactive (emission) energy transition. (See discussion of emission in Section 5.4.) In AES, the intensity of the light emitted is measured at specific wavelengths and used to determine the concentrations of the elements of interest.

Thermal excitation sources can populate a large number of different energy levels for several different elements at the same time. Consequently, all excited atoms and ions can emit characteristic radiation at nearly the same time. In general, three types of thermal sources are used in analytical atomic spectrometry to dissociate sample molecules into free atoms: flames, furnaces, and electrical discharges. The first two types are hot enough to dissociate most types of molecules into free atoms. Electrical discharges, the third type, are also called *plasmas*.

12.1.1 PLASMAS

Plasma is a state of matter usually consisting of highly ionized gas that contains an appreciable fraction of equal numbers of ions and electrons in addition to neutral atoms and molecules. Plasmas conduct electricity and are affected by magnetic fields. The plasma source has a high degree of stability to overcome interference effects. Plasma is capable of exciting several elements that are not excited by flames and has increased sensitivity to flame AES. The low detection limits, freedom from interferences, and long-line working ranges prove that it is a superior technique for AES. For more detail about plasmas, see Appendix J.

The *electrical plasmas* used in AES work are highly energetic ionized gases and are usually produced in inert gases. The plasma source for analytical AES is argon-supported inductively coupled plasma (ICP).

12.1.2 SHORT HISTORY OF AES

In the 1860s, Kirchhoff and Bunsen developed methods based on emission spectroscopy that led to the discovery of four elements: cesium (Cs), rubidium (Rb), titanium (Ti), and indium (In). At this time, the emitted lines were used in qualitative analytical work.

In the mid-twentieth century, quantitative emission spectroscopy was the tool used to determine trace concentrations for a wide range of elements, but sample preparation techniques were very difficult and time consuming.

The atomic spectra emitted from flames had the advantage of being simpler and easier. This technique, called *flame emission spectrometry* (also known as *flame photometry*) is used to determine alkali metals and other easily excitable elements. Swedish agronomist Lundegardth is credited with initiating the modern era of flame photometry in the late 1920s. This technique is commonly used in clinical laboratories for determining sodium and potassium levels in biological materials.

In the 1960s and 1970s, *flame atomic absorption* (FAA) was the preferred technique for the determination of trace metals. FAA offers high precision and moderate detection limits. *Electrothermal atomization*, or *graphite furnace atomic absorption spectrophotometry* (GrAAS), on the other hand, offers high sensitivity and lower detection limits, but poorer precision and a higher level of matrix interferences. However, most of these interferences have been reduced or eliminated (see Section 9.4). Both FAA and GrAAS techniques are widely used today and provide excellent means of trace element analysis. However, most atomic absorption instruments are limited in that they measure only one element at a time. Instrument setup or operating conditions may require changing hollow cathode lamps or using different furnace parameters for each different element to be determined. Because of the limited calibration range in AAS techniques, the need for sample dilution is much greater than in AES techniques.

The first report (Greenfield et al.) about the use of an *atmospheric pressure inductively coupled plasma* (ICP) for element analysis via AES was published in England in 1964.

At the same time, Velmer Fassel and colleagues at Iowa State University refined the technique and made it practical for laboratory use. By 1973, ICP was promoted as the most popular technique in analytical emission spectrometry because of its low detection limits, long linear working ranges, and freedom from interference.

12.2 GENERAL CHARACTERISTICS OF ICP-AES

Emission spectroscopy using ICP is a rapid, sensitive, and convenient method for the determination of elements, including metals, in solution. All matrices, including groundwater, aqueous samples, extracts, wastes, soils, sludges, sediments, and other solid wastes require digestion prior to analysis. (Sample preparation procedures are discussed in Chapter 15.) Routine determination of 70 elements can be made by ICP-AES at concentration levels below 1 mg/l. Table 12.1 lists recommended wavelengths and corresponding estimated detection limits. The detection limits are provided as a guide for instrument limits. In reality, method detection limits are sample dependent and vary according to the sample matrix. Detection limits, sensitivity, and optimum ranges of metals vary by matrix and instrument model.

12.2.1 GENERAL DISCUSSION

The ICP method measures element-emitted light by optical spectrometry. Samples are nebulized and the resulting aerosol is transported to the plasma torch. Element-specific, atomic-line emission spectra are produced by radio-frequency inductively coupled plasma. The spectra are dispersed by a grating spectrometer, and the line intensities are monitored by photomultiplier tubes. The background must be measured adjacent to analyte lines on samples during analysis.

An ICP source consists of a *flowing stream of argon gas* ionized by an applied *radio frequency field* that typically oscillates at 27.1 MHz. This field is inductively coupled to the ionized gas by a water-cooled coil surrounding a *quartz torch* that supports and confines the plasma (see Section 12.2.3). A *sample aerosol* is generated in an appropriate *nebulizer* and *spray chamber* and enters the plasma through an *injector tube* located in the torch (Section 12.3.1). The sample aerosol is injected directly into the ICP, subjecting the constituent atoms to temperatures of about 6000 to 8000 K. Because this procedure results in almost complete dissociation of molecules, significant reduction in chemical interferences is achieved. The high temperature of the plasma excites element-specific *atomic-line-emission spectra*. The spectra are dispersed by a *grating spectrometer*, and the intensities of the lines are monitored by *photomultiplier tubes*.

TABLE 12.1
Recommended Wavelengths and Estimated Instrumental Detection Limits for ICP

Element	Wavelength (nm)[a]	Estimated DL[b] (μg/l)
Aluminum	308.215	45
Antimony	206.833	32
Arsenic	193.696	53
Barium	455.403	2
Beryllium	313.042	0.3
Boron	249.773	5
Cadmium	226.502	4
Calcium	317.933	10
Chromium	267.716	7
Cobalt	228.616	7
Copper	324.754	6
Iron	259.940	7
Lead	220.353	42
Magnesium	279.079	30
Manganese	257.610	2
Molybdenum	202.030	8
Nickel	231.604	15
Potassium	766.491	[c]
Selenium	196.026	75
Silicon	288.158	58
Silver	328.068	7
Sodium	588.995	29
Thallium	190.864	40
Vanadium	292.402	8
Zinc	213.856	2

[a] The wavelengths listed are recommended because of their sensitivity and overall acceptance. Other wavelengths may be substituted if they can provide the needed sensitivity and are treated with the same corrective techniques for spectral interference. In time, other elements may be added as more information becomes available and as required.

[b] The estimated detection limits are provided as a guide for an instrument limit. In reality, method detection limits are sample dependent.

[c] Highly dependent on operating conditions and plasma position.

The ICP provides an optically "thin" source that is not subject to self-absorption except in very high concentrations. Thus, *linear dynamic ranges* of four to six orders of magnitude are observed for many elements. The efficient excitation provided by the ICP results in low concentrations. Coupled with the extended dynamic range, such efficiency permits effective multielement determination of metals.

12.2.2 Performance Characteristics

The ICP-AES technique is applicable to the determination of a large number of elements at microgram-per-liter (ppb) levels. For precise quantitation, the element's concentration should be 50 to 100 times higher than the detection limit. ICP-AES analysis is not recommended for low-level concentration elements or elements that are naturally entrained into the plasma from sources other than the analyzed sample, such as traces of argon and CO_2 from argon gas, H_2 and O_2 when water is the solvent, C from organic solvent, and H_2, O_2, and N_2 from air. ICP-AES should also not be used to determine elements whose atoms have very high excitation energy requirements, such as fluorine, chlorine, the noble gases, and synthetic elements. Table 12.2 lists elements by suggested and alternate wavelengths, estimated detection limits, calibration concentrations, and typical upper limits for linear calibration.

One advantage of ICP-AES is its long linear dynamic range. (Linear dynamic range is discussed in Section 7.5.1.) This range makes possible instrument calibration to a one- or two-point curve. Another advantage is that less sample dilution is necessary. With this technique, operators can determine a large number of elements over a wide range of concentrations, and many elements can be determined in the same analytical run. The precision and accuracy of ICP-AES results are sufficient for

TABLE 12.2
Suggested Wavelengths, Estimated Detection Levels, Alternate Wavelengths, Calibration Concentrations, and Upper Limits

Element	Suggested Wavelength (nm)	Estimated Detection (μg/l)	Alternate Wavelength (nm)	Calibration Concentration[a] (mg/l)	Upper Limit Concentration (mg/l)
Al	308.22	40	237.32	10.0	100
Sb	206.83	30	217.58	10.0	100
As	193.70	50	189.04[b]	10.0	100
Ba	455.40	2	493.41	1.0	50
Be	313.04	0.3	234.86	1.0	10
B	249.74	5	249.68	1.0	50
Cd	226.50	4	214.44	2.0	50
Ca	317.93	10	315.89	10.0	100
Cr	267.72	7	206.15	5.0	50
Co	228.62	7	230.79	2.0	50
Cu	324.75	6	219.96	1.0	50
Fe	259.94	7	238.20	10.0	100
Pb	220.35	40	217.00	10.0	100
Li	670.78	4[c]	—	5.0	100
Mg	279.08	30	279.55	10.0	100
Mn	257.61	2	294.92	2.0	50
Mo	202.03	8	203.84	10.0	100
Ni	231.60	15	221.65	2.0	50
K	766.49	100[c]	769.90	10.0	100
Se	196.03	75	203.99	5.0	100
SiO_2	212.41	20	251.61	21.4	100
Ag	328.07	7	338.29	2.0	50
Na	589.00	30[c]	589.59	10.0	100
Sr	407.77	0.5	421.55	1.0	50
Tl	190.86[b]	40	377.57	10.0	100
V	292.40	8	—	1.0	50
Zn	213.86	2	206.20	5.0	100

[a] Other wavelengths may be substituted if they provide the needed sensitivity and are corrected for spectral interference.

[b] Available with vacuum or inert gas purged optical path.

[c] Sensitive to operating conditions.

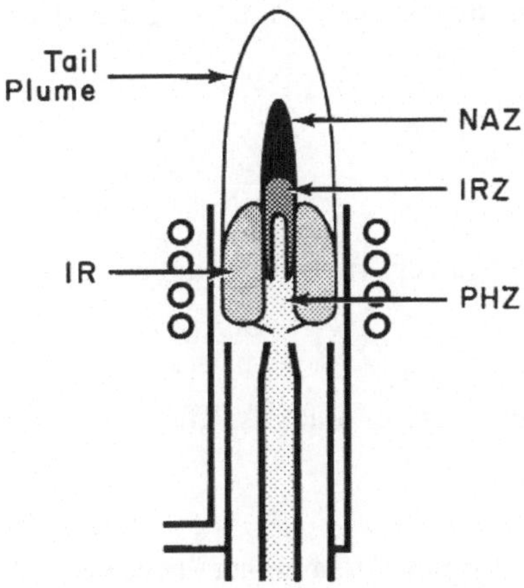

FIGURE 12.1 ICP zones: IR, induction region; PHZ, preheating zone; IRZ, initial radiation zone; and NAZ, normal analytical zone.

most analytical work. Compared to other analytical atomic spectrometry techniques, ICP-AES is subject to the lowest number of interferences. Interferences are discussed in Section 12.4.

12.2.3 ICP Discharge

Argon gas is directed through a torch consisting of three concentric tubes made of quartz or another suitable material. A copper coil, called the *load coil*, surrounds the top end of the torch and is connected to a *radio-frequency (RF) generator*. When RF power (typically 700–1500 W) is applied to the load coil, an alternating current moves back and forth (oscillates) within the coil at a rate corresponding to the frequency of the generator (27–40 MHz). Oscillation of the current in the coil causes RF electric and magnetic fields to be set up in the area at the top of the torch. With argon gas being swirled through the torch, a spark is applied to the gas causing electrons to be stripped from argon atoms. These electrons are then caught up in the magnetic field and accelerated by it. Adding energy to the electrons by the use of a coil in this manner is known as *inductive coupling*. These high-energy electrons collide with other argon atoms and produce more electrons. This collisional ionization of the argon gas continues and breaks down the gas into a plasma consisting of argon atoms, electrons, and ions, forming the *inductively coupled plasma discharge*. This ICP discharge is then sustained by the torch and load coil, as *RF energy* is continually transferred to it during the inductive coupling process.

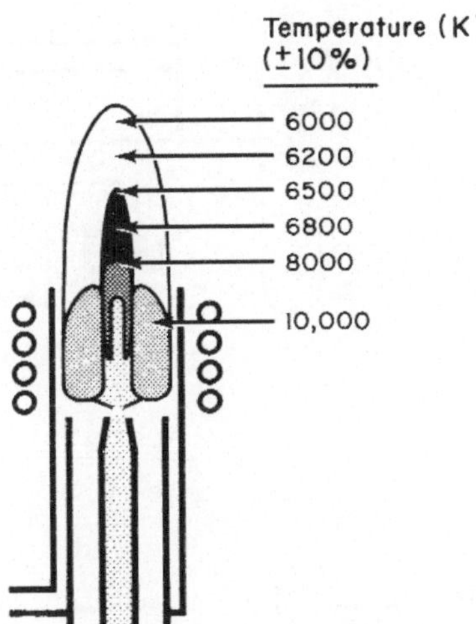

FIGURE 12.2 Temperature regions of typical ICP discharge.

The ICP discharge is very intense, brilliant white, and teardrop-shaped. Figures 12.1 and 12.2 illustrate the ICP zones and plasma temperature regions, respectively.

12.2.3.1 Plasma Functions

12.2.3.1.1 Desolvation

The first function of the high-temperature ICP is to remove the solvent from the sample droplet or desolvate, leaving the sample as microscopic salt particles.

12.2.3.1.2 Vaporization and Atomization

The salt particles decompose into gas molecules and then dissociate into atoms. These processes occur in the *preliminary zone* (PHZ) of the ICP (see Figure 12.1).

12.2.3.1.3 Excitation and Ionization

As discussed previously, an electron of any atom or ion can be promoted to a higher energy level by an excitation process during which it emits characteristic radiation. This process occurs in the *initial radiation zone* (IRZ) and in the *normal analytical zone* (NAZ) (see Figure 12.2).

12.2.3.1.4 Emission Measurement

The light emitted by the excited atoms and ions is measured in the NAZ region of the plasma. The emitted light of diverse wavelengths is measured with a polychromator and detected by a photomultiplier tube. The wavelengths are separated by a monochromator.

12.3 ICP-AES INSTRUMENTATION

In ICP-AES, the sample is usually transported into the instrument in the form steam from a liquid sample. The liquid is converted into an aerosol and transported to the plasma where it is vaporized, atomized, and excited or ionized. The emitted radiation is collected and measured. The major components and layout of a typical ICP-AES instrument are illustrated in Figure 12.3.

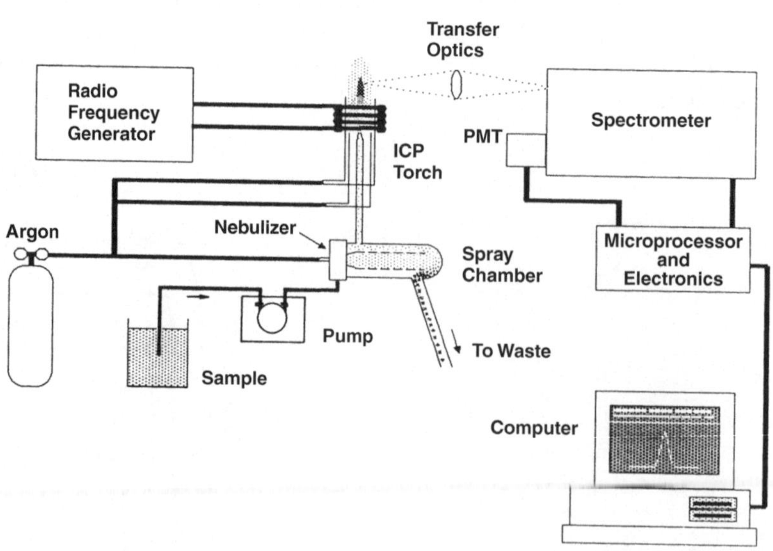

FIGURE 12.3 Major components and layout of a typical ICP-AES instrument.

12.3.1 Sample Introduction

12.3.1.1 Nebulizers

Nebulizers convert a liquid into an aerosol that can be transported to the plasma, where it is desolved, vaporized, atomized, ionized, and excited. The type of nebulizer used depends on the samples to be analyzed as well as the equipment.

12.3.1.2 Pumps

The sample solution is pumped to the nebulizer; with the help of a series of rollers, the solution is pushed through the tubing. The tubing is made of materials that are not affected by acidic solutions, organic solvents, and hydrogen fluoride. The instrument's operating manual includes instructions for the use of proper tubing. The peristalting pump tubing is the only part of an ICP system that typically requires frequent replacement. The tubing should be checked daily for wear, which is indicated by permanent depressions that can be detected by running one's fingers over the tubing.

12.3.1.3 Spray Chambers

Between the nebulizer and torch is the spray chamber, as seen in Figure 12.3. The chamber removes large droplets from the aerosol before it enters the plasma and smoothes out pulses. The diameter of the slow droplets entering the plasma should be about 10 μm or smaller. These droplets constitute about 1% to 5% of the sample, and the remaining 95% to 99% of the sample is drained into a waste container.

12.3.1.4 Drains

The drain carries the excess sample from the spray chamber to the waste container and provides the backpressure necessary to force the sample aerosol carrying the gas flow through the torch injector tube and into the plasma discharge. If the drain system does not drain evenly or it allows bubbles to pass through, the injection of the sample to the plasma will be disrupted and noisy emission signals can result.

12.3.2 Emission Production

12.3.2.1 Torches

The torches contain three concentric tubes for *argon gas flow* and *aerosol injection* (see Sections 12.2.1 and 12.3.1). The spacing between the two outer tubes is very narrow so that the gas flows between them at high velocity and in a spiral movement, thereby keeping the quartz walls of the tubes cool. For this reason, the argon gas flow is also called the *coolant flow* or *plasma flow* (because the gas flow makes the plasma). In argon ICPs, it is known as *plasma gas flow*, and the flow rate is 7 to 15 l/min.

The gas flow carrying the sample aerosol is injected into the plasma through a central tube called the *injector*. Because this flow carries the sample to the plasma, it is called the *sample flow*. When used as the nebulization gas, it is called the *nebulizer flow*. The flow rate is usually 1 l/min, and is known as the *auxiliary flow*. The three flows are illustrated in Figure 12.4.

The most popular torches can be fitted with various injector tubes, including corrosion-resistant ceramic injectors, injectors for analyzing organic solvents, and injectors for introducing samples containing highly dissolved solids.

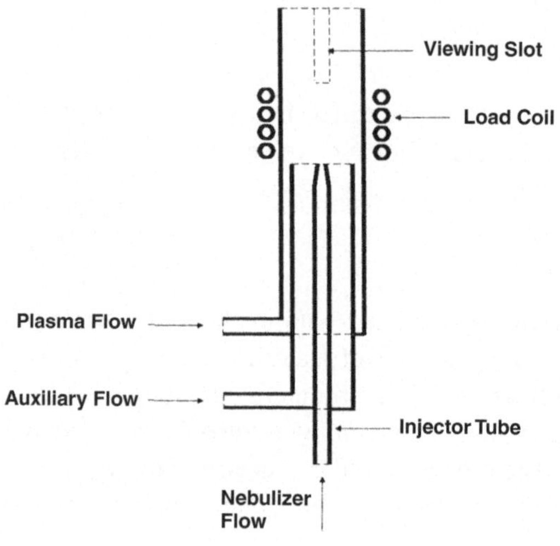

FIGURE 12.4 Schematic of a torch used for ICP-AES.

12.3.2.2 Radio-Frequency (RF) Generators

The RF generator provides the power (generally 600–1800 W) for the plasma torch. Heat is transferred to the plasma gas through a *load coil* surrounding the top of the torch. The load coil is usually made of copper tubing, and during operation it is cooled by water or gas. Most ICP-AES generators operate at a frequency of 27 to 56 MHz.

12.3.3 Collection and Detection of Emissions

12.3.3.1 Transfer Optics

The emission radiation from the normal analytical zone (NAZ) of the plasma is collected by a *focusing optic*, such as a convex lens or a concave mirror, which transfers it onto the *entrance slit* of the wavelength-dispersing device.

12.3.3.2 Wavelength-Dispersive Device

The collected emission radiation is then differentiated by elements, accomplished with a *diffraction-grating-based dispersive device*. (Diffraction grating is discussed in Section 6.2.1.) This device is simply a mirror with closely spaced lines etched into its surface, with a density of 600 to 4200 lines per millimeter. When light strikes such a grating, the light is diffracted at an angle, which is dependent on the wavelength of light and the line density of the grating. The longer the wavelength and the higher the line density, the higher the diffraction angle will be. The grating is incorporated in a *spectrometer*. The spectrometer generates the light beam, disperses it according to wavelengths selected by the grating, and focuses them to the appropriate *exit slits*. At this point, the wavelengths are passed to the detector. A *polychromator* is a device comprised of several exit slits and detectors in the same spectrometer. When only one exit slit and detector are used, the device is called a *monochromator*. Both devices can be used for multielement analysis in ICP-AES instruments. Most of the analytical emission lines in ICP-AES are in the 190 to 450 nm region. With these wavelengths, electromagnetic radiation is absorbed by oxygen molecules; therefore, air should removed from the spectrometer by purging it with nitrogen gas or by using a vacuum system.

12.3.3.4 Detectors

The detector measures the intensity of the emission line. The photomultiplier tube (see Section 7.3.5) — the most widely used detector in ICP-AES — consists of a vacuum tube containing a *photocathode* that ejects electrons when struck by light. These electrons travel to a *dynode* that produces one to five secondary electrons for every electron striking its wall. The secondary electrons strike another diode, producing new electrons, and so on. A typical photomultiplier tube contains 9 to 16 dynode stages. The anode in the tube collects the electrons from the last dynode. As many as 10^6 secondary electrons are produced from a single photon striking the photocathode in the tube. The electrical current at the anode is measured as the intensity of the radiation reaches the phototube. Figure 12.5 illustrates how the signal produced by a photon in a photomultiplier tube is measured.

12.3.4 Signal Processing and Instrument Control

12.3.4.1 Signal Processing

The electrical current measured at the anode of the photomultiplier tube is converted to information that can be passed on to a computer or immediately accessed by the analyst.

12.3.4.2 Computers and Processors

The incorporated computer is an important part of an ICP-AES instrument. Every commercial ICP-AES instrument available today uses some type of computer to control the spectrometer and to collect, manipulate, and report analytical data. The amount of control over other functions of the instrument varies widely from model to model.

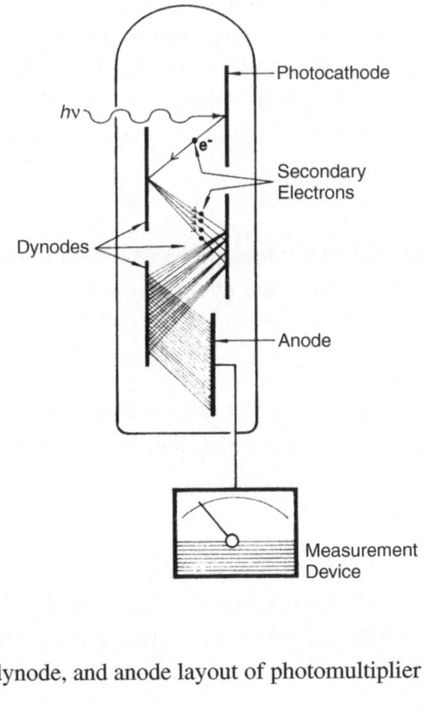

FIGURE 12.5 Photocathode, dynode, and anode layout of photomultiplier tube.

12.3.5 Accessories for ICP-AES Instruments

12.3.5.1 Autosamplers

Typical autosamplers have a capacity of 40 to 60 samples, but some models can hold 100 samples. Ideally, the analyst should be able to load the autosampler with standards and samples, start the analysis, walk away, and return to find the analysis completed.

12.3.5.2 Sample Introduction Accessories

Sample introduction accessories are widely used with ICP-AES instruments. These accessories are available directly from the instrument manufacturer or can be constructed in the laboratory.

In the *hydride generation* technique, the sample in dilute acid is mixed with a reducing agent, usually a solution of sodium borohydride in diluted sodium hydroxide. The reaction of the sodium borohydride with the acid produces atomic hydrogen. Atomic hydrogen then reacts with Hg, Sb, As, Bi, Ge, Pb, Se, Te, and Sn in the solution to form volatile hydrides of these elements. These gaseous compounds are separated from the rest of the reduction mixture and transported to the plasma. The detection limits may increase by a factor of up to 1000 by using this technique.

Another technique for ICP-AES sample introduction is a *graphite furnace* or other electrothermal device to vaporize a small portion of a liquid or solid sample. In this technique, the sample introduction system is replaced by a graphite furnace (see Section 9.2.4). The vapor of the sample goes to the center of the ICP discharge in the ICP torch.

12.3.6 Instrument Care and Maintenance

12.3.6.1 Sample Introduction and ICP Torch

Keeping the torch and sample introduction system clean and free from obstructions is important in ensuring a smooth, uncontaminated flow of sample to the plasma. Run a blank solution for several minutes after an analysis is completed or before the instrument is shut down for the day. After running a sample with a complex matrix, the sample introduction system requires a thorough cleaning. Check for depressions or flat spots on the tubing. Manually stretch new tubes before placing them on the peristaltic pump head. Make sure that the tubing is appropriate for the sample type.

12.3.6.2 Nebulizer

Make sure that the nebulizer is not clogged or leaking. When checking the aerosol for a uniform spray pattern, be sure to use deionized water and wear eye protection.

12.3.6.3 Drain System

The drain system should be filled with liquid to the level that will provide the proper backpressure for the nebulizer gas flow. Waste from the spray chamber should flow smoothly.

12.3.6.4 Torch

Check for leaks caused by damaged quartz tubes. Deposits on the torch should be removed. Check and clean the clogged injector after analyzing samples with high levels of particulates or dissolved solids. When analyzing organic-based samples, check and remove carbon deposits from the torch and injector.

12.3.6.5 RF Generator

The RF load coil should be checked for corrosion or leakage. The high-voltage wires and other parts of the ignition system must be checked and replaced if corroded. The power amplifier tubes must be checked, but replacement should be performed only by professionals. Always be sure that the laboratory exhaust venting system for the ICP torch box is functioning properly before igniting the plasma. Harmful ozone, toxic combustion products, and metal fumes may accumulate in the laboratory if not vented properly.

12.3.6.6 Spectrometer

Windows should be regularly inspected and carefully cleaned or replaced as necessary. Periodically check wavelength calibration as described in Section 6.7.

12.3.6.7 Computer

Conduct regular routine computer maintenance, such as cleaning disk drives and air filters. If data files are stored on the hard disk, "clean up" the data file directories by erasing files or transferring them to disks.

12.3.7 Verification of Instrument Performance

Several tests are available to verify that the instrument is working properly. Some of these tests should run on a daily basis, and some should be used as diagnostic tests to verify problems indicated by erratic results. Before running tests, wait for the instrument to warm up properly. The warm-up usually takes 30 to 60 min.

12.3.7.1 Bullet Test

The visual bullet test should be performed on a daily basis. A solution of yttrium or sodium in a concentration of 1000 mg/l or more is introduced into the system. The emission should produce a so-called "bullet" in the center of the ICP discharge. The mere presence of the bullet indicates that the sample aerosol is reaching the plasma, while the vertical position of the bullet in the discharge is an indicator of the gas flow and RF power settings.

12.3.7.2 Signal Intensity

The number of emission counts (called *signal intensity*) for an element with known concentration is frequently measured, because the emission count for a given concentration may vary from day to day.

12.3.7.3 Background Equivalent Concentration (BEC)

The BEC is an indicator of relative sensitivity for an emission line. A BEC of a higher-than-normal value often indicates problems with the efficiency of the sample introduction system, although it can be due to a number of causes.

12.3.7.4 Precision

The precision of an argon emission line is sometimes used as a diagnostic test for the RF generator. Precision is discussed in Section 13.9.

12.3.7.5 Detection Limits

Detection limits may also be used for diagnostic purposes. Detection limits and measurements are described in Section 13.8. The measured detection limits alone do not serve as an indicator of an instrument's performance, unless the measurement is combined with a series of other, more specific tests.

12.3.7.6 Wavelengths

Because UV/Vis spectrometers are subject to drift, ensure that the spectrometer is calibrated properly in terms of wavelength prior to ICP analysis. In some instruments, calibration is performed by the instrument software at the beginning of the analysis, but some instruments require manual checking.

12.4 INTERFERENCES IN ICP-AES

Most interferences are of spectral origin. Other types of interference are often the result of high concentrations of certain elements or compounds in the sample matrix and can be easily compensated for in most cases.

12.4.1 Spectral Interferences and Corrections

Light emission from spectral sources other than the element of interest may increase the apparent signal intensity. Spectral interferences are caused by (1) overlap of a spectral line from another element; (2) unresolved overlap of molecular spectra; (3) background contribution; and (4) stray light from the line emission of high-concentration elements.

12.4.1.2 Spectral or Background Interferences

The following interferences are well known to users of the ICP-AES technique:

- Simple and sloping background shift
- Direct spectral overlap
- Complex background shift

12.4.1.2.1 Correction of Spectral Interferences

- *Alternate analytical wavelengths:* Avoid line overlaps by selecting alternate analytical wavelengths.
- *Interelement correction:* Measure the emission intensity of the interfering element at another wavelength and calculate a correction factor (Section 12.6.8). This factor should apply to determine the correct result. This technique is often useful in correcting the simple, sloping, and complex background shifts. Analyte concentration equivalents arising from interference at the 100-mg/l level are presented in Table 12.3. The interference is expressed as analyte concentration arising from 100 mg/l of interference element. For example, assume that As is to be determined (at 193.696 nm) in a sample containing approximately 10 mg/l of Al. According to Table 12.3, 100 mg/l of Al would yield a false signal for As equivalent to approximately 1.3 mg/l. Therefore, the presence of 10 mg/l of Al would result in a false signal for As equivalent to approximately 0.13 mg/l. The interference effects must be evaluated for each individual instrument.

TABLE 12.3
Analyte Concentration Equivalents Arising from Interference at the 100-mg/l Level

		Interference[a,b]									
Metal	**Wavelength (nm)**	**Al**	**Ca**	**Cr**	**Cu**	**Fe**	**Mg**	**Mn**	**Ni**	**Tl**	**V**
Al	308.215	—	—	—	—	—	—	0.21	—	—	1.4
Sb	206.833	0.47	—	2.9	—	0.08	—	—	—	0.25	0.45
As	193.696	1.3	—	0.44	—	—	—	—	—	—	1.1
Ba	455.403	—	—	—	—	—	—	—	—	—	—
Be	313.042	—	—	—	—	—	—	—	—	0.04	0.05
B	249.773	0.04	—	—	—	0.32	—	—	—	—	—
Cd	226.502	—	—	—	—	0.03	—	—	0.02	—	—
Ca	317.933	—	—	0.08	—	0.01	0.01	0.04	—	0.03	0.03
Cr	267.716	—	—	—	—	0.003	—	0.04	—	—	0.04
Co	228.616	—	—	0.03	—	0.005	—	—	0.03	0.15	—
Cu	324.754	—	—	—	—	0.003	—	—	0.05	—	0.02
Fe	259.940	—	—	—	—	—	—	0.12	—	—	—
Pb	220.253	0.17	—	—	—	—	—	—	—	—	—
Mg	279.079	—	0.02	0.11	—	0.13	—	0.25	—	0.07	0.12
Mn	257.610	0.005	—	0.01	—	0.002	0.002	—	—	—	—
Mo	202.030	0.05	—	—	—	0.03	—	—	—	—	—
Ni	231.604	—	—	—	—	—	—	—	—	—	—
Se	196.026	0.23	—	—	—	0.09	—	—	—	—	—
Si	288.158	—	—	0.07	—	—	—	—	—	—	0.01
Na	588.995	—	—	—	—	—	—	—	—	0.08	—
Tl	190.864	0.30	—	—	—	—	—	—	—	—	—
V	292.402	—	—	0.05	—	0.005	—	—	—	0.02	—
Zn	213.856	—	—	—	0.14	—	—	—	0.29	—	—

[a] Dashes indicate that no interference was observed even when interferences were introduced at the following levels: Al, 1000 mg/l; Mg, 1000 mg/l; Ca, 1000 mg/l; Mn, 200 mg/l; Cr, 200 mg/l; Tl, 200 mg/l; Cu, 200 mg/l; V, 200 mg/l; and Fe, 1000 mg/l.

[b] The figures recorded as analyte concentrations are not observed concentrations. To obtain those figures, add the listed concentrations to the interference figure.

12.4.2 Nonspectral Interference

12.4.2.1 Physical Interference

Physical interference is associated with nebulization and transportation processes. Changes in the physical properties of samples, such as viscosity and surface tension, in highly dissolved solids and high-acid concentrations can cause significant errors. Physical interferences may be compensated with sample dilution or by the standard addition technique (see Section 7.7.1). Samples consisting of highly dissolved solids may cause salt buildup at the tip of nebulizer. Using prehumidified argon for each sample nebulization is helpful.

12.4.2.2 Chemical Interferences

Chemical interference is caused by molecular compound formation, ionization effects, and sample vaporization. Chemical interference is highly dependent on the sample matrix and the element of interest. These conditions are easily minimized by careful selection of operating conditions. Similar to physical interference, chemical interference may be compensated by using matrix-matched standards or by using the standard additions method.

12.5 REAGENTS AND STANDARDS

12.5.1 Chemicals, Standards, and Reagents

- Chemicals and reagents should be ultra-high-purity grades, except as noted.
- Dry all salts at 105°C for 1 h and store in a desiccator before weighing.
- Use deionized water prepared by passing it through at least two stages of a deionization process for preparing standards, reagents, and dilutions. Criteria and checks of laboratory pure-water quality are covered in Section 13.4.

12.5.2 Acids

- Hydrochloric acid, HCl concentrate, and 1+1
- Nitric acid, HNO_3 concentrate, and 1+1

Add 500 ml of HNO_3 concentrate to 400 ml of water and dilute to 1 liter.

12.5.3 Standard Stock Solutions

Standard stock solutions can be purchased or prepared from chemicals or metals. See Appendix H for recipes of these stock solutions. Store metal stock solutions at room temperature with a record of arrival, date opened, and expiration date (see Section 13.6.1).

12.5.4 Mixed Calibration Standard Solutions

- Prepare mixed calibration standard solutions by combining appropriate volumes of the stock solutions in 100-ml volumetric flasks.
- Add 2 ml 1:1 HNO_3 and 10 ml 1:1 HCl and dilute to 100 ml with distilled water. Mix well. Store the mixed standards in FEPs (fluorocarbon or unused polyethylene bottles).

Concentrations of standards can change with aging! Verify concentrations by using quality control sampling and monitor weekly for stability. Some typical combinations of mixed standards follow, although alternative combinations are acceptable.

Mixed standard solution I:	Be, Cd, Mn, Pb, Se, and Zn
Mixed standard solution II:	Ba, Co, Cu, Fe, and V
Mixed standard solution III:	As, Mo, and Si
Mixed standard solution IV:	Al, Ca, Cr, K, Na, and Ni
Mixed standard solution V:	Ag, B, Mg, Sb, and Tl

Note: If the addition of Ag to the recommended acid combination results in an initial precipitation, add 15 ml of distilled water and warm the flask until the solution clears. Cool and dilute to 100 ml with distilled water. The Ag concentration should be limited to 2 mg/l, which is stable for 30 days. Higher concentrations of Ag require additional HCl.

12.5.5 Blanks

12.5.5.1 Calibration Blank

Measure 2 ml 1+1 HNO_3 and 10 ml 1+1 HCl and dilute to 100 ml with laboratory pure water. Prepare a sufficient quantity to be used to flush the system between standards and samples.

12.5.5.2 Method Blank

Carry a reagent blank through the entire sample preparation procedure. Prepare with the same acid contents and concentrations as the sample solutions.

12.5.6 Instrument Check Standard

Prepare the standard by combining compatible elements at concentrations equivalent to the midpoint of respective calibration curves.

12.5.7 Interference Check Solution

This solution is prepared to contain known concentrations of interfering elements that will provide an adequate test for correction factors (Section 12.6.8). Spike the sample with the element of interest at the approximate concentration of ten times the instrument detection limits.

12.5.8 Quality Control Standards

Obtain a certified aqueous reference standard from an outside source and prepare according to the instructions provided by the supplier. Use the same acid matrix as the calibration standards.

12.5.9 Method Quality Control Sample

Carry quality control sample (Section 12.5.8) through the sample preparation procedure.

12.6 PROCEDURE

12.6.1 Sample Preparation

Preparation depends on the physical and chemical characteristics of the samples. Sample preparation methodology is discussed in Chapter 15.

12.6.2 Instrument Setup and Operation

1. The instrument should be warmed up for at least 30 min.
2. Set up the instrument with proper parameters. Because of differences among types and models of instrumentation, follow the manufacturer's instructions. Program the instrument using the computer software provided with the instrument. Establish instrument detection limits, optimum background correction positions, linear dynamic range, and interferences for each analytical line.
3. Before making analytical measurements, take the necessary steps to determine that the instrument is set up and functioning properly. (Instrument maintenance and performance verification are discussed in Sections 12.3.6 and 12.3.7.)
4. Calibrate the instrument, using the typical mixed standard solutions. Flush the system with the calibration blank (Section 12.5.5.1) between each standard. For concentrations greater than 500 μg/l, an extended flush time of 1 to 2 min is recommended.
5. Before analyzing samples, reanalyze the highest mixed calibration standard as if it is a sample. Concentration values obtained should not deviate by more than ±5% from the actual value or from the established control limit, whichever is lower.
6. Flush the system with the calibration blank for at least 1 min. Run the quality control sample (Section 12.5.8). The concentration value should not deviate more than ±5% of the original value.

7. Begin each sample run with the calibration blank (Section 12.5.5.1), and then analyze the method blank (Section 12.5.5.2). This permits a check for contamination of sample preparation reagents and procedures.
8. Flush the system with the calibration blank (Section 12.5.5.1) for at least 1 min before the analysis of each sample. Analyze samples while alternating them with a calibration blank. If carryover is observed, repeat rinsing until proper blank values are obtained.
9. Analyze the quality control check standard (highest calibration standard) and quality control sample (Section 12.5.8) once per ten samples. If agreement is not within ±5% of the expected values, terminate analysis of samples, correct the problem, recalibrate the instrument, and analyze the quality control sample again to confirm proper recalibration. Reanalyze one or more of the samples analyzed just before termination of the analytical run. Results should agree to within ±5%; otherwise, all samples analyzed after the last acceptable quality control test must be reanalyzed.
10. Analyze the quality control sample (Section 12.5.8) during each run. Use this analysis to verify accuracy and stability of the calibration standards. If any result is not within ±5% of the certified value, prepare new calibration standards, and recalibrate the instrument. If this does not solve the problem, prepare a new stock solution and new standards, and recalibrate the instrument again.
11. Analyze the method quality control sample (Section 12.5.9) with every run. Results deviating more than ±5% of the certified value indicate losses or contamination during preparation.
12. When analyzing a new or unusual sample matrix, verify that positive or negative nonlinear interferences do not exist. If the element is present above a 1 mg/l concentration, dilute the sample with a calibration blank. Results from the analysis of dilution should be within ±5% of the original result. If the result is below 1 mg/l or not detected, spike the digested sample with 1 mg/l. Recovery should be within 95 and 105%.

12.6.3 Instrument Calibration

Set up the instrument as described in Section 12.6.2, items 1 through 3. Calibrate the instrument according to the manufacturer's recommended procedure using the typical mixed standard solutions described in Section 12.5.4. Flush the system with the calibration blank (Section 12.5.5.1) between each standard. Aspirate each standard or blank for a minimum of 5 sec after reaching the plasma but before beginning signal integration. Rinse with the calibration blank for at least 60 sec between each standard to eliminate any carryover from the previous standard. For boron concentrations greater than 500 mg/l, extended flush times of 1 to 2 min may be required.

12.6.4 Sample Analysis

1. Before analyzing samples, analyze the instrument check standard (Section 12.5.6). Concentration values obtained should not deviate by more than ±5% from the actual values or the established control limits, whichever is lower. If they do, follow the recommendations of the instrument manufacturer to correct for this condition.
2. Flush the system with the calibration blank (Section 12.5.5.1) solution for at least 1 min.
3. Analyze the method blank (Section 12.5.5.2).
4. Analyze samples, alternating with analysis of the calibration blank (Section 12.5.5.1). Rinse for at least 60 sec with diluted acid between samples and blanks. Examine each analysis of the calibration blank to verify that no carryover has occurred. If carryover is observed, repeat the rinsing until proper blank values are obtained.
5. Make appropriate dilutions or concentrations of the sample to determine concentrations beyond the linear concentration range.

12.6.5 Instrument Quality Control

1. Analyze the instrument check standard (Section 12.5.6) once per ten samples to determine significant instrument drift. If agreement is not within ±5% of the expected value or within the established control limits, whichever is lower, terminate the analysis of samples, correct the problem, and recalibrate the instrument.
2. Confirm proper recalibration by analyzing the instrument check standard.
3. Reanalyze one or more of the samples analyzed just before termination of the analytical run. Results should agree to within ±5%; otherwise, all samples analyzed after the last acceptable instrument check standard analysis must be reanalyzed.
4. Analyze quality control sample (Section 12.5.8) with every run to verify the accuracy and stability of the calibration standard. If any result is not within ±5% of the calibrated value, prepare a new calibration standard and recalibrate the instrument. If this does not correct the problem, prepare a new stock solution and a new calibration standard and repeat calibration.

12.6.6 Method Quality Control

Analyze the method quality control sample (Section 12.5.9) during every run. Results should agree within ±5% of the certified value. Greater discrepancies may reflect losses or contamination during sample preparation.

12.6.7 Test for Matrix Interference

When analyzing a new or unusual sample matrix, verify that positive and negative nonlinear interference effects are not operative. If the element is present at a concentration above 1 mg/l, use serial dilution with a calibration blank. Results from analyses of a dilution should be within ±5% of the original result. If the result is below 1 mg/l or not detected, use a postdigestion addition equal to 1 mg/l. Recovery of the addition should be between 95% and 105% or within the established control limits of two standard deviations around the mean. If a matrix effect causes test results to fall outside the critical limits, complete the analysis after diluting the sample to eliminate the matrix effect while maintaining a detectable concentration at least twice the detection limit, or applying the standard addition method.

12.6.8 Calculations and Corrections

12.6.8.1 Blank Correction

12.6.8.1.1 Calibration Blank

See Section 12.5.5.1. Subtract the result of an adjacent calibration blank from each sample result to calculate a baseline drift correction. Make certain that the used calibration blank has not been contaminated.

12.6.8.1.2 Reagent Blank or Method Blank

See Section 12.5.5.2. Use the result of the method blank analysis to correct reagent contamination.

12.6.8.2 Dilution and Concentration Correction

If the sample was diluted or concentrated, correct the result accordingly. In the case of *dilution*, the result should be multiplied by the dilution factor (final volume/initial volume). For example, if the original result was 0.20 mg/l and the sample was diluted 5 times, the corrected result will be $0.20 \times 5 = 1.00$ mg/l.

Because of low concentration of a metal in a sample, the digestion technique is used and the sample will be *concentrated*. For example, an original 100-ml sample was cooked down to 10-ml final volume. The reading of the concentrated sample was 0.06 mg/l, so the final result is 0.06/10 = 0.006 mg/l or 6 μg/l.

12.6.8.3 Correction for Spectral Interference

Correct for spectral interference (see Section 12.4.1) by using computer software supplied by the manufacturer or by using the manual method based on interference correction factors. Determine interference correction factors by analyzing single-element stock solutions of appropriate concentrations under conditions that match as closely as possible those used in the sample analysis. Calculate the correction factors by using the following equation:

$$\text{C.F.} = A/B \tag{12.1}$$

where

A = difference between the observed concentration in the stock solution and the observed concentration in the blank.

B = actual concentration.

All results should be reported in micrograms per liter with up to three significant figures.

12.7 QUALITY CONTROL

All quality control data should be maintained and available for easy reference and inspection. For complete quality control criteria, see Chapter 13. The quality control check process for ICP analysis is discussed in Section 12.6.2, items 6 through 12.

13 Quality Control in Metals Analysis

13.1 GENERAL DISCUSSION

Quality control procedures are necessary to ensure that the obtained analytical data and reported values are correct. Validation and approval of analytical data are based on a well-designed and regularly applied quality assurance/quality control (QA/QC) program. The components of this program are divided into a management (QA) and a functional (QC) part. Quality Assurance is a laboratory operation program of specified standard procedures that are aimed at producing data of defensible quality and highly reliable reported results. Quality Control consists of a set of measures within a sample analysis methodology to ensure that the process is controlled.

Laboratory personnel develop their own QA/QC program, which should be delineated in a QA/QC manual. The program should be strictly enforced, continually reviewed, and updated as needed. Large laboratories typically have a QC control officer or group independent of laboratory management charged with oversight of the QA/QC program.

Quality assessment is the mechanism for verifying that a laboratory system is operating within acceptable limits. This mechanism consists of all activities aimed at providing assurance that the overall quality control job is being done effectively. Each laboratory develops its own QA program, which should be delineated in a QA/QC manual and should be comprehensive enough to apply to most operations. This program should be continually reviewed and updated as needed. The laboratory should also have a separate QA "project plan" for each project. Understanding the QA program helps analysts to be responsible participants in the laboratory system in order to produce defendable, precise, and accurate analytical data.

13.2 GLASSWARE USED IN METALS ANALYSIS

The mainstay of a modern analytical laboratory is a highly resistant borosilicate glass, such as *Pyrex* or *Kimax*. *Corning*-brand glassware is highly resistant to alkalis and practically boron free. *Raysorb*- or *Lo-actinic*-brand glassware is recommended for light-sensitive solutions. Stoppers and cups should be carefully selected: *Do not use metal caps or rubber-stoppered bottles for metal solutions!*

13.2.1 VOLUMETRIC GLASSWARE

Volumetric glassware is used for accurate volume measurements; therefore, it should meet the *Class A glassware* specification and be permanently marked with "A" and the temperature at which calibration was made. Carefully check glassware for *TC* (to contain) or *TD* (to deliver) marks, and use them accordingly. Not suitable, incorrectly used, or improperly cleaned glassware endangers the quality of analytical results.

13.2.2 Cleaning Glassware

When cleaning glassware for metals analysis, use the following guidelines. Always use appropriate detergent, such as *Liquinox, Alconox*, or an equivalent.

1. Remove all labels or marks from the glassware (using acetone is acceptable).
2. Wash with hot soapy water. Use appropriate detergent. Use brush to scrub inside the glassware. (Do not use a brush with any metal parts.)
3. Rinse thoroughly with hot tap water.
4. Rinse thoroughly with distilled water.
5. Rinse with 1:1 HCl.
6. Rinse with 10% HNO_3.
7. Rinse with laboratory-grade water.
8. Volumetric class A glassware should not be dried by heating!
9. Store glassware to protect from contamination, dust, breakage, and chipping.

Store glassware in an area separate from the metals analysis work area to avoid contamination.

13.3 CHEMICALS

Carefully select the grade of the chemical that meets the requirements of the work to be done. Always recheck the label of the chemical that you are using. The use of a wrong chemical can cause an explosion or ruin the analytical work. Check the information carefully on the container of the chemical: name, formula, formula weight, percentage of impurities, analytical grade, health hazards, safety codes, and expiration date. Store in chemical storage room. All chemicals used for Hg analysis must be Hg-free.

13.4 LABORATORY-PURE WATER

13.4.1 Quality of Laboratory-Pure Water

One of the most important aspects of chemical analysis is the quality of the laboratory-pure water or reagent-grade water to be used in the preparation of standard solutions, reagents, dilutions, and blank analysis.

13.4.1.1 Distilled Water

Distillation is the procedure in which the liquid is vaporized, recondensed, and collected. Distilled water quality depends on the type of still and the quality of the feed water. Deionized feed water is preferred.

13.4.1.2 Demineralized or Deionized Water

This type of water is purified in a mixed-bed exchanger. Commercial resin-purification trains that produce superior water quality are available.

13.4.1.3 Redistilled Water

This type of water is prepared by redistilling single-distilled water from an all-borosilicate-glass apparatus.

13.4.1.4 Reagent Water

Reagent water is a sample that conforms to ASTM grades II, III, or IV (see Section 13.4.2).

13.4.1.5 Analyte-Free Water

This type of water is free of the substance analyzed.

13.4.1.6 Reagent-Grade Water

This is the highest-quality, laboratory-pure water. It is prepared by passing distilled water through an activated carbon cartridge to remove dissolved organic materials. It is then passed through two deionized cartridges to remove dissolved inorganic substances. Finally, it is passed through membrane filters to remove microorganisms and any particulate matter with a diameter as large as 0.22 µm. This kind of high-quality water is commercially available and used in AA, GC work, and tissue culturing, among other things.

13.4.2 Types of Laboratory-Pure Water

The type of laboratory-pure water used depends on the analytical work. For metals analysis, the criterion is analyte-free water with an ASTM-grade type II classification. ASTM International (ASTM, formerly the American Society for Testing and Materials) specifies the various grades of laboratory-pure water.

13.4.2.1 Type I Water

Type I water has no detectable concentration of the compound or element to be analyzed at the detection limit of the analytical method. Use type I water in test methods requiring minimum interference and bias and maximum precision. It is prepared by distillation, deionization, or reverse osmosis treatment of feed water, followed by polishing with a mixed-bed deionizer and passage through a 0.2-µm-pore-size membrane filter. type I water cannot be stored without significant degradation; therefore, produce it continuously and use immediately after processing.

13.4.2.2 Type II Water

Type II is the same as type I water but without passage through a 0.2-µm-membrane filter. This type is used in tests in which the presence of bacteria can be tolerated. Type II water is recommended for metals analysis. It can be stored, but keep storage to a minimum. Store only in materials that protect the water from contamination, such as Teflon or glass for organic analysis and plastic for metals analysis.

13.4.2.3 Type III Water

Type III water may be used for washing glassware, preliminary rinsing of glassware, and as feed water for preparation of higher-quality water. Storage is similar to type II water.

The quality of laboratory-pure water should be checked regularly. Parameters and monitoring frequency for quality checks are listed in Table 13.1; an example of recommended documentation of quality checks is presented in Table 13.2.

13.5 FIELD QUALITY CONTROL

The quality of data resulting from sampling activities depends on the following major activities:

- Collecting representative samples
- Use of appropriate equipment
- Proper sample handling and preservation
- Proper chain-of-custody and sample identification procedure
- Proper QA and QC in the field

13.5.1 Field QA/QC Program

The field QA/QC program consists of the following areas and corresponding documentation:

1. Sample collection methodology called "field standard operation procedure" (FSOP) with special procedures: Each method must be accompanied by method numbers, method reference, method detection limits, and accepted limits for precision and accuracy. These methods should be approved by the Environmental Protection Agency (EPA) and DEP.
2. Field QC requirements
3. Procedures to record and process data
4. Procedures to review and reduce data based on QC results
5. Processes to validate field measurement data for reporting purposes
6. Procedures to calibrate and maintain field instruments and equipment
7. Qualification and training of sampling personnel to attain proficiency in the following areas:
   - Determination of the best representative sample site
   - Use of proper sampling techniques by choosing grab or composite sampling, selection of the appropriate equipment, use of proper sample preservation, and sample identification
   - Use of appropriate data recording techniques and reporting form
   - Calibration and maintenance of field instruments and equipment
   - Use of QC samples such as duplicate, split, and spiked samples
   - After the training program, the fresh-sample collector must be involved in sampling activities under the direction of a more experienced person for at least 1 month prior to assuming field responsibility; special training workshops are available for training of sampling personnel.

13.5.2 Criteria for Field QC Checks

Sampling operations must also be supported by a well-designed and reliable quality assurance program, including QC checks.

13.5.2.1 Equipment Blanks

Equipment blanks are used to detect contamination from sampling equipment. At least one equipment blank should be collected for every 20 samples per parameter group and for each matrix. Each type of equipment used in sampling must be accompanied by an equipment blank. This blank is prepared in the field before sampling begins by using the precleaned equipment and filling the appropriate container with analyte-free water. Preservation and documentation should be the same as for the collected samples. If equipment is cleaned on site, then additional equipment blanks should be collected for each equipment group.

13.5.2.2 Field Blanks

Field blanks are collected at the end of the sampling event. Fill an appropriate sample container with analyte-free water and preserve and document in the same manner as the collected samples.

TABLE 13.1
Quality Check of Laboratory-Pure Water

Parameter	Monitoring Frequency	Limit
Conductivity	D	1–2 µmhos/cm
pH	D	5.5–7.5 unit
Total organic carbon	A	<1.0 mg/l
Trace metal, single (Cd, Cr, Cu, Ni, Pb, Zn)	A	<0.05 mg/l
Trace metal, total (Cd, Cr, Cu, Ni, Pb, Zn)	A	<1.0 mg/l
Ammonia, as NH_3–N	M	<0.1 mg/l
Free chlorine, Cl_2	M	<0.1 mg/l
Heterotrophic count		
Fresh water	M	<1000 cnt/ml
Stored water	M	<10,000 cnt/ml
Water suitability test	A	Ration: 0.8–3.0

Note: A = annually; M = month; D = daily; cnt = count (bacterial).

TABLE 13.2
Documentation of Laboratory-Pure Water Quality

Date	pH	Cond. (µmhos/cm)	TOC	Cd	Cr	Cu	Ni	Pb	Zn	NH_3 as N	Het. Bact. (c/ml)	In.

Note: Parameters are reported in mg/l, except as noted.
Cond. = conductivity, µmhos/cm; TOC = total organic carbon; Cd = cadmium; Cr = chromium; Cu = copper; Ni = nickel; Pb = lead; Zn = zinc; NH_3–N = nitrogen ammonia; Het. Bact. = heterotrophic bacteria count, count/ml; In. = initial of the logger.

13.5.2.3 Trip Blank

The purpose of trip blanks is to verify contamination that may occur during sample collection and transportation, as a result of improperly cleaned sampling containers, contaminated reagents, airborne contamination during transportation, and so on. Trip blanks are blanks of analyte-free water prepared by the laboratory and transported to the field that remain unopened during the sampling and are then transported back to the laboratory with the collected samples. These blanks should be properly labeled and documented. Trip blanks are usually collected with volatile organic compound (VOC) samples.

13.5.2.4 Duplicates

Duplicates are samples collected at the same time from the same source (called field duplicates) or aliquots of the same sample that are prepared and analyzed at the same time (laboratory duplicates). Duplicate samples are analyzed to calculate measurement precision. During each independent sampling event, at least one sample or 10% of the samples, whichever is greater, must be collected for duplicate analysis. This requirement applies to each parameter group and each matrix sampled.

13.5.2.5 Field-Spiked Samples

Field-spiked samples are environmental samples that contain specific added concentrations of various parameters of interest. Spiked samples are used to measure the performance of the complete analytical system, including interference from the sample matrix. Field preparation and transportation to the laboratory of a spiked sample should be similar to other samples, and the spiked sample should be marked as *FSp* (field-spiked sample). If spiked duplicates are collected, they are identified as FSp1 and FSp2. Spiked samples are selected according to a specific requirement, previous evaluation of the sample site, or an on-site inspection.

13.5.2.6 Split Samples

Split samples are replicas of the same sample. Split samples are given to two independent laboratories for analysis.

13.6 INSTRUMENT CALIBRATION

Calibration and standardization of analytical systems are necessary to ensure that the produced data are accurate.

13.6.1 Calibration Stock and Standard Solutions

13.6.1.1 Calibration Stock Solutions

Calibration stock solutions are either commercially available or "house prepared" by the laboratory. Records should be maintained for both types of stock solutions as illustrated in Figures 13.1 and 13.2. Calibration stock solutions used for metals analysis are stored at room temperature. In the case of purchased stock, the applicable holding time is indicated by the given expiration date. House-prepared stock should be renewed when instrument response is not satisfactory.

13.6.1.2 Calibration Standard Solutions

Calibration standard solutions are prepared from calibration stock solutions by appropriate dilution. A log form for preparation of calibration standards is illustrated in Figure 13.3. The form should contain the concentration of the stock solution, dilution technique for the desired concentration, date and

Name and concentration: __________
Test for use: __________
Source: __________
Lot number: __________
Date received: __________
Date opened: __________
Expiration date: __________
Manufacturer's certification: __________
Storage: __________
Date of disposal: __________
Mode of disposal: __________

Remarks: __________

Signature of logger __________ Date __________

FIGURE 13.1 Documentation log form for purchased calibration stock and standard solutions.

Name and concentration: __________
Test of use: __________
Source of chemical used:
Name and formula: __________
Grade: __________
Lot number: __________
Source: __________
Date received: __________
Date opened: __________
Preparation procedure:
Temperature and drying time of chemical prior to preparation: __________
Volume prepared: __________
Quantity of chemical used: __________

Remarks: __________

Date: __________ Signature: __________
Date: __________ Approval by supervisor: __________

FIGURE 13.2 Documentation log form for preparation of calibration stock solution.

signature of the preparer, and, if applicable, holding time and mode of storage. Calibration standards for metals analysis should be preserved with 0.5% HNO_3. Concentrations such as 100 ppm and 10 ppm standards may be stored at room temperature up to 1 month but should be replaced when readings of concentration values show decline. Documentation for storage and preparation of the standards must be available. Every time calibration is performed, it is recommended that working standards be prepared fresh.

13.6.2 Calibration Check Solutions

13.6.2.1 Continuing Calibration Standard (CCS)

The CCSs are used to ensure calibration accuracy during every analytical run. The CCS represents the value of the midpoint initial calibration standard. It must run immediately after the standard curve is established, during the analytical batch analysis (at a frequency of 5%), and after the last sample is analyzed. The deviation from the original value should be within ±5%.

13.6.2.2 Calibration Verification Standard (CVS)

The CVS is a known-value standard used to verify that the standards and calibration are accurate and to confirm the calibration curve. It should be certified (purchased from the EPA or another source) or independently prepared by a source other than the source of the calibration standards. The value is accepted within a ±10% deviation from 100% recovery. The log form for preparation of the CVS is illustrated in Figure 13.4.

Test: ____________________

Optimum Calibration Range: ____________________

Stock Solution, Concentration: ____________________

Final Volume of Standards, ml: ____________________

Concentration of Standards	Volume Stock Used
________	________
________	________
________	________
________	________
________	________

Concentration of Continuing Calibration Standard: ____________________

____

Date of Preparation: ____________________

Expiration Date: ____________________

Storage Description (if applicable): ____________________

____

____

Date: ____________ Signature: ____________

Date: ____________ Approval by Supervisor: ____________

FIGURE 13.3 Documentation log form for preparation of calibration standards.

13.6.2.3 Preparation Blank (Prep Blank) and Laboratory Control Standard (LCS)

When samples go through pretreatment (digestion, extraction, filtration, etc.) prior to analysis, pretreatments should be verified and the effects of sample preparation monitored. To support these measures, a blank and standards should be prepared and analyzed together with the sample. The *preparation blank* is a volume of analyte-free water processed through the sample preparation procedure. The LCS is the same concentration as the CVS, except that it is carried through the preparation and analysis procedure in the same way as the samples. Acceptable recovery is a ±15% deviation from 100% recovery.

13.6.3 Initial and Continuing Calibration

Calibrations are performed at the beginning of the analysis to ensure that the instrument is working properly. This initial calibration has to be proven during the analytical process by continuing calibration. When continuing calibration fails to meet acceptance criteria, the initial calibration should be repeated. Calibrations must be performed according to specifications of a particular analytical method.

13.6.3.1 Initial Calibration

An initial calibration is determined for each parameter tested and is based on the instrument responses for different concentration ranges of the calibration standards. The number and optimum concentration range of calibration standards used in a particular analytical method are specified in

Name and Concentration ____________________

Storage ____________________

Outside Source

Manufacturer: ____________________

Certification available: ____________________

Date received: ____________________

Expiration date: ____________________

Storage: ____________________

In-House Prepared

Name, formula, and grade of the chemical used: ____________________

____

Lot number: ____________________

Date received: ____________________

Date opened: ____________________

Expiration date: ____________________

Preparation procedure: ____________________

____

____

Signature of Logger: ____________________ Date: ____________________

Approval by Supervisor: ____________________ Date: ____________________

FIGURE 13.4 Documentation log form for preparation of calibration verification standards or quality control check standards.

the approved methodology and described by the laboratory's standard operation procedure (SOP). If this information is not available, then a minimum of one blank and three standards must be utilized for calibration. Instrument calibration varies according to the type and model of the equipment. Detailed operation and calibration procedures for each instrument are available in laboratory SOPs and the manufacturer's instruction. Initial calibration is based on the instrument response for different concentrations of calibration standards against calibration blanks.

Standards are prepared by analysts or highly reliable suppliers. The concentration of the standards should be bracketed in the optimum concentration range given by the analytical method. The number of standards is recommended by the method or instrument manufacturer. When the number of standards is not known, a three-standard calibration is satisfactory.

The concentration of the "high standard" is the upper level of the optimum range, the concentration of the "middle-point standard" is half of the highest standard, and the value of the "low-level standard" is five times lower than the highest standard. The response of the instrument should be linear with the concentration of the introduced standards. The concentration of the standards and the response of the instrument are plotted on a calibration curve (see Section 6.6), or the instrument software automatically prepares the curve. After the calibration curve is prepared, it should be approved through calculation of the corresponding correlation coefficient via linear regression. The coefficient should be greater than 0.9998; otherwise, a new calibration must be performed. Calculation results should be available for inspection at any time.

13.6.3.2 Continuing Calibration

At a 5% frequency, or once per analytical batch, the initial calibration must be approved by continuing calibration. (Samples analyzed under the same method sequence and using reagents of the same lot number — i.e., samples comprising a single analytical batch — should have similar matrices.) Continuing calibration includes analysis of the calibration blank, CCS, and CVS, with the above-mentioned frequency and acceptance levels.

13.6.4 Accepted Calibration

Preparation of the calibration curve with the accepted correlation coefficient, the CCS within the proper range, and the CVS with the correct percentage of recovery are the criteria of the appropriate initial calibration for the instrument and analytical system. If these criteria are not met, the initial calibration should be repeated and samples analyzed before the criteria failure must be analyzed again.

13.6.5 Outline of Calibration Procedure

1. Establish the calibration curve by using a blank and a number of calibration standards in the optimum range as dictated by the analytical method.
2. Determine acceptance or rejection of the curve by calculating the correlation coefficient.
3. Run the CCS to verify the curve. The result must be within ±5% of the verified value.
4. Run the CVS. As stated in Section 13.6.2.2, it must be supplied by a source different from that of the calibration standards. The initial calibration is accepted if the result is within ±10% of the true value.
5. If the analytical method includes sample pretreatment (e.g., digestion), a preparation blank and a treated CVS (called the LCS) should be incorporated into the analytical run. Results within ±15% of the true value are accepted.
6. Analyze ten samples (including reagent blanks, spikes, and duplicates).
7. After the analysis of the ten samples, check the curve again with the CCS. It must be within ±5% of the verified value.

8. The accepted curve should be certified again with the CVS. If the result is within ±10% of the true value, continue measuring the samples.
9. If the CCS or CVS fails to meet the criteria, the run must be stopped and a new initial calibration performed. Samples analyzed before the failed standards were discovered must be analyzed again. If verification of the initial calibration is satisfactory, proceed with analysis and repeat the same verification at a 5% frequency.

13.6.6 Special Calibration Criteria in Metals Analysis

13.6.6.1 Atomic Absorption Spectrophotometer (AAS)

Calibration is based on a three-point standard curve (low, middle, and high level) in the optimum linear range as specified per analyte in the respective methodology. This is performed every time the instrument is used, or after failure of continuing calibration standards. Calculation of the calibration curve is done by computer software using linear regression. The curve is checked to ensure linear correlation of greater than 0.9998. Complete the initial calibration by checking the curve with CCS and CVS, and using continuing calibration check in the same manner as outlined above in Section 13.6.5, numbers 3 through 9.

13.6.6.2 Inductively Coupled Plasma (ICP) Analyzer

By monitoring several wavelengths, either all at once or in a programmed sequence, many elements can be determined in a single automated analysis. The ICP analyzer offers a significant speed advantage for determination of metals in environmental samples. Calibration is performed with the use of mixed, multielement calibration standards. The number of calibration standards and the calibration technique are defined by the instrument manufacturer. Initial calibration and checking with continuing calibration are the same as described in Section 13.6.5.

13.6.7 Summary of Definitions Related to Calibration

13.6.7.1 Optimum Concentration Range

The optimum concentration range is the range in which the calibration curve remains linear. This range is defined by the analytical method and varies according to instrumentation.

13.6.7.2 Calibration Stock Solution

Calibration stock solution is a high-concentration solution of the analyte. It is commercially available or prepared by the laboratory.

13.6.7.3 Intermediate Standard Solution

This standard is diluted from the stock solution (Section 13.6.7.2) to obtain a concentration suitable to dilute the calibration standards.

13.6.7.4 Calibration Standards

Calibration standards are solutions of precise concentrations prepared by diluting the calibration stock or the intermediate standard solution. The number of standards is determined by the analytical method and the instrument, while the concentration is determined by the given optimum range.

13.6.7.5 Mid-Range Standard

The mid-range standard is defined as a standard in the middle of the linear range of the established calibration curve or a standard concentration in the middle of the expected sample concentration range, depending on the type of determination to be performed.

13.6.7.6 Calibration Blank

The calibration blank is a volume of analyte-free water used to zero the instrument. It is also run at the end of the analytical work to check whether contamination or drift occurred.

13.6.7.7 Preparation Blank

The preparation blank is a volume of analyte-free water processed through the sample preparation procedure.

13.6.7.8 Continuing Calibration Standard (CCS)

The CCS is used to ensure calibration accuracy during each analytical run. The CCS is analyzed at the beginning and end of the analysis and after every ten samples. The CCS concentration must be at or near the mid-range level.

13.6.7.9 Calibration Verification Standard (CVS) or QC Check Standard

The CVS is an independently prepared standard with a concentration near the mid-level standard and is prepared by a source other than the calibration standards source. It is analyzed at the beginning and end at a frequency of 10% during the analysis.

13.6.7.10 Laboratory Control Standard (LCS)

The LCS is a CVS carried through the entire sample preparation and analysis process. It is used to monitor the effects of sample preparation.

13.6.7.11 Sensitivity Check Standard

This standard is used for optimizing an AA spectrophotometer. The concentration of the standard varies according to the analytical method and is also stated in the manufacturer's instructions. With this standard, the instrument must provide an absorbance reading at or near 0.200.

13.6.7.12 QC Check Samples

These samples, which are obtained from an independent source, contain a known value of the analyte. Check samples are analyzed along with a sample set of similar matrix, and the obtained results are used to determine the accuracy of laboratory performance.

13.7 INSTRUMENT PERFORMANCE CHECK

13.7.1 Atomic Absorption Spectrophotometer (AAS)

13.7.1.1 Flame and Graphite

The AAS performance checks are performed every time a different metal is analyzed as part of an analytical procedure. The performance check provides a reading of the instrument's optimal operation, and thus is an indicator of deterioration in the lamps or the spectrophotometer. Performance is measured via a *sensitivity check standard*, a concentration of a particular metal dictated by the analytical method. The absorbance of this standard should be 0.200. If it differs by more than ±10%, the instrument is not performing correctly and must be corrected. Sensitivity standard concentrations for the FAAS and GrAAS techniques are presented in Tables 8.3 and 9.4, respectively.

13.7.2 Inductively Coupled Plasma Analyzer (ICP)

While ICP-AES analysis is often simple to perform, obtaining good analytical result requires attention to details that could influence analysis outcomes. Besides preparing samples carefully and selecting the proper hardware and operating parameters, the analyst should ensure that the instrument is properly set up and maintained. Specific information on maintenance and performance tests can usually be found in manufacturer's operator manuals. Guidelines for the care and maintenance of these instruments are discussed in Section 12.3.6.

Checking the performance of the instrument before beginning an analysis is recommended. In some cases, such as certain environmental analyses, the analysis protocol may specify performance tests that must be run before, during, and after an analysis. Such protocols are often quite rigorous, as the validity of the data produced must be able to stand up in a court of law if challenged. General tests that are often used to verify that an ICP-AES instrument is performing properly are listed below. Some of these tests are quite simple and should be performed on a daily basis. Other tests are more rigorous and can be used as diagnostics when malfunctions are suspected.

13.7.2.1 Warm-Up Time

Before analytical work begins, the instrument should be warmed up. The recommended warm-up time is 30 to 60 min.

13.7.2.2 Visual Bullet Test

This test can be performed on a daily basis. The analyst introduces 1000 mg/l or more of an element whose atomic emission produces a well-defined "bullet" in the center of the ICP discharge. The mere presence of the bullet indicates that the sample aerosol is reaching the plasma, while the vertical position of the bullet in the discharge is an indicator of the gas flow and RF power settings being used. Sodium and yttrium are often used in this test.

13.7.2.3 Signal Intensity

The number of emission counts for a given concentration of an element is often used as a quick instrument diagnostic. This test is usually more useful for trend analysis than as an absolute indicator of performance because the emission counts may vary somewhat from day to day.

13.7.2.4 Background Equivalent Concentration (BEC)

The BEC is an indicator of relative sensitivity for an emission line. A BEC value higher than normal often indicates problems with the efficiency of the sample introduction system.

13.7.2.5 Precision

The precision of an argon emission line is used as a diagnostic test for the RF generator. Five to ten measurements for a strong emission line are used to calculate precision, and this figure may be expressed as %CV (correlation of variation) or %RSD (relative standard deviation).

13.7.2.6 Detection Limits

The causes of measured detection limits outside expected range are numerous. Running this test along with other specific tests is recommended.

13.7.2.7 Ion/Atom Ratio

The ratio of emission intensity of an ion line to emission intensity of an atom line is an indicator of relative excitation in the plasma.

13.7.2.8 Wavelength/Peak Alignment

Prior to sample analysis, make sure that the spectrometer is calibrated properly in terms of wavelength. Calibration is performed manually or automatically by the instrument software at the beginning of an analysis.

13.8 LABORATORY QC CHECKS

Laboratory QC checks include all practices and activities that ensure accuracy and precision of the analytical measurements. QC checks serve as tools for approval of generated analytical data. Laboratory staff should follow all method-specific QC requirements. Guidelines for basic controls of chemical measurements are summarized below.

13.8.1 Blanks

13.8.1.1 Calibration Blank

A calibration blank is used to establish the analytical curve. It is prepared by measuring 2 ml of 1:1 HNO_3 and 10 ml of 1:1 HCl into a 100-ml volumetric flask (see Section 13.2 for preparation of glassware for metals analysis) and dilute to the mark with analyte-free water. Prepare a sufficient quantity to flush the system between running standards and samples.

13.8.1.2 Reagent Blank or Method Blank

A reagent blank is used to correct possible contamination caused by reagents used in processing samples prior to sample analysis. If the analytical method includes any pretreatment (digestion, distillation, etc.) one analyte-free water blank must be processed and analyzed the same way as the samples.

13.8.2 Duplicate Samples

Duplicate samples are collected at the same time and from the same source as analytical samples but are submitted and analyzed as separate samples. The analyst does not know that they are duplicates. Duplicate samples are used to determine precision of analytical performance. Analyze one duplicate in every 20 samples.

13.8.2.1 Field Duplicates

Duplicate field samples are collected at the same time from the same source and are submitted and analyzed as separate samples.

13.8.2.2 Laboratory Duplicates

Laboratory duplicates are aliquots of the analytical sample that are prepared and analyzed at the same time.

13.8.3 Spikes

13.8.3.1 Spiked Sample or Matrix Spike

Specific concentrations of the parameters of interest are added to spiked samples. Spiked samples are used to measure the performance of the complete analytical system, including chemical interferences from the sample matrix. A small quantity of a known concentration of analyte stock solution is added to the sample (or aliquot). Always use highly concentrated stock solutions, so that a small quantity of spike added to the sample does not change sample volume. Always add spike solution before sample preparation.

Water samples are thoroughly mixed by shaking the sample container before measuring out the sample aliquot. The spike addition should produce a minimum level of ten times and a maximum level of 100 times of the instrument detection limit (IDL).

Use the following formula to calculate the volume of the spike addition, concentration of the spike solution, desired spike concentration, and the volume of the sample spiked:

$$c_1 v_1 = c_2 v_2 \quad (13.1)$$

where

$$c_1 = (c_2 v_2)/v_1$$

$$v_1 = (c_2 v_2)/c_1$$

$$c_2 = (c_1 v_1)/v_2$$

$$v_2 = (c_1 v_1)/c_2$$

where

c_1 = concentration of the spike stock solution.
v_1 = the unknown.
c_2 = desired spike concentration.
v_2 = volume of the sample before spiking.

If the analytical set involves multiple matrices, matrix spikes should be prepared for all matrix types. Analyze one per set at a rate equivalent to 5% of all samples.

Calculate the percent recovery on the spike ($\%R_{sp}$) by using the following formula:

$$\%R_{sp} = (\text{spiked sample value} - \text{sample value}) \times 100/\text{added spike value} \quad (13.2)$$

13.8.3.2 Reagent Water Spike

Analyte-free water is spiked with the analyte and is prepared in the same way as the samples. A reagent water spike is used to monitor analytical method effectiveness. Run at 5% frequency. Concentration and preparation are the same as described for matrix spike (Section 13.8.3.1). The recommended level of the spike addition is five times the method quantification limit.

13.8.3.3 Duplicate Matrix Spike Sample

Duplicates of matrix spike samples (instead of duplicate samples) may be used to check the precision of the analytical system.

13.8.4 Calibration Check Standards

Calibration check standards are also part of proper calibration criteria. See Section 13.6.

13.8.5 Blind QC Check Samples

Qualiy Control check samples (also known as reference materials) are obtained from an independent source with known analytical levels. The analyst is not informed of the true concentration value of the sample. QC check samples measure and validate the analytical system and analyst performance. Accuracy is expressed as percent recovery (see Formula 13.6). Blind check samples are run on a semiannual basis and are part of the QA report.

13.8.6 Performance Evaluation Samples

Reference materials obtained from an independent source for which the level(s) of analytes have been validated are used in performance evaluations. Performance testing evaluates a laboratory's ability to produce a specified quality of data and is used in obtaining certification and contract qualification.

Evaluation of results is an essential part of performance testing. Each participant should be informed of his or her performance in relation to an acceptable standard. Information on each analyte should include the true value, acceptance range, reported value, and the evaluation as "accepted," "not acceptable," or "check for error." Laboratory managers conduct overviews of results, initiate corrective action as needed, and document all steps of remedial actions.

13.8.7 Interference Check

Verify interelement and background correction factors at the beginning and end of the analytical run or twice during every 8-h work shift, whichever is more frequent. This verification is accomplished by analyzing the interference check sample. Review interelement correction!

13.9 DETECTION LIMITS

Detection limits are the smallest concentration of the analyte of interest that can be measured within a specific probability of significance.

13.9.1 Method Detection Limit (MDL)

The MDL is the smallest concentration of the analyte of interest that can be measured and reported with 98% confidence that the concentration is greater than zero. MDL determination steps are outlined below.

1. Estimate the MDL by using the minimum detection concentration described by the analytical method.
2. Prepare 1 liter of standard by adding the analyte to analyte-free water to create a concentration close to the estimated MDL.

3. Analyze seven portions of the solution and calculate the standard deviation (SD) of the results.
4. In a Student's t table (see Table 13.3), select the t value for 6 degrees of freedom (df) at the 98% level (df $= n - 1$, or $7 - 1 = 6$ in this case). The value of t at the 98% level is 3.14.
5. The calculation of the MDL follows:

$$\text{MDL} = \text{SD} \times 3.14 \tag{13.3}$$

In this formula, SD is the standard deviation calculated in step 3.

13.9.2 Instrument Detection Limit (IDL)

The IDL is the concentration of the analyte that produces a signal greater than five times the signal-to-noise ratio of the instrument. An operating analytical instrument usually produces a signal greater than three standard deviations of the mean noise. IDL determination follows:

1. Analyze analyte-free water seven times and record instrument signals.
2. Calculate the standard deviation of the seven instrument responses.
3. Calculate the IDL:

TABLE 13.3
Student's *t* Table

df	80% $t_{.90}$	90% $t_{.95}$	95% $t_{.975}$	98% $t_{.99}$	99% $t_{.996}$	99.73% $t_{.9985}$
1	3.078	6.314	12.706	31.821	63.657	235.80
2	1.886	2.920	4.303	6.965	9.925	19.207
3	1.638	2.353	3.182	4.541	5.841	9.219
4	1.533	2.132	2.776	3.747	4.604	6.620
5	1.476	2.015	5.571	3.365	4.032	5.507
6	1.440	1.943	2.447	3.143	3.707	4.904
7	1.415	1.895	2.365	2.998	3.499	4.530
8	1.397	1.860	2.306	2.896	3.355	4.277
9	1.383	1.833	2.262	2.821	3.250	4.094
10	1.372	1.812	2.228	2.764	3.169	3.975
11	1.363	1.796	2.201	2.718	3.106	3.850
12	1.356	1.782	2.179	2.681	3.055	3.764
13	1.350	1.771	2.160	2.650	3.012	3.694
14	1.345	1.761	2.145	2.624	2.977	3.636
15	1.341	1.753	2.131	2.602	2.947	3.586
16	1.337	1.746	2.120	2.583	2.921	3.544
17	1.333	1.740	2.110	2.567	2.898	3.507
18	1.330	1.734	2.101	2.552	2.878	3.475
19	1.328	1.729	2.093	2.539	2.861	3.447
20	1.325	1.725	2.086	2.528	2.845	3.422
25	1.316	1.708	2.060	2.485	2.787	3.330
30	1.310	1.697	2.042	2.457	2.750	3.270
40	1.303	1.684	2.021	2.423	2.704	3.199
60	1.296	1.671	2.000	2.390	2.660	3.310
	1.282	1.645	1.960	2.326	2.576	3.000

$$IDL = SD \times 3 \quad (13.4)$$

In this formula, SD is the standard deviation of the average signal of the instrument in steps 1 and 2.

13.9.3 Practical Quantitation Limit (PQL)

The PQL is the smallest concentration of the analyte of interest that can be reported with a specific degree of confidence. This limit has been proposed as the lowest level achievable among laboratories within specified limits during routine laboratory operations. The PQL is significant because different laboratories will produce different MDLs even though they are using the same analytical procedures, instruments, and sample matrices. The concentration related to the PQL is about five times the MDL and represents a practical and routinely achievable detection limit with a relatively high certainty that any reported value is reliable. To determine the PQL, take the standard deviation (SD) from the determination of MDL and multiply by 10. The ten standard deviations correspond to an uncertainty factor of ±30% in the measured value at the 98% confidence limit.

$$PQL = SD \times 10 \quad (13.5)$$

where SD is the standard deviation of MDL (see Section 13.9.1).

13.10 ACCURACY AND PRECISION

13.10.1 General Discussion

Satisfactory analytical data are accurate and precise.

13.10.1.1 Accuracy

Accuracy is the degree of agreement of a measured value with the true or expected value of the quantity of interest. Accuracy is measured and expressed as percent recovery (%R) and calculated according to the following formula:

$$\%R = \text{analytical value} \times 100/\text{true value} \quad (13.6)$$

13.10.1.2 Precision

Precision is the degree of mutual agreement among individual measurements as the result of repeated applications under the same conditions. Precision measures the variation among measurements and is expressed in different ways.

13.10.1.3 Standard Deviation (SD)

Calculate by using a calculator or the following formula:

$$SD = E(x - X)^2/n - 1 \quad (13.7)$$

where

E = sum
X = mean of measurements
x = sum of measurements
n = number of measurements

13.10.1.4 Relative Standard Deviation (RSD)

The RSD is derived in four steps:

1. Calculate standard deviation (SD).
2. Calculate the mean of the measurements (x).
3. Calculate the coefficient of variant (CV) (standard deviation divided by the mean):

$$CV = SD / x \tag{13.8}$$

4. Calculate the RSD:

$$RSD = CV \times 100 \tag{13.9}$$

13.10.1.5 Relative Percent Difference (RPD)

The RPD is the difference between duplicate values divided by the average of the duplicate values and multiplied by 100:

$$RPD = (A - B)/[(A + B)/2] \times 100 \tag{13.10}$$

or

$$RPD = (A - B)/(A + B) \times 200 \tag{13.11}$$

13.10.2 Quality Control Delineation for Accuracy and Precision

Calculated accuracy (%R) and precision (SD, RSD, or RPD) values are used in determination of QC limits for each parameter.

13.10.2.1 Accuracy Control Limits

1. Collect 20 data points for the parameter of interest.
2. Calculate the mean (x) and the standard deviation (SD).
3. Calculate the accuracy of QC limits as %R.

Warning limits (WLs)	$= x \pm 2$ SD
Upper warning limit (UWL)	$= x + 2$ SD
Lower warning limit (LWL)	$= x - 2$ SD
Control limits (CLs)	$= x \pm 3$ SD
Upper control limit (UCL)	$= x + 3$ SD
Lower control limit (LCL)	$= x - 3$ SD

When %R data are outside the warning limits, the analytical system is critical (approaching an out-of-control situation) and may require corrective action. Data falling outside the control limits indicate an out-of-control system; analysis must be stopped and corrective action taken. Samples analyzed after the failed QC check sample should be analyzed again.

13.10.2.2 Precision Control Limits

1. Collect 20 data points of expressed precision (see Section 13.9.) for the parameter of interest.

2. Calculate the mean (x) and the standard deviation (SD).
3. Calculate the QC limits for precision. Because the minimum value for precision is zero, the lower limits should always be zero.

Warning limits (WLs)	$= 0 - (x + 2\ \text{SD})$
Upper warning limit (UWL)	$= x + 2\ \text{SD}$
Lower warning limit (LWL)	$= 0$
Control limits (CLs)	$= 0 - (x + 3\ \text{SD})$
Upper control limit (UCL)	$= x + 3\ \text{SD}$
Lower control limit (LCL)	$= 0$

Any value falling above the warning limit should be interpreted as a signal that the system is critical and may indicate the need for corrective action. Data falling outside the control limits indicate that the system is out of control.

13.10.2.3 Confidence Limit or "Target Limit" for Precision and Accuracy

At the completion of any analysis, the calculated precision and accuracy values are documented on the related QC charts and recorded on summary log forms (illustrated in Tables 13.4, 13.5, and 13.6). Based on these collected statistical results, QA "target limits" should be established for each analyte per matrix. These limits serve as a confidence range for precision and accuracy values. This calculation consists of the following steps:

1. Collect a minimum of 20 data points for precision determination *or* 20 data points for accuracy determination.
2. Calculate the mean of these values (x) and the corresponding standard deviation (SD).
3 Calculate the confidence interval (CI) according to the following formula:

$$\text{CI} = x \pm [(t \times \text{SD})/n] \tag{13.12}$$

The value of t (Student's t) depends on the degrees of freedom ($\text{df} = n - 1$) for the 98% confidence level (see Table 13.3). For example, assume that the average confidence-level value of the 20 data points for $\%R$ is 98% and the standard deviation is 5.26:

$$\begin{aligned} x &= 98\% \\ \text{SD} &= 5.26 \\ n &= 20 \\ \text{df} &= n - 1 = 19 \\ t &= 2.593 \end{aligned}$$

$$\text{CI} = 98 + (2.539 \times 5.26)/4.47 = 101$$
$$\text{CI} = 98 - (2.539 \times 5.26)/4.47 = 95$$

The confidence interval or "target limit" for accuracy is 95–101%.

13.10.3 Quality Control Charts

Control charts are statistical tools for monitoring the performance of a particular task on a continuing basis and have become very useful in demonstrating statistical control. Both accuracy and precision control charts should contain the name of the laboratory, the test parameter, analytical method (method number with reference), range, and calculated standard deviation. All initial data used to

TABLE 13.4
Monitoring Form for Precision (RPD) Values

ANALYTE ______________________ **METHOD** ______________________ **MATRIX** ______________________
Unit, values are expressed: ____

Date	Sample value (A)	Duplicate value (B)	RPD %	Control limit	Remark	Sign.

Note: RPD = relative percent difference. RPD = $[(A - B)/(A + B)/2)] \times 100$ or $[(A - B/A + B)] \times 200$.

TABLE 13.5
Monitoring Form for Accuracy (% Recovery) Values

ANALYTE ______________________ **METHOD** ______________________ **MATRIX** ______________________
Unit, values are expressed: ____

Date	QC check sample true value	QC check sample measured value	% Recovery	Control limit	Remark	Sign.

Note: % Recovery (%R) = (measured value/true value) × 100.

TABLE 13.6
Monitoring Form for Spike Recovery (%R_{sp}) Values

ANALYTE ____________ **METHOD** ____________ **MATRIX** ____________
Unit, values are expressed: ____________

Date	Sample value (SV)	Spike added (SA)	Spiked sample value (SSV)	% Recovery	Control limit	Remark	Sign.

Note: %R_{sp} = [(SSV − SV)/SA] × 100.

prepare the control limits are established by the analytical methodology and should be less than 5% outside the upper and lower warning limits.

Quality Control charts are used on a daily basis. Each analytical method has established accuracy and precision control limits according to the control charts that are used to determine the acceptability of data on a continuous basis. The response to an out-of-control event must be immediate, and all conditions and corrective actions must be documented. This report includes the out-of-control test parameter, date, description of the QC problem, and necessary corrective action. Once the out-of-control condition has been corrected, QC requirements are reapplied until satisfactory data points have been plotted on the control chart.

13.10.3.1 Accuracy Control Chart

The accuracy control chart monitors the percent recovery with the standard deviation as the limiting control. The control chart is prepared for each test parameter after 20 determinations have been performed. The mean is plotted with the warning control limits of ±2 SDs (standard deviations) and upper and lower control limits of ±3 SDs. In other words, the upper and lower warning limits are defined as the limits that would encompass 95% of the measured values for percent recovery, and the upper and lower control limits are defined as the limits that would encompass 99% of the measured values of the percent recovery.

13.10.3.2 Evaluation of Conditions in Accuracy Control Charts

The evaluation of possible conditions on accuracy control charts is illustrated in Figure 13.5.

- *Condition is satisfactory:* Data are variable, showing no trends and remaining within the warning limits.
- *Condition is critical*: One or more points are outside the UWL and LWL; seven successive points are in the same direction are causing either an upward or downward trend; ten successive points are on the same side of the average value (x) of the chart.
- *Condition is out of control:* One or more points are outside the UCL and LCL.

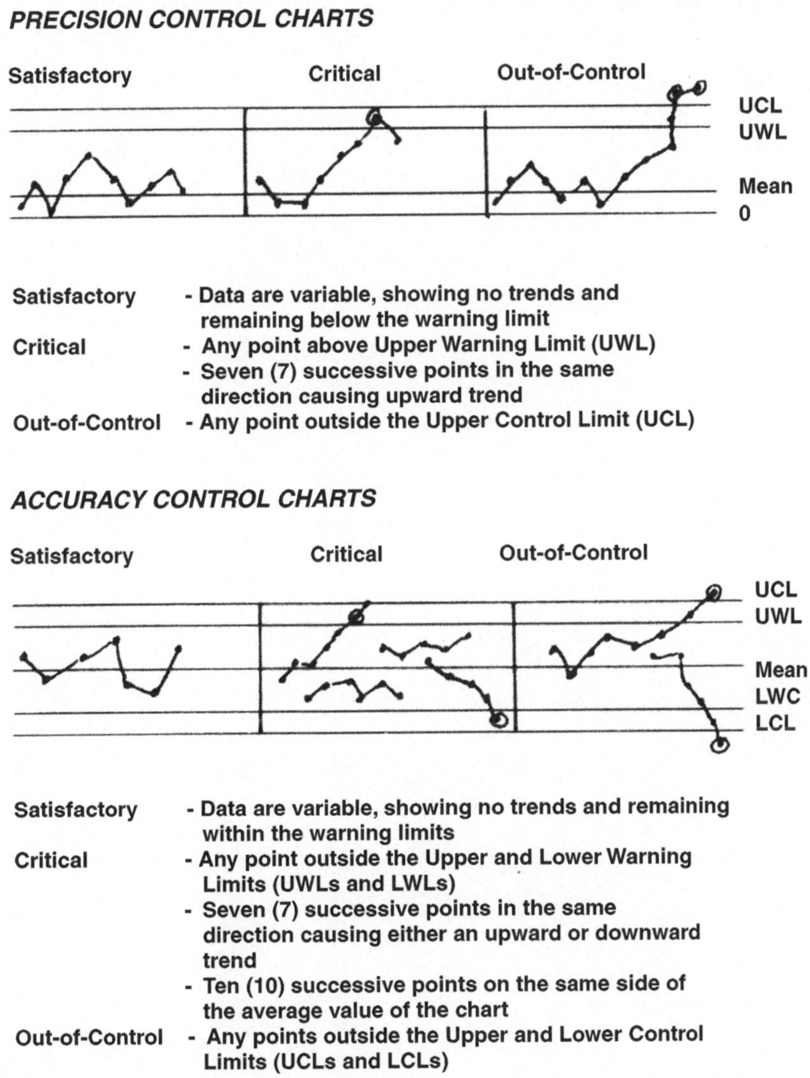

FIGURE 13.5 Interpretation of quality control charts.

13.10.3.3 Precision Control Chart

The precision control chart monitors the repeatability of a measurement system, irrespective of accuracy. It is based on the term that was selected to express precision values. These charts are based on results from duplicate samples after accumulating 20 data points. Precision control charts have only upper warning and upper control limits. The lower limit for both warning and control limits is zero.

13.10.3.4 Evaluation of Conditions in Precision Control Charts

The interpretation of conditions on precision control charts is illustrated in Figure 13.5.

- *Condition is satisfactory*: Data are variable, showing no trends and remaining below the warning limits.
- *Condition is critical*: Any one or more points are above the warning limit (WL); seven successive points are in the same direction, causing an upward trend.
- *Condition is out of control*: One or more points are beyond the control limit.

14 Sample Collection for Metals Analysis

14.1 GENERAL CONSIDERATIONS IN SAMPLING

The quality of any analytical system depends primarily on the sample analyzed. A sample must be representative of the environmental system from which it is taken so that chemical analysis results, in turn, represent the system.

14.1.1 FACTORS AND REQUIREMENTS OF SAMPLING PROGRAM TO BE CONSIDERED

- Parameters of interest with method number and references
- Duration of survey
- Frequency of sampling
- Number of samples
- Sample matrices
- Sample source
- Site identification
- Grab or composite samples (see Section 14.1.4)
- Manual and automatic sampling (see Section 14.1.5)
- Field measurements
- Quality control (QC) requirements

14.1.2 PREPARATION FOR SAMPLE COLLECTION

- Understand the sampling plan; all information should be written and discussed with field personnel.
- Prepare, clean, and calibrate sampling equipment so that it is ready to use.
- Check, calibrate, and prepare equipment for field tests.
- Prepare sample containers.
- Prepare preservative and dispose in a safe container.
- Collect labels and markers, field notebook, pH paper, and small disposable cups to check pH of preserved samples.
- Prepare all blanks.
- Check all calibration standards and expiration dates for freshness. If necessary, prepare new ones.
- Check QC samples for availability, and check dates for freshness. If necessary, prepare new ones. Determine whether sample should be spiked and discuss concentration of the spikes. Calculate the volume of the added spike stock solution for each spiked parameter.

- Collect pipets with suitable volumes and pipet bulbs.
- Collect empty bottles for splits, duplicates, and so on.
- Collect glassware for field tests and check cleanliness of glassware.
- Make sure that thermometers are stored in protective carriers to avoid breakage.
- Check spike stock solutions and check dates for freshness. If necessary, prepare new ones.
- Collect soap for cleaning sampling equipment and for washing hands, paper towels, soft tissues, and bottles with DI (deionized) water.

14.1.3 Prefield Procedures

Several prefield procedures must be considered prior to the sampling activities:

- Selection of proper sampling equipment (preferred materials for sampling and purging equipment for metals analysis include Teflon, polypropylene or polyethylene, and stainless steel)
- Decontamination of sampling equipment
- Selection of sample bottles
- Preservative preparation
- Preparation and calibration of field analytical instruments
- Preparation of sample labels, chain-of-custody forms, field notebook, waterproof ink, and so on

14.1.4 Types of Samples

14.1.4.1 Grab or Individual Samples

Samples collected at a particular time and place are called *grab* or *individual samples*. This type of sample represents conditions at the time it was collected. Therefore, a grab sample should not be used as a basis for decision making about pollution abatement. However, some sources are quite stable in composition, thus single-grab samples would be considered representative.

14.1.4.2 Composite Samples

If results for an entire source system are to be reported, a series of small samples are collected in a single container and blended for analysis. The mixing process averages variations in sample composition and minimizes analytical effort and expense. These types of samples are called *composite samples*. When a time factor is being taken into consideration, grab samples are collected at suitable intervals according to expected changes. Composite samples reflect average characteristics during the sampling period, and in most cases a 24-h period is standard. Subsample volume should be constant and at least 200 ml.

14.1.5 Manual and Automated Sample Collection

14.1.5.1 Manual Sample Collection

When collecting samples for immediate field tests or when automatic samplers are not available, collect samples directly into a sample container. If a sample cannot be placed directly into the container, an intermediate vessel should be used. The intermediate container must be as clean as the sample container and must be made from the required material for parameter of interest. The sample is collected by lowering a properly cleaned device on a rope, pole, or chain into the sample medium. In some cases, using a power or hand-operated pump is necessary to withdraw the sample. When collecting samples for metals analysis, rinsing the sampling device three times is sufficient, except if the bottles are prepreserved.

14.1.5.2 Automated Sample Collection

A wide variety of automatic samplers are commercially available. When sampling a large number of locations, automatic samplers are more practical, help reduce human error, and are able to keep the samples cool to 4°C during the time spent gathering samples. Automatic samplers, however, are expensive.

14.1.6 General Rules in Sampling

- Samples must be collected from the least to the most contaminated sampling locations within the site.
- Disposable latex gloves must be worn while sampling, and new, unused gloves must be used for each separate sampling point.
- For compositing or mixing samples for metals analysis, use a stainless steel or Teflon bowl.
- Keep in mind that the order of sample collection is as follows:
  1. Volatile organic compounds (VOCs)
  2. Extractable organics
  3. Total metals
  4. Dissolved metals
  5. Microbiologicals
  6. Inorganic nonmetals

- For aqueous matrices, sampling equipment and containers are rinsed with the sample fluid before the actual sample is taken, with the exception of prepreserved containers.
- A step-by-step, written sampling procedure should be available. The procedure should contain all sample collection activities.

14.1.7 Proper Material for Sampling Devices

Devices used for collecting samples for metals analysis should be made of plastic, stainless steel, or Teflon.

14.1.8 Errors Introduced during Sampling

Serious errors that may be introduced during sampling and storage are the contamination resulting from improperly cleaned sampling devices and sample containers and loss of metals by absorption or precipitation in sample containers because of failure to acidify the sample properly.

14.1.9 Waste Disposal in the Field

Wastes generated during sampling are separated into specialized and properly labeled waste containers. Laboratory- and field-generated wastes are disposed of by certified waste management companies. The certificate and contract of this company should be recorded.

14.2 AUTOMATIC SAMPLERS

14.2.1 Proper Operation of Automatic Samplers

To ensure proper operation of automatic samplers and thus the collection of representative samples, correct maintenance and calibration must be followed:

- A maintenance log containing all repair information should be available.
- Prior to each field trip, check the sampler for correct operation (proper working order, batteries, desiccant, etc.).

- Before sampling, check the constant pumping volume.
- After returning from the field, check operation of sampler and repair if necessary.

14.2.2 Preparation of Sampling Equipment

Step-by-step cleaning procedures (called *decon* for decontamination) should be performed. These procedures derive from specific regulations and must be available in written form. Equipment should be cleaned before sampling and in the field between samples. At the end of the field trip, sampling equipment must be labeled as "rinsed, ready for house cleaning." After being sufficiently cleaned in the laboratory, the equipment should be labeled as "in-house cleaned, ready for field," accompanied by the date and the signature of the cleaner. Both house and field cleaning should be documented properly. Detergents specified for cleaning include *Alconox* (or equivalent) with 5% phosphate, or *Liquinox* (or equivalent), which is free of phosphates and ammonia.

The purity and reliability of the analyte-free water used for rinsing and blank preparation are shown in results of tests performed on the blank.

14.2.2.1 In-House Cleaning of Sampling Equipment

1. Wash with hot, soapy tap water and scrub with a brush.
2. Rinse thoroughly with hot tap water.
3. Rinse with 10 to 15% nitric acid (HNO_3). Acid rinses should never be applied to stainless steel or metallic equipment.
4. Rinse thoroughly with deionized water.
5. Rinse thoroughly with pesticide-grade isopropanol.
6. Rinse thoroughly with analyte-free water.
7. Air dry completely.
8. Wrap in aluminum foil for storage and transportation.

14.2.2.2 Field Cleaning of Sampling Equipment

Use the same procedure as in-house cleaning procedure, with the exception of hot water wash and rinse. Laboratory-pure water rinse is recommended, but optional. Rinsing with sample water is acceptable when proper cleaning of the equipment is impossible. It should be disposed of until effective cleaning is possible.

14.3 SAMPLE CONTAINERS

14.3.1 Preferred Sample Containers

- Preferred sample containers for metals analysis are polyethylene bottles with tight, screw-type lids.
- Borosilicate glass containers also may be used, but avoid soft glass bottles for samples containing metals in the microgram-per-liter (ppb) range.
- Store samples for silver analysis in light-absorbing containers.
- Sample containers may be cleaned in-house or in the field or purchased from commercial vendors as precleaned containers. The cleaning grades must meet EPA analyte-specific requirements. All records for these containers (lot numbers, certification statements, date of receipt, etc.) and their uses must be documented.

14.3.2 Proper Cleaning of Sample Containers

1. The soap should be metal-free *Acationox* or equivalent.
2. Wash bottles and caps in hot, soapy water and rinse liberally with tap water until soapsuds are gone.
3. Rinse bottles and caps with 1+1 HCl, followed by tap water rinse.
4. Rinse bottles and caps with 1+1 HNO_3.
5. Rinse three times with liberal amounts of laboratory-pure water.
6. Drain and cap tightly until used.

14.4 SAMPLE PRESERVATION

Sample preservation is necessary for all samples (40 CFR, Part 136). Sample preservation may be accomplished by using ready, prepreserved bottles obtained from the laboratory, but additional preservatives must be available in the field if the measured pH of the preserved sample indicates that additional preservative is necessary.

If the sample is preserved in the field, the following protocols should be practiced:

- Preservative should be prepared from reagent-grade chemical.
- Fresh preservative should be used in each sampling trip.
- Preservatives transported to the field should be stored in properly cleaned plastic or Teflon containers to avoid breakage.
- Chemicals should be segregated from sample containers to avoid accidental contamination.
- Preservatives should be added with a pipet or premeasured droppers.
- After preservation, the pH of the preserved sample should be measured. Transfer a small quantity from the preserved and well-mixed sample into a disposable container, and determine the pH by using a narrow-range pH paper. If the pH value indicates the addition of more preservative, the preservative should be from the same source as used in the original treatment. The amount of the additional preservative should be documented, and the additional preservative should be added to the corresponding blank as well.
- Acid preservation should be done in a well-ventilated area to avoid inhalation of acid fumes and toxic gases. Any unusual reaction should be noted!
- In the case of any acid spill, wipe up immediately and flush the area with a great amount of water.

14.5 SPECIAL SAMPLING PROCEDURES

Before collecting a sample, decide on the metal fraction to be analyzed: dissolved (filterable), suspended (nonfilterable), or total metals. This decision will determine whether the sample is acidified with or without filtration.

14.5.1 Total Metals

Total metals are defined as the concentration of metals in an unfiltered sample or the sum of the concentrations of metals in both the dissolved and suspended fractions. Preserve the sample with 3 ml of 1+1 HNO_3 or 1.5 ml of concentrated HNO_3 per liter. Samples with high buffer capacity and high alkaline samples may require more acid, as indicated by pH measurement. Samples should be transported to the laboratory without cooling.

14.5.2 Dissolved Metals

Dissolved metals are defined as the concentration of metals determined in the sample after it is filtered through a 0.45-μm filter. Samples must be filtered through a 0.45-μm filter prior to preservation. Filter paper should be acid washed and dried before use. After the sample is filtered, the filtrate will be the sample for dissolved metals and acidified in the same way as for total metals.

14.5.3 Suspended Metals

Suspended metals are defined as the concentration of metals determined in the portion of the sample that is retained in a 0.45-μm filter. Unpreserved samples are filtered through a 0.45-μm filter, as mentioned above for the sample collection of dissolved metals, and the filter paper is retained for further analysis of the suspended or unfilterable metals. The filter paper containing the suspended matter is transferred to the laboratory for determination of suspended metals. Samples should be filtered in the field, or immediately after transport to the laboratory. In the latter case, preserve the filtrate.

14.5.4 Sample Collection of Hexavalent Chromium

Materials containing hexavalent chromium (Cr^{6+}) are sampled separately from other metals. Do not add acid preservation to this sample; transport it to the laboratory for analysis as soon as possible. During transportation and storage, samples should be kept at 4°C.

"No preservative added" should be clearly written on the sample label in the request for this type of sample. *Holding time is 24 h for these samples.*

14.6 HOLDING TIME

Holding time for most preserved samples is 6 months. For mercury (Hg) determination, holding time of the preserved sample is 28 days. Samples collected without preservation for the determination of hexavalent chromium (Cr^{+6}) can be held for only 24 h. A sample holding-time log is illustrated in Figure 14.1.

14.7 FIELD RECORDS

Field records are taken for all data generated during sample collection. These records are kept in a chain-of-custody form (Figure 14.2), sample label (Figure 14.3), field notebook (Figure 14.4), sample field log (Figure 14.5), preservative preparation log (Figure 14.6), and QC sample and spike preparation log (Figure 14.7).

14.7.1 Chain-of-Custody

All sampling events should be documented and recorded on a chain-of-custody form. This practice ensures that the sample is collected, transferred, stored, analyzed, and destroyed only by authorized personnel. Each custodian or sampler must sign, record, and date the transfer. The form includes the name of the sampling project; collector's signature; sampling location; sampling site; sampling point, date, and time; type of sample; number of containers; and analysis required. The chain-of-custody form is illustrated in Figure 14.2.

			Holding Time (days)			Day of Preparation				Storage	
Sample ID	Matrix	Analysis Required	prep	anal	dispo	rec	prep	anal	dispo	Sample Prepared	Sign

Sample ID = sample identification number; prep = prepared; anal = analysis; dispo = disposal; rec = received; sign = signature of logger.

Holding Time Explanation:

prep = number of days between the date sample received and the date sample prepared
anal = number of days between the date sample prepared and the date of actual analysis
dispo = number of days between the date sample received and the date sample disposed

Storage Designations:

R. T. = room temperature in designated area
Ref. O. = refrigerator, designated for organic samples
Ref. I. = refrigerator, designated for inorganic samples
Fr. = freezer, designated for special samples

FIGURE 14.1 Sample holding-time log.

14.7.2 Sample Label

A sample label (Figure 14.3) should be affixed to all sample containers and serves as an important part of sample identification. The label should be waterproof, and all information should be written in waterproof ink.

14.7.3 Field Notebook

The field notebook (Figure 14.4) is specially designed for fieldwork, with waterproof paper and a hard cover. All field records should be written in waterproof ink. Errors in documents should be deleted by a single-line cross-through, accompanied by the date and initial of the person making the correction.

Field ID: ____________________ **Site Name:** ____________________

Date Sampled Received: ____________________ **Address:** ____________________

Sampler(s): ____________________ **Laboratory:** ____________________

Sample Container Description

Sample Identity	Date Sampled										Total	Remarks

Total Number of Containers ______

Relinquished By: ________ **Organization:** ________ **Received By:** ________ **Organization:** ________

Date: ________ **Time:** ________ **Date:** ________ **Time:** ________

Relinquished By: ________ **Organization:** ________ **Received By:** ________ **Organization:** ________

Date: ________ **Time:** ________ **Date:** ________ **Time:** ________

Delivery Method: ____________________ **(attach shipping bills, if any)**

Use extra sheets if necessary.

FIGURE 14.2 Chain-of-custody form.

Field Sequence No. ______________________________

Field Sample No. ____________ **Date** ____________ **Time** ____________

Sample Location ______________________________

Sample Source ______________________________

Preservative Used ______________________________

Analyses Required ______________________________

____

Collected by ______________________________

Remarks ______________________________

____

____

Final pH Checked ______________________________

Additional Preservative Used (If Applicable) ______________________________

____

FIGURE 14.3 Sample label.

Date ____________ **Time** ____________

Sampler's name ____________ **Signature** ____________

Other field people ______________________________

____

____

Sample location ______________________________

____

Sample type **grab** ________

composite ____________ **Compositing time** ____________**hr**

Time interval ____________**min**

Subsample volume ____________**ml**

Sq. No.	FID	Preserv. container	Analysis required	pH	T(°C)	Cond. (μmhos /cm)	DO (ppm)	Cl_2	Comment

Sq. No. = sample sequence number; FID = field identification number; Cond. = conductivity; DO = dissolved oxygen; Cl_2 = chlorine, residual; ppm = parts per million (mg/l).

Field conditions:
pH check:
Additional preservative used:
Other observations:

FIGURE 14.4 Field notebook.

Purpose of Analysis:____________________**Sample Field ID:**______________________________

Type of Sample:____________ **Sampler:** ________________________ **Date/Time:**________________________

Sample Site Number	Sample Source Description	Bottle Type	Bottle No.	Preservative	Analysis Required							
Remarks: * Field Measurements												

FIGURE 14.5 Sample field log.

14.7.4 Sample Field Log and Preservative Preparation Log

Other field records are the sample field log (Figure 14.5) and preservative preparation log (Figure 14.6). Immediately after sampling while still at the sampling point, the correct preservation and proper identification of samples (chain-of-custody and submittal forms, sample labels, etc.) should be checked.

14.7.5 Information Available in Field Records

The following information should be available in field records:

- Name of sample collector and field personnel
- Date and time of sampling
- Field conditions (weather, important information about the sample site)
- Description of sample location (address, exact sampling points)
- Sample type (grab, composite). If composite sample, record the time intervals, duration of sampling, and volume of subsamples
- Requested analytical parameters, type and number of containers, preservation technique
- Preservative preparation

Preservative ______________________________
Preparation Procedure ______________________________
Date Prepared ______________________________
Date of Expiration ______________________________
Analyte Preserved ______________________________
Information Related to the Chemical Used:
  Name, formula, and grade of the chemical
  ____
  ____
  Source of the chemical (name of manufacturer)
  ____
  Lot no. of the chemical ______________________________
  Date chemical received ______________________________
  Date container was opened ______________________________
  Expiration date ______________________________
  Storage of the chemical ______________________________
  ____
Check of Preservative ______________________________
____

______________________ ______________________
Preparer **Supervisor**

FIGURE 14.6 Preservative preparation log.

Analyte spiked ______________________________
Field no. of sample spiked ______________________________
Sample volume spiked ______________________________
Value of spike added ______________________________
Concentration of spike stock solution ______________________________
Volume of spike stock solution added ______________________________
Source of spike stock solution:
  Commercial source
    Manufacturer: ______________________________
    Lot no.: ______________________________
    Date received: ______________________________
    Date expired: ______________________________
  Laboratory prepared
    Date of preparation: ______________________________
    Expiration date: ______________________________
Date spike sample prepared ______________________________
Signature of field personnel ______________________________

FIGURE 14.7 Field sample spike preparation log.

- How pH was checked on the preserved sample and the value of the measured pH; if additional preservative was used to obtain the correct pH, how many extra milliliters were added, and how was the blank prepared with the additional preservative
- Sequential order of the samples taken; each sample should be accompanied by a sequence number and a field identification number
- If duplicate samples are taken, properly identified as FD_1 and FD_2
- If split samples are taken, correctly identified as FS_1 and FS_2
- Information about the preparation and true value of field quality control samples
- Spiked samples marked as FSp_1 and FSp^2 (if duplicates are taken)
- Field measurement data (temperature, pH, etc.)
- List of purging and sampling equipment used
- Documentation for monitoring wells:
  - Well-casing composition and diameter
  - Depth of water table and well
  - Total volume of water purged
  - Calculation used for volume purged
  - Date and time well was purged
  - Measurements to monitor stabilization of wells: purging should continue until measurements (temperature, pH, conductivity) are stable. If no measurements are taken, at least five well volumes must be purged before sample collection can begin
- Documentation for surface waters (depth at which samples were taken)
- Documentation for wastewater effluent:
  - If composite samples were taken, beginning and ending times of composition
  - Duration of compositing
  - Volume of subsamples
- Documentation for soil and sediments (depth at which samples were taken)
- Documentation for drum sampling:
  - Type of drum and description of contents
  - If stratified, layer(s) sampled
- How samples are transported to the laboratory (packing, cooling, separated, etc.)
- Sample transmittal form (typically, chain-of-custody form), which must include the following information:
  - Site name and address
  - Date and time of sample collection
  - Name of sampler
  - Complete identification of samples, such as field identification number, number of samples, date and time sample collected, requested analysis, preservation, and comments about the sample

Failure to fill out these records properly could result in data invalidation.

14.8 FIELD QUALITY CONTROL

The quality of data resulting from sampling activities is measured by quality control (QC) procedures. The goals of QA/QC in sample collection are to prove the validity of data derived from field measurements and to prevent improper sampling techniques and inadequacies in sample preservation, identification, and transportation. Field QA/QC is described in Section 13.5.

14.8.1 General Requirements of Field QA/QC Program

- Availability of field standard operating procedure (FSOP)
- Documentation of calibration and maintenance of field instruments and equipment
- Qualification and training of field personnel
- QC check criteria
- Validation of field measurements
- Written statement about packing and transfer of collected samples to laboratory

14.8.2 Field Quality Control Check Criteria

14.8.2.1 Equipment Blanks

Equipment blanks are used to detect contamination from sampling equipment. This blank is prepared in the field before sampling begins by using the precleaned equipment and filling the proper sample container with analyte-free water. Preservation and documentation of the blank are the same as for collected samples. If equipment is cleaned on site, then additional equipment blanks should be collected for each equipment group. Each sample matrix should be accompanied by separate equipment blanks.

14.8.2.2 Field Blanks

Field blanks are collected at the end of sample collection by filling the sample container with analyte-free water and are preserved and documented in the same way as the collected samples.

14.8.2.3 Trip Blanks

Trip blanks are collected to verify contaminations that may occur during sample collection and transportation (improperly cleaned sample containers, contaminated reagents, contamination during transportation, etc.). Trip blanks are blanks of analyte-free water prepared in the laboratory and transported to the field. They remain unopened during the sampling event and are transported back to the laboratory with the collected samples. Trip blanks should be properly labeled and documented!

14.8.2.4 Duplicates

Duplicates are samples collected at the same time from the same source. During one sample collection event, at least one sample or 10% of the collected samples (whichever is greater) should be duplicated.

14.8.2.5 Split Samples

Split samples are replicas of the same sample that are given to two independent laboratories for analysis.

14.8.2.6 Field Spiked Samples

Spiked samples are used to measure the performance of the complete analytical system, including interference from the sample matrix. Spiked samples are environmental samples with the addition of known concentrations of the analyte of interest. Field preparation, preservation, and documentation should be the same as for the collected samples. Selection of spiked and split samples may be dictated by the requirements of the site, a previsit evaluation, or on-site inspection.

14.8.2.7 Validation of Field Measurements

Field measurements are validated through *precision* (based on duplicate samples) and *accuracy* (based on measurement of the known value of a QC sample). Calculations of precision and accuracy are discussed in Section 13.9.

14.9 SAMPLE COLLECTION FROM DIFFERENT MATRICES

14.9.1 GROUNDWATER SAMPLING

Groundwater, the base flow of all perennial flows, accounts for over 90% of the world's freshwater resources. Groundwater is the primary source of drinking water; about 50% of the U.S. population uses groundwater. In many instances of groundwater contamination, the ability to predict how the contaminant plume will behave in the future can only be done on the basis of an extensive drilling and sampling program. The most frequently used approach in groundwater quality monitoring is to collect and analyze water samples from *monitoring wells*. The purpose of a monitoring well is to determine hydrogeologic properties, provide a facility for collecting water samples, and monitor the movement of the contamination plume. Critical factors include the number and location of monitoring wells and the depth at which samples are taken.

14.9.1.1 Well Purging Prior to Sampling

- Prior to sampling, an adequate amount of stagnant well water must be removed so that the collected water sample will be representative of groundwater conditions.
- For most wells, removing three to five well volumes is adequate, or until the values of temperature, pH, and conductivity measurements of the water are stabilized.
- Wells should be sampled within 6 h of purging.

14.9.1.2 Water Level Measurement

Water levels are measured by using electronic tape or chalked tape, among other techniques. When *electrical devices* are used, a light or ammeter indicates a closed circuit when the probe touches the water. *Depth markers* are commonly attached to the cable by the manufacturer at about 5-ft (1.5-m) intervals. When using a steel tape, a lead weight is attached to the bottom. The lower end of the tape is wiped dry and coated with carpenter's chalk before measurement. The tape is dropped into the well, and after withdrawal the wetted line of the tape can be read on the chalked section. The reading is subtracted from the foot mark held at the measuring point; the difference is the water level depth.

14.9.1.3 Sampling Technique Using a Bailer

The *bailer* is the most common sampling equipment used for collecting samples from groundwater. Bailers are constructed in a wide variety of diameters and in a wide variety of materials. They are easy to transport, easy to clean, and inexpensive. The disadvantage of the bailer is that atmospheric oxygen may be introduced when the sample is transferred into the sampling bottle. Select the material for the bailer that is convenient for collecting metal-analysis samples (stainless steel, plastic, or Teflon). As with all sampling equipment, the bailer must be scrupulously clean. The bailer is illustrated in Figure 14.8. Wear latex rubber gloves to avoid sample contamination.

1. Lower the bailer slowly into the well. As the bailer moves slowly down through the water in the well, the check valve remains open, allowing the water to pass through the bailer.
2. At the desired depth, stop lowering the bailer.

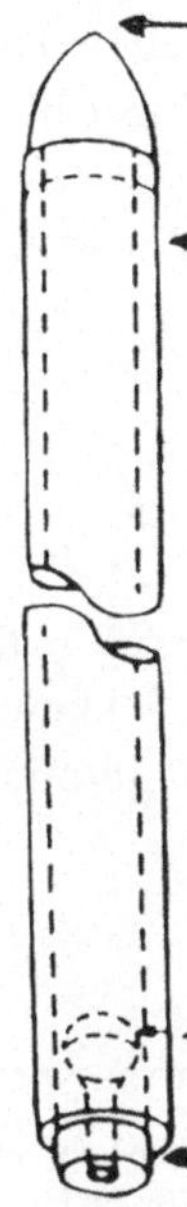

FIGURE 14.8 Teflon bailer. The top of the bailer is open and the bottom contains a sample ball-and-seat check valve arrangement. As the bailer moves down through the water in the well, the check valve remains open, and the water passes through the bailer. As the bailer is lifted, the weight of water inside the bailer causes the ball valve to seat, thus trapping the sample inside.

3. As the bailer is lifted, the weight of the water inside the bailer will close the valve, trapping the sample inside.
4. When the bailer reaches the surface, the sample is transported to the sampling bottle. (As mentioned previously, the preferred material of sample containers for metals analysis is polyethylene.)
5. Remove the cap from the bottle and rinse the bottle with the sample. Do not rinse the bottle if it is a prepreserved container!
6. Fill the bottle with the sample, but do not fill to the top. Leave space for the addition of preservative and mixing.
7. With a pipet or a premeasured dropper, add 3 ml of 1+1 HNO_3 or 1.5 ml concentrated HNO_3 per liter of sample to take the sample pH to less than 2. If prepreserved bottles are used, do not add acid!
8. Mix sample well; pour a small amount of sample into a small, disposable container; and check the pH with narrow range pH paper. If the pH is not less than 2, add more preservative until the desired pH is achieved.
9. Record the volume used, the concentration of the preservative, and the measured pH of the preserved sample in the field notebook and on the sample label.
10. The corresponding equipment blank should contain the same amount of preservative as the sample. Samples with additional preservative should have a separate blank with the same amount of acid as in the sample.
11. Samples for hexavalent chromium (Cr^{6+}) do not need preservative. Carefully select the sampling container. Do not use prepreserved sampling bottle! Complete sample label and transfer to the laboratory as soon as possible.

12. If the analysis request is for suspended and dissolved metal determination, the sample should be filtered prior to preservation and treated as discussed in Section 14.5.2.
13. Quality control requirements are dependent on the project plan. Field quality control checks are listed in Section 14.8.2.
14. Affix the sample labels, fill out the chain-of-custody form, and record all sampling data in the field notebook, as discussed in Section 14.7.1.

14.9.2 Drinking Water Sampling

14.9.2.1 Sampling Potable Well Water

When sampling drinking water from residential, private potable wells, the wells must be purged as described in Section 14.9.1. If the capacity of the pressure tank is not known, purge the well for about 15 to 20 min. After purging, reduce the flow to approximately 500 ml/min. Take samples as discussed in Section 14.9.1.3, steps 5 to 14.

14.9.2.2 Sampling from Distribution System

Samples should be collected in areas free from excessive dust, rain, snow, or other sources of contamination. If samples are collected from faucets, the faucet should be clean and free from possible contamination. The faucet should also be flushed thoroughly, generally 2 to 3 min, but sometimes a longer flush is needed, such as when sampling to test for lead pollution. After flushing the water, adjust the flow so that it does not splash against the walls of bathtubs, sinks, or other surfaces. Then collect samples as discussed in Section 14.9.1.3, steps 6 to 15.

14.9.3 Sampling Surface Waters

Selection of sample sites depends on the nature of the sampling project, type of samples, and whether the sites are permanent monitoring stations established by the Surface Water Improvement and Management (SWIM) program.

14.9.3.1 General Rules in Surface Water Sampling

- When gathering samples from a powerboat, samples must be taken from the bow, away and upwind from the outboard gasoline engine.
- Both water and sediment samples should be collected from downstream to upstream.
- When water and sediment samples are taken from the same area, water samples must be collected first.
- Care should be taken not to disturb sediment when taking water samples.
- Do not take samples at or near dams, piers, or bridges, because the unnatural water flow may disturb the representativeness of the sample.

14.9.3.2 Grab Samples

Grab samples are taken by using unpreserved containers.

1. Submerge the container in the water.
2. Invert the bottle so that the neck is upright and pointing to the water flow. Fill the bottle and return it to the surface.
3. Pour out a small quantity of the sample to leave space for adding preservative and mixing.
4. Preserve samples and follow activities as discussed in Section 14.9.

5. Another grab sampling method is to use a *pole-mounted flask* and follow steps 1 to 4 above. This kind of equipment must be constructed of material that does not interfere with the sampled parameters. (If the pole is copper, of course, the sample would not be acceptable for copper testing.) When using a pole sampler, samples can be taken from a bridge, a boat, or from the shore.
6. Composite samples are taken when a given depth interval is desired for the sample. Care should be taken that all subsamples are of equal volume (Section 14.1.4).
7. A *peristaltic pump* may also used to take grab or composite samples.

14.9.3.3 Samples Taken at Different Depths in Same Sample Location

Samples are taken just below the surface, at mid-depth, and just above the bottom. For this kind of sampling, a *depth-specific sampler*, such as the Kemmerer sampler, a pump, or a bailer may be used. Each kind of sampling equipment should be selected based on the proper material for the parameter of interest. For example, for metal testing, the equipment should be made of stainless steel, plastic, or Teflon. The Kemmerer sampler is illustrated in Figure 14.9. Collect and preserve samples as in Section 14.9.3.

14.9.3.4 Sediment Samples

Sediment samples are usually taken as a part of surface water samples. Equipment used for sediment sampling varies according to sample location, water depth, sediment grain size, and water velocity. The most common sampling equipment types are *scoops, corers, Eckman sampler* (for sand, silt, and mud sediments), and the *Peterson and Ponar sampler* (for hard, rocky sediments). The Eckman bottom grab sampler is illustrated in Figure 14.10.

14.9.4 Sampling Waste Water

Requirements for the analyzed parameters from both municipal and industrial waste waters are regulated by the National Pollutant Discharge Elimination System (NPDES) permit program.

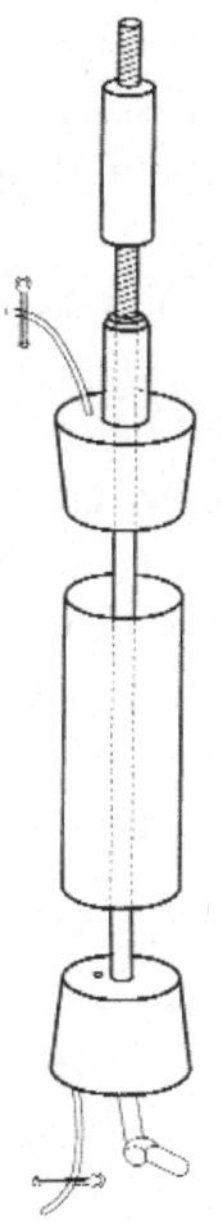

FIGURE 14.9 Modified Kemmerer sampler.

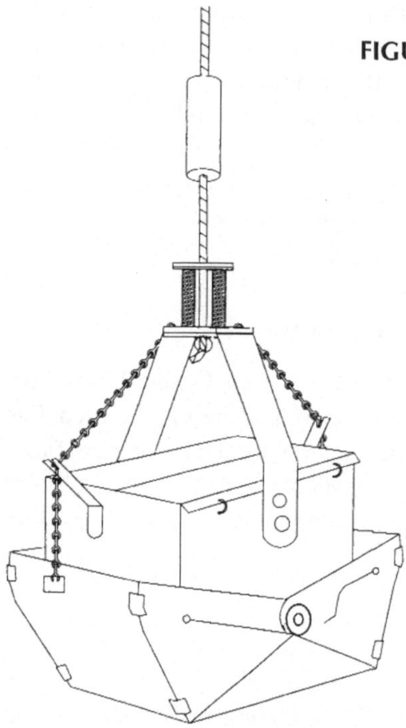

FIGURE 14.10 Eckman bottom-grab sampler.

These permits specify the types and amounts of pollutants that may be discharged and are intended to ensure that the effluent content remains within the limits of the relevant groundwater or surfacewater standard. Sampling locations should be described in the permit or the project plan.

14.9.4.1 General Criteria for Collecting Representative Samples

- Sampling time should not exceed 15 min.
- Samples must be taken where wastewater flow is properly mixed.
- The most representative sample from effluent should be taken downstream from the wastewater stream before it enters the disposal site (surface water, wetland, deep-well injection, etc.)
- The best point for sampling from influents is from the turbulent flow where the sample is well mixed.
- When taking a sample, the container should be inverted and submerged below the waste-stream surface and filled. Do not use prepreserved sample bottles! Follow up with activities discussed in Section 14.4.3, steps 6 to 14.
- When collecting composite samples, the compositing interval and duration as well the sampling point should be described in the project plan. Use of automatic samplers is practical.

14.9.5 Sampling Agricultural Discharges

Agricultural discharges can be categorized into three types:

1. Concentrated animal waste or manure
2. Runoff from an agricultural watershed
3. Irrigation return flow

The frequency and location of sampling, the number of samples, and the required parameters must be followed as stated in the discharge permit.

14.9.6 Collecting Domestic Sludge

All samples for sludge classification should be representative and taken after final sludge treatment but before disposal. The preferred container for metals analysis is plastic, and the sample is preserved with HNO_3.

14.9.7 Collecting Soil Samples

All samples for soil analysis should be representative of the area to be sampled. If the area where the sample is taken shows natural disturbances (e.g., dead vegetation or discoloration), the sample must be accompanied by a sample from a uncontaminated area. For metals analysis, the preferred container is plastic. The material of the sampling equipment is dictated by the need to avoid contaminating the selected analyte group in the sample.

14.9.7.1 General Rules for Soil Sampling

1. Wear natural latex rubber gloves.
2. Select the appropriate sampling devices.
3. Select the appropriate sample container.
4. Take the soil sample, mix well in a stainless steel plate, and transfer into the sample container with minimal headspace.
5. Clean sample container exterior if necessary, and label it.
6. Fill out chain-of-custody form and field notebook.
7. Place the sample container into a plastic bag.
8. When samples are collected from a large area, composition of soil samples is recommended to reduce the number of samples. Composition of the samples should be handled in stainless steel or glass containers. The origin and sample size of each subsample must be documented in the field notebook.

14.9.7.2 Surface Soil Sampling

Before taking the sample, remove dirt, leaves, and grass from the soil surface, and take the sample with a stainless steel spoon or scoop.

14.9.7.3 Shallow Subsurface Soil Sampling

Dig a hole with a stainless steel shovel, bucket, or auger to the desired depth. To avoid the collapse of the hole, insert a rigid PVC support into the hole, and after sampling, remove the support.

14.9.7.4 Deep Subsurface Soil Sampling

The sample is taken from a hole more than 15 ft below the surface. Various types of sampling equipment are available for this kind of sample collection. For rocks and hard soils, the head of the sampling device has a small diamond bit to cut through the hard surfaces as the drilling rod is rotated.

14.9.8 Sampling Hazardous Wastes

To ensure representative sample collection, a sampling plan should be available to determine the correct number of samples taken with the appropriate frequency. Hazardous material samples can be gas, liquid, solid, paste, sludge, or some combination. Therefore, methods and equipment vary according to sample makeup. Reaction of the sample with sunlight or temperature should also be taken into account. Be sure that the sampling equipment and sample containers do not react with the waste sample!

The most common sampler is the *colivasa,* designed to sample free-flowing liquids from drums, open tanks, and pits, among other sources. It consists of a metal, glass, or plastic tube equipped with an end closure that can be opened and closed. Open and lower the colivasa into the waste and let the tube fill. Lock the stopper and withdraw from the waste. Wipe the exterior with a disposable cloth and transfer the sample to the sample container. The colivasa sampler is illustrated in Figure 14.11.

Another type of liquid sampler is the *weighted bottle.* It is a glass or plastic bottle with a sinker, stopper, and a line that is used to lower, raise, and open the bottle. Lower the bottle into the sample, let it fill (when bubbling stops, the bottle is filled), raise, and use the bottle itself as a sample container.

The *dipper,* a beaker on the end of a long pole, is similar to the weighted bottle. Dippers are useful for sampling liquids and free-flowing slurries.

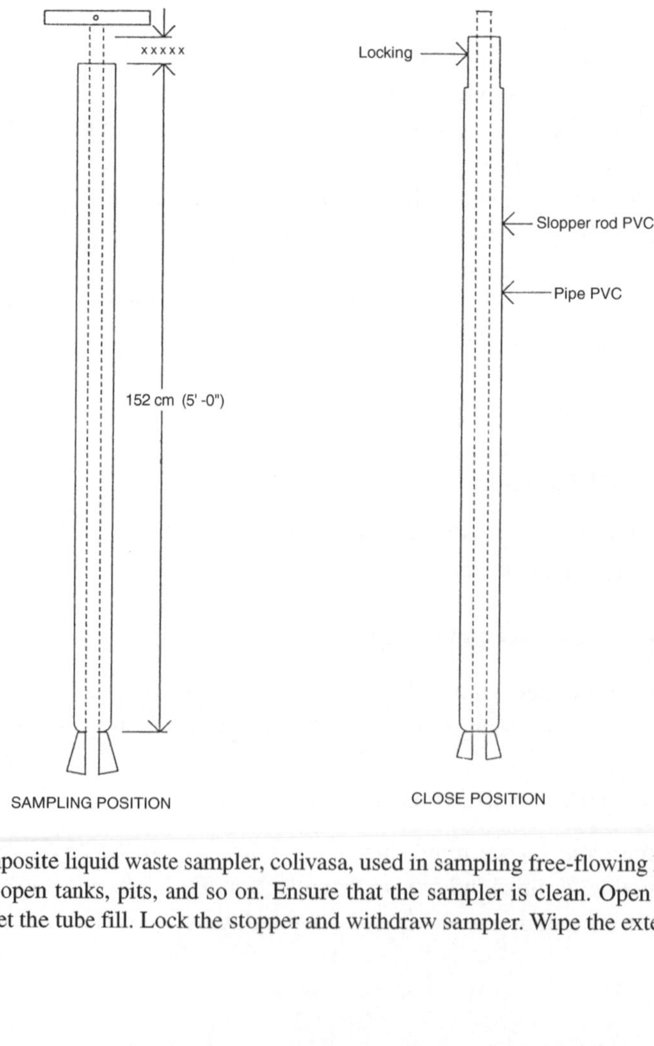

FIGURE 14.11 Composite liquid waste sampler, colivasa, used in sampling free-flowing liquids and slurries from drums, shallow open tanks, pits, and so on. Ensure that the sampler is clean. Open and lower into the sample material and let the tube fill. Lock the stopper and withdraw sampler. Wipe the exterior with a disposable cloth.

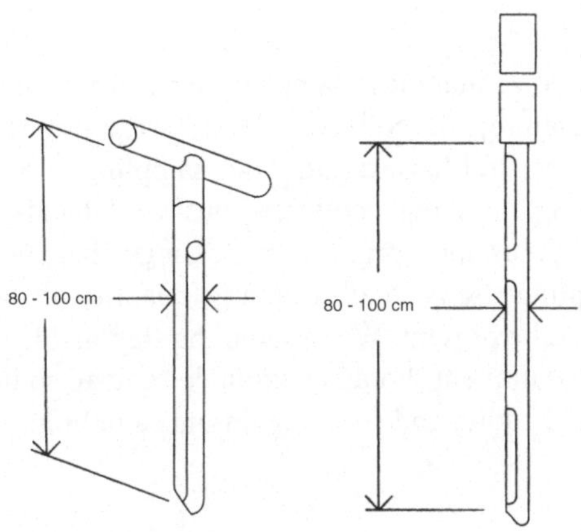

FIGURE 14.12 Sampling trier (left), used in sticky solids and loose soils, is a tube cut in half lengthwise with a sharpened top that allows cutting into the sample material. Insert clean trier into the sample, cut the core, remove with concave side up, and transfer sample to container. Thief sampler (right), used in any bulk material, is especially useful in sampling grain-like material. It consists of two slotted concentric tubes, usually made of brass or stainless steel. Insert clean, closed thief into sample. Wiggle sampler to let material enter the slots. Close, withdraw, and remove inner tube; transfer sample to the sample container.

The *trier*, used for sampling sticky solids and loosened soils, is a tube with a sharpened tip. Insert the trier into the waste, cut the core, remove with concave side up, and transfer the sample into the sample container. The sampling trier is illustrated in Figure 14.12.

For bulk material, the best sampler is the *sampling thief*, and for hard and packed solids, a convenient sampler is the *auger. Scoops and shovels* are also useful for sampling granular or powdery materials.

The material of the sample container should be chosen so that it will not react with the sample. The container should be resistant to leakage and breakage and the appropriate size for the sample. Wide-mouth plastic containers with tight, screw-type lids are desirable if the sample is not used for organic analyte determination. After the sample is taken, clean the container exterior, label properly, and place in a plastic bag for transport to the laboratory. When the nature of the hazardous material is known, a safety label should also be affixed to the sample container. Common safety labels are shown in Figure 14.13.

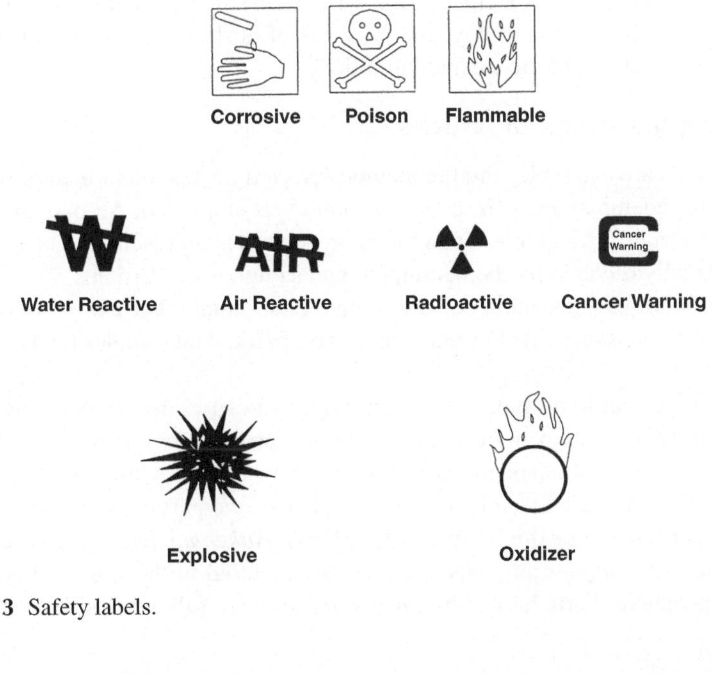

FIGURE 14.13 Safety labels.

14.9.8.1 Safety Concerns

The person collecting hazardous waste samples must realize that these samples are hazardous materials and should be handled with extreme care! Therefore, the collector should wear protective gloves, face aspirator, and special safety clothing, shoes, and hazard hat. After sampling, clothes, shoes, and hat must be removed. Each time after sampling, the sample collector must wash hands and exposed portions of the body; in some cases, a full shower is appropriate. A fire extinguisher should be available if the material is flammable and the sampling site is small. If a large quantity of flammable material is at the sampling site, a fire truck should be present. When sampling flammable materials with a high vapor pressure or low flashpoint, all equipment should be grounded and all sources of ignition should be prohibited. As a general rule, avoid eating and smoking during sampling.

14.9.9 Sampling Fish Tissues

Fish tissues are usually analyzed for metals and organic pollutants. Equipment used for sampling should be scrupulously cleaned and decontaminated properly. First, wash equipment with laboratory detergent, rinse with deionized water, isopropylalcohol, and finally, analyte-free water. The dried equipment should be stored in aluminum foil until use. The captured fish is placed in wet ice in a cooler. In the laboratory, weigh and prepare the tissue. The fillet should not be skinned. The fish tissue should be wrapped in aluminum foil and kept in wet ice for 24 h, or frozen for longer storage.

14.9.10 Collecting Air Samples

Dust, silica, and other suspended particles in the air are measured by gravimetry. The filter in the cassette should be weighted before and after sampling for accurate mass determination of deposited particles. For metals analysis, the metal dusts deposited on the filter must be acid digested and analyzed via atomic absorption spectrometry or inductively coupled plasma spectrometry.

The primary concern of the sampler must be directed to the collection of representative samples and the homogeneity of the air mixtures employed to calibrate both the collection and the analytical systems. The concentration of the contaminant at a specific location is influenced by the source of contaminant, airflow direction and velocity (due to wind or thermal gradients), density of the contaminants, intensity of sunlight, time of day, and presence of obstructions, such as trees, buildings, and machinery (which produce turbulence and humidity).

14.9.10.1 Sampling and Storage of Particles

Many sampling methods are available, and the method selected depends on the purpose of the sampling. For example, for chemical analysis, a *Hi-Vol sampler* is employed. All parts of the airstream must be sampled and properly weighted so that the entire stream is represented. The size and type of the filter paper are usually dictated by the instrument and sample site. Particles can react with filter paper, evaporate, and sublimate, depending on the nature of the sample. The analyst must always consider the method prior to the sampling. For many purposes, particulate samples tend to keep well for a long period of time.

Site selection is important in all types of air sampling, but especially for particles because they are much less uniformly dispersed in ambient air as well as in process equipment. Particles of all sizes are continually emitted into the atmosphere. *Large particles* fall rapidly, while smaller sizes fall more slowly. The height of the source, wind velocity and turbulence, and particle size distribution will determine how fast the particles settle out. *Very small particles* (*Attken nuclei*) tend to become attached to larger particles. *Aerosol samples* are dispersions of any material in the solid or liquid phase in a gas stream or the atmosphere. Particles can be categorized into the following size groupings:

Settleable particles, larger than 30 μm in diameter
Suspended particles, smaller than 30 μm in diameter
Condensation or *Attken nuclei*, 0.01 to 0.1 μm in diameter
Agglomerates, several small particles attracted by a large particle or attracted to each other
Fine particles, particles less than 2.5 μm in diameter
Coarse particles, particles greater than 2.5 μm in diameter

14.9.10.2 Isokinetic Sampling of Particles

The momentum of a particle is mass × velocity. Particles of different sizes are displaced by different amounts. Isokinetic sampling refers to taking a sample under conditions in which there is no change in momentum. This is accomplished by using a thin-walled tube aligned with the stream flow and drawing the sample into it and at the same linear velocity as the stream flow at that point. Particles of all sizes can thus be collected efficiently.

14.9.10.3 General Rules for Particulate Sampling

A 24-h sampling has become a standard practice. General rules for particulate sampling follow:

- Take the sample at the point of major interest.
- Do not place the sampler directly downwind from a major point source.
- Place the sampler about 1.45 m above ground level.
- Locate downwind from major obstacles at a distance of about ten times their height.
- Take several samples at different locations in the area of interest.
- Sample during the time of day of greatest interest, or take a 24-h sampling.

The objective is to collect a sample that is representative of the material emitted. The specific points of sampling are generally determined by discussion with plant engineers or others who understand the process or the source of emission. A site visit is generally required for final selection. Particulate sampling should be carried out with probes inserted in the duct at each end of the points-of-flow measurement. Care must be exercised to ensure that particles may be vaporized. If the ambient temperature is too low, water or other vapors form mist that will collect with the solids and plug up the filter, leading to bad results.

For each sample, the following data must be attached:

- Date and time of collection
- Sample location
- Sample flow rate
- Sample pressure
- Sample temperature
- Dew point
- Plant operating condition
- Sampler's name

15 Sample Preparation for Metals Analysis

15.1 GENERAL DISCUSSION

Samples containing particulate or organic material generally require pretreatment before analysis.

15.1.1 Sample Pretreatment for Total Metals

Colorless, transparent samples (primarily drinking water) containing a turbidity of less than 1 NTU (nephelometric turbidity unit), and single-phase samples can be analyzed directly via atomic absorption spectroscopy or inductively coupled plasma spectroscopy for total metals without digestion. For further verification or to determine changes in existing matrices, compare digested and undigested samples to ensure comparable results. Digest all other samples before determination of total metals.

Take care not to introduce metals into samples during preliminary treatment. During pretreatment, avoid contact with rubber, metal base paints, cigarette smoke, paper tissues, and all metal products, including those made of stainless steel, galvanized metal, and brass. Conventional fume hoods can contribute significantly to sample contamination, particularly during acid digestion in open containers. Plastic pipet tips are often contaminated with copper, iron, zinc, and cadmium. Before use, soak pipets in 2N HCl or HNO_3 for several days and rinse with deionized (DI) water. Check reagent-grade acids used for preservation, extraction, and digestion for purity. If excessive metal concentrations are found, purify the acids by distillation or use ultra-pure acids. Carry blanks through all digestion and filtration steps and apply necessary corrections to the results.

15.1.2 Sample Pretreatment for Dissolved Metals

To analyze for dissolved metals, filter the sample (as in Section 15.1.4), acidify the filtrate, and analyze directly.

15.1.3 Sample Pretreatment for Suspended Metals

To determine suspended metals, filter the sample, digest the filter and the material on it, and analyze.

15.1.4 Preliminary Filtration of Samples

If dissolved or suspended metals are to be determined, filter the sample at the time of collection. Use a preconditioned, plastic, vacuum (or pressure) device equipped with a filter support made of plastic or TFE. The filter should have the following characteristics: prewashed, ungridded, 0.45-µm membrane, and made of polycarbonate or cellulose acetate. Before use, filter a blank consisting of reagent water to ensure freedom from contamination. Precondition the filter and filter device by rinsing with 50-ml of DI water. If the filter blank contains significant metal concentrations, soak membrane filters in approximately 0.5N HCl or 1+1 HNO_3 and rinse with water before use. If the filter is to be digested for suspended metals, record the sample volume filtered, and analyze a digested filter as a blank. Before filtering, centrifuge highly turbid samples in acid-washed TFE or a high-density plastic tube to reduce loading on filters. Stirred-pressure filter units foul less readily than vacuum filters. Filter at a pressure of 70 to 130 kPa (kiloPascal; 1 atm (atmosphere) = 100 kPa).

After filtration, acidify filtrate to pH 2 with HNO_3 concentrate and analyze directly. If a precipitate forms on acidification, digest acidified filtrate before analysis. Retain filter and digest it for direct determination of suspended metals.

15.1.5 Sample Pretreatment for Acid-Extractable Metals

To determine acid-extractable metals, extract metals as indicated below and analyze extract. Extractable metals are lightly absorbed on particulate material. Because some sample digestion may be unavoidable, use rigidly controlled conditions to obtain meaningful and reproducible results. Maintain constant sample volume and contact time. Express results as extractable metals and specify extraction procedure.

At the time of collection, acidify the entire sample with 5 ml of HNO_3 concentrate per liter of sample. Extract metals as follows:

1. Mix sample well.
2. Transfer 100 ml of the sample to a beaker or flask.
3. Add 5 ml of 1+1 HCl.
4. Heat for 15 min over a steam bath.
5. Filter through a membrane filter, adjust filtrate volume to 100 ml with laboratory-pure water, and analyze.

15.2 DIGESTION PROCEDURES FOR METALS

15.2.1 Introduction

To reduce interference by organic matter and to convert metal associated with particulate to a form (usually a free metal) that can be determined by AAS or ICP, use one of the digestion techniques presented below. Use the least rigorous digestion method required to provide complete and consistent recovery compatible with the analytical method and the metal being analyzed.

HNO_3 will digest most samples adequately. Nitrate is an acceptable matrix for both FAAS and GrAAS. Some samples may required the addition of perchloric, hydrochloric, or sulfuric acid for complete digestion. These acids may interfere in the analysis of some metals and all provide a poor matrix for GrAA analysis. Confirm metal recovery for each digestion and analytical procedure used. Table 15.1 lists the acids used in conjunction with HNO_3. As a general rule:

- HNO_3 alone is adequate for clean samples or easily oxidized materials.
- HNO_3–H_2SO_4 or HNO_3–HCl digestion is adequate for readily oxidizable organic matter.
- HNO_3–$HClO_4$ or HNO_3–$HClO_4$–HF digestion is necessary for difficult-to-oxidize organic matter or minerals.

Report the digestion technique used. Because acid digestion techniques do not normally achieve total digestion, the microwave digestion procedure may be used as an alternate. The microwave method is a closed-vessel procedure, and thus typically provides improved precision when compared with the hot-plate technique. Suggested sample volumes for digesting are presented in Table 15.2. Larger samples require additional acid.

TABLE 15.1
Acids Used in Conjunction with HNO_3 for Sample Preparation

Acid	Recommended For	May Be Helpful for	Not Recommended for
HCl	—	Sb, Ru, Sn Th, Pb	—
H_2SO_4	Ti	—	As, Pb, Ba
$HClO_4$	—	Organic materials	—
HF	—	Siliceous materials	—

TABLE 15.2
Suggested Sample Volumes for Digestion

Estimated Metal Concentration (mg/l)	Sample Volume (ml)
<1	**1000**
1–10	**100**
10–100	**10**
100–1000	**1**

15.2.2 Nitric Acid Digestion

1. Transfer a suitable volume (50 to 100 ml) of well-mixed sample into a 125-ml flask or beaker.
2. Add 5 ml of HNO_3 concentrate and a few boiling chips or glass beads. Bring to a slow boil and evaporate on a hot plate to the lowest volume (10–20 ml) before precipitation occurs.
3. Continue heating and adding acid until digestion is complete, as indicated by a light-colored, clear solution. Do not let sample dry during digestion.
4. Wash down flask or beaker walls with DI water and filter, if necessary. Transfer filtrate to a 100-ml volumetric flask. Cool and dilute to the mark, and mix thoroughly.

Alternatively, take a larger sample volume using the procedure for concentration.

15.2.3 Nitric Acid–Hydrochloric Acid Digestion

1. Transfer a suitable volume of the well-mixed, acid-preserved sample appropriate for the expected metal concentrations to a flask or beaker.
2. Add 3 ml of HNO_3 concentrate. Heat on a hot plate and evaporate to less than 5 ml, making certain that the sample does not boil and is not allowed to dry.
3. Cool and add another 5 ml of acid. Cover container with a watch glass and return to a hot plate. Increase temperature of the hot plate so that a gentle reflux action occurs.
4. Continue heating, adding additional acid as necessary until digestion is complete, indicated by a light-colored and clear digestate.
5. Evaporate to less than 5 ml and cool. Add 10 ml of 1+1 HCl and 15 ml of DI water per 100-ml total volume. Heat for an additional 15 min to dissolve any precipitate or residue.
6. Cool, wash down beaker walls and watch glass with DI water, and filter to remove insoluble material that could clog the nebulizer. Alternatively, centrifuge or let settle overnight.
7. Adjust to a predetermined volume based on expected metal concentrations.

15.2.4 Nitric Acid–Sulfuric Acid Digestion

1. Transfer a suitable volume of the well-mixed, acid-preserved sample into a flask or beaker.
2. If sample is not already acidified, acidify to methylorange endpoint with H_2SO_4 concentrate and add 5 ml of HNO_3 concentrate and a few boiling chips or glass beads.
3. Bring to slow boil on a hot plate and evaporate to 15 to 20 ml.
4. Add 5 ml of HNO_3 concentrate and 10 ml of H_2SO_4. Evaporate on a hot plate until dense white fumes of SO_3 begin to form. If the solution does not clear, add 10 ml of HNO_3 concentrate and repeat evaporation until fumes of SO_3 begin to form. Heat to remove all HNO_3 before continuing treatment. All HNO_3 will be removed when the solution is clear and no brownish fumes are evident. Do not let sample dry during digestion.
5. Cool and dilute to about 50 ml with DI water. Heat to almost boiling to dissolve slowly soluble salts. Filter if necessary.
6. Complete procedure by transferring filtrate into a volumetric flask and dilute to the mark. Mix thoroughly.

15.2.5 Nitric Acid–Perchloric Acid Digestion

1. Mix sample and transfer a suitable volume into a flask or beaker.
2. If sample is not already acidified, acidify to methylorange endpoint with HNO_3 concentrate, add an additional 5 ml of HNO_3 concentrate and a few boiling chips or glass beads, and evaporate on a hot plate to 15 to 20 ml.
3. Add 10 ml each of HNO_3 concentrate and $HClO_4$, cooling flask or beaker between additions. Evaporate gently on a hot plate until dense white fumes of $HClO_4$ begin to appear.
4. If solution is not clear, cover container with a watch glass and keep solution at boiling temperature (but no higher), boiling until it clears.

If Pb is to be determined in the presence of high amounts of sulfate (e.g., determination of Pb in power-plant fly ash in samples), dissolve $PbSO_4$ precipitate as follows:

1. Add 50 ml of ammonium acetate solution to flask or beaker in which digestion was carried out, and heat to incipient boiling. Rotate container occasionally to wet all interior surfaces and dissolve any deposited residue.
2. Using a preconditioning plastic filtering device with either vacuum or pressure and containing a filter support of plastic or TFE, filter the sample through a prewashed ungridded 0.45-mm membrane filter as described in Section 15.1.4.
3. Transfer filtrate to a 100-ml volumetric flask, cool, dilute to the mark, mix thoroughly, and set aside for determination of Pb.

Caution: Heated mixtures of perchloric acid ($HClO_4$) and organic matter may violently explode. Avoid this hazard by taking the following precautions:

- Do not add $HClO_4$ to a hot solution containing organic matter. (Always pretreat samples containing organic matter with HNO_3 before adding $HClO_4$!)
- Avoid repeated fuming with $HClO_4$ in ordinary hoods. (For routine operations, use a water pump attached to a glass fume eradicator. Stainless-steel fume hoods with adequate water wash-down facilities are available commercially and are acceptable when using $HClO_4$.)
- Never let samples being digested with $HClO_4$ evaporate to dryness.

15.2.6 Nitric Acid–Perchloric Acid–Hydrofluoric Acid Digestion

Caution: See precautions for using $HClO_4$ in Section 15.2.5. Handle with extreme care and provide adequate ventilation, especially for the heated solution. Avoid all contact with exposed skin. Seek medical attention for hydrofluoric acid burns.

1. Mix sample and transfer a suitable volume into a 250-ml TFE beaker.
2. Add a few boiling chips and bring to a slow boil. Evaporate to 15 to 20 ml.
3. Add 12 ml of HNO_3 concentrate and evaporate to near dryness. Repeat HNO_3 addition and evaporation.
4. Cool solution and add 20 ml of $HClO_4$ and 1 ml of HF, and boil until solution is clear and white fumes of $HClO_4$ have appeared.
5. Cool, add about 50 ml of DI water, filter, and proceed as directed in Section 15.4.2, step 4.

15.2.7 Dry Ashing

Dry ashing is helpful if large amounts of organic matter are present. Dry ashing yields highly variable precision and bias depending on the sample type and metal analyte.

1. Mix sample and transfer a suitable volume into a platinum or high-silica-glass evaporating dish (Vycor, manufactured by Corning Glass Works, or equivalent).
2. Evaporate to dryness over a steam bath.
3. Transfer dish to a muffle furnace and heat sample to a white ash. If volatile elements are to be determined, keep temperature at 400 to 450°C.
4. If only Na is to be determined, create the ash sample at a temperature up to 600°C.
5. Dissolve ash in a minimum quantity of HNO_3 concentrate and warm water. Filter diluted sample and adjust to a known volume, preferably so that the final HNO_3 concentration is about 1%.
6. Use a portion of this solution for metal determination.

15.2.8 Microwave-Assisted Digestion

Caution: This method is designed for microwave digestion of waters only. It is not intended for the digestion of solids, in which high concentrations of organic compounds may result in high pressure and possibly unsafe conditions.

15.2.8.1 Requirements for Microwave Unit

Use a microwave unit with programmable power (minimum 545 W) to within ±10 W of required power, with a corrosion-resistant, well-ventilated cavity, and with all electronics protected against corrosion for safe operation. Use a unit with a rotating turntable with a minimum speed of 3 rpm to ensure homogeneous distribution of microwave radiation. Only laboratory-grade microwave equipment and closed digestion containers with pressure relief that are specifically designed for hot acid should be used.

Vessels should be constructed of perfluoroalkoxy (PFA) Teflon capable of withstanding pressures of at least 760±70 kPa (±110 psi) and capable of controlled pressure relief at the manufacturer's maximum pressure rating. Acid wash all digestion vessels and rinse with reagent water. When using a new PFA Teflon vessel or when changing between high- and low-concentration samples, clean by leaching with hot 1+1 HCl for a minimum of 2 h and then with hot 1+1 HNO_3 for a minimum of 2 h, rinse with reagent water, and dry in a clean environment.

15.2.8.2 Procedure

The following procedure is based on heating acidified samples in two stages where the first stage is to reach 160±4°C in 10 min, and the second stage is to permit a slow rise to 165–170°C during the second 10 min. A verified program that meets this temperature-time profile is 545 W for 10 min followed by 344 W for 10 min using five single-wall PFA Teflon digestion vessels. The usable number of vessels is determined by vessel design and power output.

1. Weigh entire digestion vessel assembly to 0.1 g before use and record (*A*).
2. Accurately transfer 45 ml of well-shaken sample into the digestion vessel.
3. Pipet 5 ml of HNO_3 concentrate into each vessel. Make sure that pressure-cap relief disks are inserted according to manufacturer's directions. Tighten caps to manufacturer's specification.
4. Weigh each capped vessel to the nearest 0.1 g (*B*).
5. Evenly distribute the appropriate number of vessels in the carousel.
6. Treat sample blanks, known additions, and duplicates in the same manner as samples.
7. When fewer samples than the appropriate number are digested, fill the remaining vessels with 45 ml of reagent water and 5 ml of HNO_3 concentrate to obtain the full complement of vessels for the particular program in use.
8. Place carousel in microwave and set it carefully on the turntable. Program microwave unit to heat samples to 160±4°C in 10 min; for the second stage, permit a slow rise to 165 to 170°C for 10 min. Start microwave generator, making sure that the turntable is turning and that the exhaust fan is on.
9. Upon completion of the microwave program, let vessels cool for at least 5 min in the unit before removal. Samples may then be cooled further outside the unit by removing the carousel and letting them cool on a bench or in a water bath. When cooled to room temperature, weigh (to 0.1 g) each vessel and record weight (*C*).
10. If the net weight of sample plus acid decreased by more than 10%, discard sample.
11. Complete sample preparation by carefully uncapping and venting each vessel in a fume hood. Transfer to acid-cleaned, noncontaminating plastic bottles. If the digested sample contains particulate, centrifuge at 2000 to 3000 rpm for 10 min and then filter or let settle overnight.

15.2.8.3 Calculation

Dilution correction: Multiply results by 50/45, or 1.11, to account for the dilution caused by the addition of 5 ml of acid to a 45-ml sample.

Discarding of sample: To determine if the net weight of the sample plus acid decreased by more than 10% during the digestion process, use the following calculation:

$$[(B-A)-(C-A)]/(B-A) \times 100 > 10\% \tag{15.1}$$

15.2.8.4 Quality Control (QC)

Including a QC sample in each loaded carousel is recommended. Prepare samples in batches, including preparation blanks, sample duplicates, and predigested known additions. Determine size of batch and frequency of QC samples according to the analytical method and laboratory practice. The power of the microwave unit and batch size may prevent including one or more QC samples in each carousel. Do not group QC samples together but distribute them throughout the various carousels to render the best monitoring of digestion.

15.3 ACID DIGESTION FOR TOTAL AND DISSOLVED METALS

15.3.1 INTRODUCTION

This procedure is used to prepare surfacewater and groundwater samples for analysis by flame atomic absorption spectroscopy (FAAS) or by inductively coupled plasma spectroscopy (ICP), for the following metals: Al, Sb, As*, Ba, Be, Cd, Ca, Cr, Co, CI, Fe, Pb, Mg, Mn, Ni, K, Se*, Ag, Na, Ta, V, and Zn (* = ICP only). Note that this digestion procedure may not be vigorous enough to destroy some metal complexes. Total metal samples must be acidified at the time of collection with HNO_3. For dissolved metals, all samples must be filtered through a 0.45-μm filter and the filtrate acidified with HNO_3 (see Section 15.1). (For discussion of sample preservation, see Section 14.4.)

15.3.2 PROCEDURE

1. Measure 100-ml aliquot of well-mixed sample into a beaker. (To avoid metal contamination, the cleaned beaker should be rinsed with 1+1 HNO_3.)
2. Add 2 ml of concentrated HNO_3 and 5 ml of concentrated HCl. Cover with a ribbed watch glass, and heat in a steam bath or on a hot plate at 90 to 95°C until the volume is reduced to 15 to 20 ml. *Do not boil!* Antimony (Sb) is easily lost by volatilization from HCl media.
3. Remove from hot plate and allow to cool.
4. Wash down the beaker walls and watch glass with DI water. If necessary (i.e., when suspended material appears), filter or centrifuge the sample to remove suspended materials. *Caution:* The filter and filtration apparatus should be washed with 1+1 HNO_3 before filtration.
5. Adjust the final volume to 100 ml with DI water.

Note: As and Se determination from this digestate are suitable for the ICP technique only.

15.3.3 QUALITY CONTROL (QC)

Together with the samples, at a frequency of 5% or one per analytical batch, the following should be digested as part of the quality control process:

1. 100 ml of analyte-free water, called the *preparation blank* (prep blank)
2. 100 ml of calibration verification standard (CVS), called *laboratory control sample* (LCS)
3. *Spiked sample* (spiked sample may be duplicated instead of a sample duplicate)
4. One sample *duplicate*

15.4 ACID DIGESTION OF AQUEOUS SAMPLES AND EXTRACTS FOR TOTAL METALS BY FLAME ATOMIC ABSORPTION SPECTROMETRY (FAAS) AND INDUCTIVELY COUPLED PLASMA (ICP) ANALYZER

15.4.1 INTRODUCTION

This digestion procedure is used to determine total metals in the preparation of aqueous samples, EPTOX and mobility-procedure (TCLP) extracts, and wastes that contain suspended solids for FAAS and ICP analysis. Samples may be analyzed for the following parameters: Al, As, Ba, Be, Cd, Ca, Cr, Co, Co, Fe, Pb, Mg, Mn, Mo, Ni, K, Se, Na, Tl, V, and Zn. This procedure is not suitable for samples that will be analyzed by GrAAS because HCl can cause interference during furnace atomization. Collection and preservation of samples are discussed in Section 14.4.

15.4.2 PROCEDURE

1. Transfer 100-ml representative aliquot of well-mixed sample into a beaker. (The cleaned beaker should be rinsed with 1+1 HNO_3 to avoid contamination.)
2. Add 3 ml of concentrated HNO_3.
3. Cover the beaker with a ribbed watch glass and place on a hot plate. Heat slowly, until it evaporates to about 5 ml. *Do not boil sample! Make certain that no portion of the bottom of the beaker is allowed to dry.*
4. Cool and add 3 ml of concentrated HNO_3.
5. Recover the beaker with the watch glass and return to the hot plate.
6. Increase the temperature of the hot plate so that a gentle reflux action occurs.
7. Continue heating, adding additional acid if necessary, until the sample is clear and light in color.
8. Evaporate until the volume is about 3 ml. Do not dry sample! If a sample is allowed to dry or burn, discard and redigest.
9. Cool, add about 10 ml of 1+1 HCl and warm up for 15 min to dissolve all of the precipitate and residue.
10. Wash down the beaker wall and watch glass with DI water, and filter or centrifuge, if necessary, to remove silicates and other insoluble material that could clog the nebulizer. This step may cause contamination, unless the filter and filtering apparatus are thoroughly cleaned and rinsed with diluted HNO_3.
11. Adjust the final volume to 100 ml with DI water.

15.4.3 QUALITY CONTROL (QC)

As a QC requirement, the substances listed in Section 15.3.3 should be digested together with the analytical samples at a frequency of 5% or one per analytical batch (as described in Section 15.3.3).

15.5 ACID DIGESTION OF AQUEOUS SAMPLES AND EXTRACTS FOR TOTAL METALS BY GRAPHITE FURNACE SPECTROSCOPY (GrAAS)

15.5.1 INTRODUCTION

This digestion procedure is used for the preparation of aqueous samples, EPTOX and mobility-procedure (TCLP) extracts, and wastes that contain suspended solids for analysis by GrAAS for the following metals: Be, Cd, Cr, Co, Pb, Mo, Tl, and V. (Digestion and GrAAS analysis for As, Se, and Ag are different. The digestion procedure for As and Se is described in Section 15.8, and for Ag in Section 15.9.) Aqueous samples must be acidified to a pH of less than 2 with HNO_3, while nonaqueous samples should be refrigerated as soon as possible.

15.5.2 Procedure

1. Transfer 100-ml aliquot from well-mixed sample into a beaker. (The cleaned beaker should be rinsed with 1+1 HNO_3 to avoid contamination.) Cover the beaker with a ribbed watch glass.
2. Heat on a hot plate (*do not boil!*) until the sample evaporates to a volume of 5 ml (*do not dry!*).
3. Cool and add another 3 ml of concentrated HNO_3.
4. Continue heating, and add additional acid if necessary, until the sample is clear, and light in color.
5. Evaporate to about 3 ml of volume. Do not dry!
6. Add 10 ml of DI water and heat for about 10 to 15 min.
7. Wash down the walls of the beaker and the watch glass with DI water and filter if necessary.
8. Complete to 100 ml of volume with DI water.

15.5.3 Quality Control (QC)

See Section 15.3.3.

15.6 SAMPLE PREPARATION FOR ARSENIC AND SELENIUM DETERMINATION BY GRAPHITE FURNACE SPECTROSCOPY (GrAAS)

15.6.1 Introduction

This method is used in the determination of As and Se in groundwater, wastes, extracts, and soils. Aqueous samples must be acidified to a pH of less than 2 with HNO_3 at the time of collection. Nonaqueous samples must be refrigerated and analyzed as soon as possible. Furnace parameters should be employed by following the method described in Chapter 9. The calibration curve should be calculated every hour when continuous analysis is employed.

15.6.2 Procedure for Aqueous Samples

1. Transfer 100 ml of well-mixed sample into a beaker. (The cleaned beaker should be rinsed with 1+1 HNO_3 to avoid contamination.)
2. Add 2 ml of 30% H_2O_2 and sufficiently concentrated HNO_3 to result in an acid concentration of 1% (v/v).
3. Heat for 1 h at 95°C or until the volume is slightly less than 50 ml.
4. Cool and bring back the volume to 50 ml with DI water.
5. Pipet 5 ml of this digested solution into a 10-ml volumetric flask.
6. Add 1 ml of 1% nickel nitrate ($Ni(NO_3)_2$) solution, and dilute to 10 ml of volume with DI water. The sample is now ready to inject into the furnace.

15.6.3 Procedure for Solid Samples

Solid samples digestion for arsenic and selenium analysis is the same as described in Section 15.10.

15.6.4 Quality Control (QC)

See Section 15.3.3.

15.7 SAMPLE PREPARATION FOR SILVER DETERMINATION

The digested samples are analyzed by the flame atomic absorption method and approved for determination of silver in wastes, extracts, soils, and ground waters.

1. Transfer a representative aliquot of well-mixed sample to a beaker, add 3 ml concentrated HNO_3, and cover the beaker with a watch glass.
2. Place the beaker on a hot plate and cautiously evaporate to near dryness, taking care that the sample does not boil. *Do not bake!*
3. Cool the beaker and add another 3-ml portion of concentrated HNO_3 and cover again with the watch glass and return to the hot plate. Increase the temperature, so that a gentle reflux action occurs. *Note:* If the sample contains thiosulfates, this step may result in splatter of the sample as the sample approaches dryness (as with some photographic types of waste samples).
4. Continue heating and adding additional acids until the digestate is light in color and does not change in appearance with further refluxing.
5. Evaporate to dryness and cool.
6. Add a small quantity of HNO_3, so that the final dilution contains 0.5% (v/v) acid, and warm to dissolve any precipitate.
7. Wash down the wall of the beaker and the watch glass with DI water, filter if necessary, and dilute to volume. (Volume depends on the expected concentration of the metal.)

15.8 SAMPLE PREPARATION FOR ANTIMONY DETERMINATION

For antimony (Sb) determination, the recommended sample preparation method is soft digestion, as discussed in Section 15.1. *The addition of HCl to the digestate prevents furnace analysis!*

15.9 SAMPLE PREPARATION FOR MERCURY DETERMINATION (COLD-VAPOR TECHNIQUE)

15.9.1 Preparation of Aqueous Samples

1. Transfer 100 ml (or an aliquot diluted to 100 ml) to a 300-ml BOD bottle.
2. Add 5 ml of 0.5 nitrogen sulfuric acid (H_2SO_4) and 2.5 ml of concentrated HNO_3, mixing after each addition.
3. Add 15 ml of 5% potassium permanganate ($KMnO_4$) solution to each bottle. Shake and add additional portions of $KMnO_4$ until the purple color is persistent for at least 15 min. (Be sure that the same amount of permanganate is added to the accompanied standards and blanks!)
4. Add 8 ml of 5% potassium persulfate ($K_2S_2O_8$).
5. Heat for 2 h in a water bath maintained at 95°C.

6. Cool and add 6 ml of sodium chloride–hydroxylamine sulfate or hydroxylamine hydrochloride (12 g per 100 ml of DI water) to reduce excess $KMnO_4$.
7. After a delay of at least 30 sec, add 5 ml of stannous sulfate or stannous chloride suspension (10 g per 100 ml of 0.5N H_2SO_4 by stirring continuously), and immediately attach to the aeration apparatus and analyze.
8. Blanks, calibration standards, QC check standards, and spiked samples must be treated the same way.

15.9.2 Preparation of Solid and Semisolid Samples

1. Weigh three 0.2-g portions of untreated sample and place in the bottom of a 300-ml BOD bottle.
2. Add 5 ml of DI water and 5 ml of aqua regia (3:1 HCl and HNO_3).
3. Heat for 2 min in a water bath at 95°C.
4. Cool and add 50 ml of DI water and 15 ml of 5% $KMnO_4$ solution.
5. Mix thoroughly and place on the water bath at 95°C for 30 min.
6. Cool and add 6 ml of 12% of hydroxylamine hydrochloride or sodium chloride-hydroxylamine sulfate solution to reduce excess $KMnO_4$. *Caution:* Add this material under a fume hood because chlorine (Cl2) could evolve!
7. Add 55 ml of DI water and 5 ml of 10% stannous chloride or stannous sulfate solution, immediately attach to the aeration apparatus, and analyze.
8. Calibration standards and QC checks are treated in the same way.
9. For calculations to report on the dry base, the percent moisture content of the sample should be determined as described in Appendix I.

15.10 ACID DIGESTION OF SEDIMENTS, SLUDGES, AND SOILS FOR TOTAL METALS ANALYSIS

15.10.1 Introduction

This method is an acid digestion procedure used to prepare sediments, sludges, and soil samples for analysis by FAAS, GrAAS, or ICP. Samples prepared with this method may be analyzed by ICP for all listed metals and by FAAS or GrAAS as indicated below:

FAAS: Al, Ba, Be, Cd, Ca, Cr, Co, Cu, Fe, Pb, Mg, Mn, Mo, Ni, K, Na, Tl, V, Zn
GrAAS: As, Be, Cd, Cr, Co, Fe, Mo, Se, Tl, V

A representative (wet weight) sample of 1 to 2 g is digested in HNO_3 and H_2O_2. The digestate is then refluxed with either HNO_3 or HCl. Diluted HCl is used as the final reflux acid for the ICP analysis of As and Se, and the FAAS or ICP analysis of Al, Ba, Be, Ca, Cd, Cr, Co, Cu, Fe, Mo, Pb, Ni, K, Na, Tl, N, and Zn. Dilute HNO_3 is employed as the final dilution acid for GrAAS analysis of As, Be, Cd, Cr, Co, Pb, Mo, Se, Tl, and V. A separate sample should be dried for a total solids determination. All samples must be collected as discussed in Chapter 14.

15.10.2 Procedure

1. Mix sample thoroughly to achieve homogeneity.
2. Weigh 1 to 2 g of sample into a beaker. (The cleaned beaker should be rinsed with 1+1 HNO_3 before use to avoid contamination.)
3. Add 10 ml of 1+1 HNO_3 and mix with the solid sample.

4. Cover with a watch glass, heat to about 90 to 95°C, and reflux for 10 to 15 min without boiling.
5. Cool and add 2 ml of DI water and 3 ml of 30% hydrogen peroxide (H_2O_2).
6. Put back onto the hot plate for warming until effervescence stops. (Peroxide reaction is effervescent.)
7. Continue the addition of 30% H_2O_2 in 1-ml portions, and warm until effervescence stops or until the general sample appearance is unchanged. Do not add more than a total of 10 ml of 30% H_2O_2.

15.10.2.1 Sample Preparation for Flame AA and ICP Techniques, Including ICP Analysis of As and Se

1. Add 5 ml of concentrated HCl and 10 ml of DI water and reflux for 10 to 15 min without boiling.
2. Cool and dilute to 100 ml with reagent water.
3. Due to particulate in the digestate that may clog the nebulizer, filter through with Whatman No. 41 filter paper, and fill up the filtrate to 100 ml with DI water. Alternatively, centrifuge at 2000 to 3000 rpm for 10 min. The final concentration of the diluted sample is 5% (v/v) HNO_3 and 5% (v/v) HCl.

15.10.2.2 Sample Preparation for GrAAS Technique

Use this method if the sample is prepared for GrAAS technique to determine As, Be, Cd, Cr, Co, Pb, Mo, Se, Tl, and V.

1. Same as step 1 for FAAS technique (Section 15.10.2.1), but instead of HCl, add HNO_3.
2. Cover the sample with a ribbed watch glass and continue heating the acid–peroxide digestate until the volume has been reduced to approximately 5 ml.
3. After cooling, dilute to 100 ml with reagent water.
4. Particulate in the digestate should then be removed by filtration or centrifugation or by settling as mentioned above. Final concentration of the diluted sample is 5% (v/v) HNO_3.

15.10.2.3 Calculation of Results and Reporting for Solid Samples

For solid samples, the values are reported as dry-weight milligrams per kilogram or micrograms per kilogram. Therefore, the dry solid percentage of the sample must be provided. Dry weight is calculated as discussed in Appendix I.

15.10.3 Quality Control (QC)

1. For each group of samples processed, *preparation blanks* (reagent water plus reagents) should be carried through the entire sample preparation and analytical process. These blanks are useful in determining whether samples are contaminated.
2. *Duplicate samples* should be processed on a routine basis and are used to determine precision. The sample load will dictate frequency, but 20% is recommended.
3. *Spiked samples* or *standard reference materials* must be employed to determine accuracy. A spiked sample should be included with each group of samples processed and whenever a new sample matrix is analyzed.
4. The concentration of all calibration standards should be verified against a QC check sample obtained from an outside source.

15.11 DISSOLUTION PROCEDURE FOR OILS, GREASES, AND WAXES

This method is used for the preparation of samples containing oils, greases, or waxes for analysis via AAS or ICP for the following metals: Sb, Be, Cd, Cr, Cu, Fe, Mn, Ni, and V. This method is a solvent-dissolution procedure, not a digestion. The procedure can be very useful in the analysis of crude oil, but is less effective with spent or used oil with a high amount of particulate material; most particulate material is not dissolved; therefore, the analysis is not a "total metal" determination. Because particulate material is expected to contain the highest percentage of metal, oil analysis using this method will not provide an adequate estimate of total metal concentration.

15.11.1 PROCEDURE

A representative sample is dissolved in an appropriate solvent (e.g., xylene or methyl-isobutyl-ketone). Organometallic standards are prepared using the same solvent, and the samples and standards are analyzed via AAS or ICP.

15.11.2 SAMPLE COLLECTION, PRESERVATION, AND HANDLING

1. All samples must be collected using a sampling plan that addresses considerations discussed in Chapter 14.
2. Samples should be stored in an undiluted state at room temperature.
3. Samples should be processed and analyzed as soon as possible.

15.11.3 PROCEDURE

1. Weigh 2-g representative sample of the waste or extract. Separate and weigh the phases if more than one phase is present.
2. Weigh an aliquot of the organic phase and dilute it in the appropriate solvent. Warming facilitates the subsampling of crude type oils, greases, and wax-type wastes. Xylene is usually the preferred solvent for long-chain hydrocarbons and for most analyses performed via ICP. Long-chain hydrocarbons require a minimum 1:10 dilution, and lighter oils require 1:5 dilutions if low detection limits are required.
3. All metals must be analyzed by the standard addition method (Section 7.7.1.1).
4. Organometallic standards are prepared by using the same solvent. Diluted samples and diluted organometallic standards are unstable. Once standards and samples are diluted, they should be analyzed as soon as possible.

Organometallic standards are available from Conostan Division, Conoco Specialty Products, Inc., P.O. Box 1267, Ponca City, OK 74601, and U.S. Department of Commerce, National Bureau of Standards, Washington, D.C. 20234.

15.12 SAMPLE PREPARATION FOR HEXAVALENT CHROMIUM (CHELATION/EXTRACTION)

This method is approved for determining the concentration of dissolved hexavalent chromium in EPTOX-characteristic extracts and groundwaters. It may also be applicable to certain domestic and industrial wastes, provided that no interfering substances are present (high concentrations of other metals may interfere). The procedure is applicable in the range of 0.1 to 25 μg/l Cr^{6+} concentrations.

To retard the chemical activity of Cr^{6+}, the samples and extracts should be stored at 4°C until sample preparation. Chelation and extraction should be carried out as soon as possible.

15.12.1 Reagents

15.12.1.1 Potassium Dichromate Standard Solution I

$$1 \text{ ml} = 100 \text{ µg Cr}$$

Use 0.2829 g of pure dried potassium dichromate ($K_2Cr_2O_7$). Dissolve and dilute to 1 liter with analyte-free water.

15.12.1.2 Potassium Dichromate Standard Solution II

$$1 \text{ ml} = 10 \text{ µg Cr}$$

Use 10 ml of potassium dichromate standard solution I. Dilute to 100 ml with analyte-free water.

15.12.1.3 Potassium Dichromate Standard Solution III

$$1 \text{ ml} = 0.10 \text{ µg Cr}$$

Use 10 ml of potassium dichromate standard solution II. Dilute to 1 liter with analyte-free water.

15.12.1.4 Ammonium Pyrolidine Dithiocarbamate (APDC)

Dissolve 1.0 g of APDC in DI water and dilute to 100 ml. *Prepare fresh daily*!

15.12.1.5 Bromphenol Blue Indicator Solution

Dissolve 0.1 g of bromphenol blue in 100 ml of 50% ethanol.

15.12.1.6 Methyl Isobutyl Ketone (MIBK) Analytical Reagent Grade

15.12.1.7 Sodium Hydroxide Solution, 1M

Dissolve 40 g of sodium hydroxide (NaOH) in DI water and dilute to 1 liter. Because the reaction is highly exothermic, the solution should be prepared with extreme care under a laboratory hood!

15.12.1.8 Sulfuric acid, 0.12M

Slowly add 6.5 ml of sulfuric acid (H_2SO_4) to DI water and dilute to 1 liter.

15.12.2 Chelation and Extraction

1. Prepare a blank and a sufficient number of standards (minimum of 3) at a volume of 100 ml.
2. Pipet a volume of sample of less than 2.5 µg (maximum 100 ml) into a 200-ml volumetric flask. If the sample is less than 100 ml, adjust the volume to 100 ml with analyte-free water.
3. Add two drops of bromphenol blue indicator solution to samples, blank, and standards.
4. Adjust the pH by adding drops of 1M NaOH solution until the blue color persists.
5. Add 0.12M H_2SO_4 until the blue color disappears. Then add 2 ml of sulfuric acid in excess. The pH in this point should be 2.4.

6. Add 5.0 ml of APDC solution and mix. The pH should be approximately 2.8.
7. Add 10 ml of MIBK and shake vigorously for 3 min.
8. Allow the layers to separate and add DI water until the ketone layer is completely in the neck of the flask. Determine the chromium content by aspirating the ketone layer into the flame of an atomic absorption spectrophotometer. At the same time chelate and extract also a calibration verification standard (CVS) and spike duplicate. (Run a spike duplicate sample for every ten samples.)

15.13 EXTRACTION PROCEDURE (EP) TOXICITY

15.13.1 INTRODUCTION

Extraction procedure (EP) toxicity (EPTOX test) is designed to simulate the leaching a waste undergoes when disposed of in a sanitary landfill. In this laboratory test, a representative sample of a waste is extracted with distilled water maintained at a pH of 5, using acetic acid. The extract obtained from the EP is then analyzed to determine if any of the thresholds established for the eight elements (arsenic, barium, cadmium, chromium, lead, mercury, selenium, and silver), four pesticides (Endrin, Lindane, Methoxychlor, and Toxaphene), and two herbicides (2,4,5-trichloropnenoxypropionix acid and 2,4-dichlorophenoxyacetic acid) have been exceeded. If the extract contains any one of the above substances in an amount equal to or exceeding the levels specified in 40 CFR, 261.24, the waste possesses the characteristic of EP toxicity and is a hazardous waste. Maximum concentration of contaminants for EP toxicity characteristics are discussed in Section 4.7.4 and are listed in Table 4.5. The EP method is applicable to liquid, solid, and multiphase samples.

15.13.2 SAMPLE COLLECTION, PRESERVATION, AND HANDLING

All samples must be collected using a sampling plan that addresses the considerations discussed in Chapter 14. Preservatives must not be added to samples. Samples can be refrigerated if refrigeration will not affect the integrity of the sample.

15.13.3 APPARATUS AND MATERIALS

15.13.3.1 Extractor

The extractor must provide sufficient agitation to the mixture to prevent stratification of the sample and extraction fluid and ensure that all sample surfaces are continuously brought into contact with the well-mixed extraction fluid. An example of a suitable extractor for this method is shown in Figure 15.1. Extractors are available from Associated Designs & Manufacturing Co., Alexandria, VA; Glas-Col Apparatus Co., Terre Haute, IN; Millipore, Bedford, MA; and Rexnard, Milwaukee, WI.

15.13.3.2 pH Meter

The pH meter must be accurate to 0.05 pH units with temperature compensation.

15.13.3.3 Filter Holder

The filter holder should be capable of supporting a 0.45-μm filter membrane and withstanding the pressure needed to accomplish separation. Suitable filter holders range from simple vacuum units to relatively complex systems that can exert up to 5.3 kg/cm^3 (75 psi) of pressure. The filter holders used depend on the properties of the mixture to be filtered.

15.13.3.4 Filter Membrane

Filter membranes suitable for conducting the required filtration are fabricated from a material that is not physically changed by the waste material to be filtered and does not absorb or leach the chemicals from the waste. In cases of doubt about physical effects on the filter, contact the filter manufacturer to determine if the membrane or the prefilter is adversely affected by the particular waste. If no information is available, submerge the filter in the waste's liquid phase; a filter that undergoes visible physical change after 48 h (e.g., if it curls, dissolves, shrinks, or swells) is unsuitable for use.

To test the filter for absorption or leaching, take the following steps:

1. Prepare a standard solution of the chemical species of interest.
2. Determine the concentration of the chemical in the standard.
3. Filter the standard and reanalyze. If the concentration of the filtrate differs from the original standard, the filter membrane leaches or absorbs one or more of the chemical species and it is not usable for this test method.

15.13.4 Reagents

15.13.4.1 Acetic Acid (0.5N)

Add 57 ml of concentrated glacial acetic acid (17.5N) to 1000 ml of laboratory water and dilute to 2 liters. The glacial acetic acid should be of high purity and monitored for impurities.

15.13.5 Procedure

15.13.5.1 Liquid or Multiphase Samples

1. Weigh filter membrane and prefilter to ±0.01 g. Handle membrane and prefilters with blunt, curved-tip forceps or vacuum tweezers, or by applying suction with a pipet.
2. Assemble filter holder, membranes, and prefilters following the manufacturer's instructions. Place the 0.45-µm membranes on the support screen and add prefilters in ascending order of pore size. Do not prewet filter membrane.
3. Weigh out a representative subsample of the waste (100 g minimum).
4. Allow slurries to stand to permit the solid phase to settle. Wastes that settle slowly may be centrifuged prior to filtration.
5. Wet the filter with a small portion of the liquid phase of the sample or with the extraction mixture. Transfer the remaining material to the filter holder and apply vacuum or gentle pressure (10–15 psi) until all liquid passes through the filter. Stop filtration when air or pressurizing gas moves through the membrane. If this point is not reached under vacuum or gentle pressure, slowly increase the pressure in 10-psi increments to 75 psi. Halt filtration when liquid flow stops. This liquid will constitute part or all of the extract. The liquid should be refrigerated until analysis.
6. Remove the solid phase and filter and, while not allowing them to dry, weigh to 0.01 g.

$$\text{Wet weight of the residue (g)} = B - A \tag{15.2}$$

where

B = weight of the solid phase and filter (g).
A = weight of the filters (g).

7. If the solid content is less than 0.5% of the waste, discard the solid. The extract is prepared

and analyzed. Determine the exact percentage of solids by drying the filter and residue at 80°C and then calculate as follows:

$$\% \text{ solids} = (C - A) / D \times 100 \tag{15.3}$$

where

C = dry weight of the filter and residue.
A = weight of the filters.
D = initial weight of the waste.

Do not extract solid material that has been dried at 80°C!

8. If the sample contains less than 0.5% solids, use the wet weight of the solid phase obtained in step 6 to calculate the amount of liquid and acid employed for extraction. If the waste does not contain free liquids, 100 g of the material should be subjected to the extraction procedure.
9. Place the appropriate amount of material into the extractor (see step 8).
10. Add reagent-grade water at 16 times the weight of material. Begin agitation and measure the pH of the solution in the extractor. If the pH is more than 5.0, the pH of the solution should be decreased to 5.0±0.2 by adding 0.5N acetic acid. If the pH is less than 5.0, no acetic acid should be added. Monitor the pH during extraction and, if the pH rises above 5.2, more 0.5N acetic acid should be added to bring the pH down to 5.0±0.2. The pH of the solution should be checked and adjusted at 15-, 30-, and 60-min intervals. This adjustment procedure should be continued for at least 6 h. However, in no event should the aggregate amount of acid added to the solution exceed 4 ml of acid per gram of solid. The mixture should be agitated for 24 h and maintained at 22 to 40°C (68–104°F) during this time.
11. If at the end of the extraction period the pH of the solution is not below 5.2 pH and the maximum amount of acid has not been added, adjust the pH of the solution to 5.0±0.2 and continue the extraction for an additional 4 h, during which the pH should be checked at 1-h intervals.
12. At the end of the extraction period, calculate the amount of the reagent-grade water added to the extractor:

$$V = [(20)(W) - 16(W)] - Ac \tag{15.4}$$

where

V = ml of reagent-grade water added.
W = weight of solid extracted.
Ac = ml of 0.5N acetic acid added during extraction.

13. Allow the extracted material to stand to permit the solid phase to settle. Wastes that are slow to settle may be centrifuged prior to filtration.
14. Set up filtration apparatus as in step 2. Wet the filter with the extraction mixture. Transfer material to the filter holder and apply vacuum or gentle pressure (10–15 psi) until all liquid passes through the filter. Stop filtration when air or pressurizing gas moves through the membrane. Slowly increase the pressure in 10-psi increments to 75 psi. Halt filtration when liquid flow stops.
15. Combine filtrate with the remaining liquid from step 5 or the waste itself if it contains less than 0.5% of solid (step 7).
16. The extract is prepared and analyzed. If the extract includes two phases, concentration of

contaminants is determined by using a simple weighted average. For example, assume that the extract contains 50 ml of oil and 1000 ml of aqueous phase. Determine the contaminant concentration for each phase and calculate the final concentration according to the following formula:

$$[(50)(\text{concentrate in oil})] + [(1000)(\text{concentrate in aqueous phase})]/1050 \qquad (15.5)$$

15.13.5.2 Solid Samples

If a representative sample of the waste contains more than 0.5% of solids, the solid phase of the sample is ground to pass through a 0.5-mm sieve and extracted with reagent-grade water maintained at a pH of 5.0±0.2, with acetic acid. Follow the procedure in Section 15.3.5.1, steps 8 through 16.

15.13.6 Quality Control (QC)

All QC measures described in Chapter 13 should be followed. Employ a minimum of one blank per sample batch to determine if contamination effects are occurring.

15.14 EXTRACTION PROCEDURE FOR OILY WASTES

15.14.1 Introduction

This method is used to determine the mobile metal concentration in oily wastes. The method is applicable to EPA-defined separator sludges, rag oil emulsions, and other oil wastes derived from petroleum refining.

1. The sample is separated into solid and liquid components by filtration.
2. The solid phase is placed in a Soxhlet extractor charged with tetrahydrofuran (THF) and extracted.
3. When the THF is recovered, the extractor is then charged with toluene and the sample is re-extracted.
4. The EP extraction method (Section 15.13) is run on the dry solid residue.
5. The original liquid, combined extracts, and EP leachate are analyzed for the EP toxicity metals (As, Ba, Cd, Cr, Pb, Hg, Se, and Ag).

15.14.2 Interferences

Matrix interferences should be extracted from the sample. The extent of these interferences varies considerably from waste to waste, depending on the nature and diversity of the particular refinery waste being analyzed.

15.14.3 Apparatus and Materials

Soxhlet extraction apparatus (see Appendix K)
Vacuum pump or other source of vacuum
Buchner funnel 12
Electric heating mantle
Paper extraction thimble
Filter paper
Evaporating flask, 250 ml
Apparatus and materials listed in Section 15.3.3

15.14.4 REAGENTS

15.14.4.1 Tetrahydrofurane, ACS Reagent Grade

15.14.4.2 Toluene, ACS Reagent Grade

15.14.4.3 Acetic Acid, 0.5N

Preparation is discussed in Section 15.13.4.1.

15.14.5 SAMPLING

Samples must be collected in glass containers having a total volume of at least 150 ml. No solid material should interfere with sealing the sample container. Sampling devices should be wiped clean with paper towels or absorbent cloth, rinsed with a small amount of hexane followed by acetone rinse, and dried between samples. Alternatively, samples can be taken with disposable sampling devices in beakers.

15.14.6 PROCEDURE

1. Separate the sample (minimum of 100 g) into solid and liquid components using the filtration steps in Section 15.13.5.1, steps 1 through 6.
2. Determine the quantity of liquid (milliliters) and the concentration of the species of interest in the liquid phase (milligrams per liter) using appropriate analytical methods.
3. Place the solid phase into a Soxhlet extractor, charge the concentration flask with 300 ml of tetrahydrofuran, and extract for 3 h.
4. Remove the flask containing tetrahydrofuran and replace it with one containing toluene.
5. Extract the solid for a second time for 3 h with toluene.
6. Combine the tetrahydrofurane and the toluene extracts.
7. Determine the quantity of liquid (milliliters) and the concentration of the species of interest in the combined extract (milligrams per liter).
8. Take the solid material remaining in the Soxhlet thimble and dry it at 100°C for 30 min.
9. Run the EP procedure (Section 15.13) on the dried solid.
10. Calculate the mobile metal concentration in milligrams per liter using the following formula:

$$1000 \times (Q_1 + Q_2 + Q_3) / (L_2 - L_1) \quad (15.6)$$

where

Q_1 = amount of metal in initial liquid phase of sample (amount of liquid × concentration of metal in milligrams; see step 2).

Q_2 = amount of metal in combined organic extracts of sample (mg) (see step 7).

Q_3 = amount of metal in EP extract of solid (amount of extract × concentration of metal in milligrams; see step 9).

L_1 = amount of initial liquid in milligrams (see step 2).

L_2 = amount of liquid in extraction procedure (20 × weight of dried solid).

15.14.7 Quality Control (QC)

Laboratory duplicates should be analyzed to validate analytical precision. QC samples should be carried through all stages of sample preparation and measurement and analyzed to validate the sensitivity and accuracy of the analysis.

15.15 DOCUMENTATION DURING SAMPLE PREPARATION

All numerical data relating to preparation processes and that require further calculations should be briefly documented (volume or weight of the samples used for preparation, dilution factors, concentrations, reagent preparations, pH checks, etc.). Each container carrying the pretreated sample should be properly identified with sample ID, date of preparation, and all pertinent information (volume, weight, dilution, concentration, etc.) related to the preparation procedure. A sample preparation log sheet is illustrated in Table 15.3.

15.16 DISPOSAL OF SAMPLES, DIGESTATES, EXTRACTS, AND OTHER WASTES

Samples and pretreated samples in the form of extracts and digestates should be stored properly until the end of recommended holding times. Refrigerators and separate storage areas must be designated for this purpose. Regular water samples may be disposed of into the sewer system, with the exception of hazardous wastes. Hazardous laboratory wastes are stored in special containers until collected and transported by a professional waste disposal company. Such vessels should be marked clearly according to the nature of the waste, such as "acid wastes," "organic solvents," "mercury waste," and so on. Disposal of samples and treated products should be documented as shown in Table 15.4.

TABLE 15.3
Sample Preparation Log Sheet (per Analyte Group)

Sample I.D. Number	Matrix	Test for	Method No. Analysis	Method No. Prep.	Date Sample Rec'd.	Date Sample Prep.	Sample Size		Final Volume (ml)	Signature
							(ml)	(g)		

TABLE 15.4
Disposal Log Form for Digestates and Extracts

Sample ID	Date Sample		Mode of		Date of dispo	Sign
	rec	prep	prep	dispo		

Sample ID = sample identification number; rec = received; prep = prepared; dispo = disposed; Sign = signature of logger.

Mode of preparation:
D = digested; E = extracted; Dist = distilled.

Mode of disposal:	Storage:
Ac.W. = acid waste container	Designated area
B.W. = basic waste container	Designated area
Org.S. = organic solvent container	"Flammable" cabinet
Hg.W. = mercury waste container	Designated area
CN = cyanide waste container	Designated area

16 Converting Raw Data into Reportable Form

Raw data are generated by the analytical process, which includes QC checks. *Reportable data* or final results are generated from raw data by mathematical or statistical calculations. Final results may be produced by direct readings from the instrument or calculations based on readings or instrument output. Before starting calculations, make sure that all readings or outputs are correct and the selected formulas used in calculations are appropriate. Formulas and calculations should be recorded in the laboratory notebook or on the work sheet. In all records, calculations are entered in ink and mistakes are never erased; just cross through the error with a single line, and date and sign. Generation of raw data and all related calculations and complete record keeping are the responsibility of the analyst.

16.1 RESPONSIBILITIES OF THE ANALYST

The following list contains the duties of the analyst from the selection of proper samples for a particular analysis through analytical performance, including QC checks, documentation, calculations, recognition and correction of problems, and giving the accurate, defendable result.

- Analytical work should be planned and organized so that laboratory time is used efficiently.
- The analyst should be familiar with the test method described in the laboratory SOP (standard operating procedure).
- If samples require pretreatment (drying, digestion, extraction, etc.), start as soon as possible because pretreatments are usually time-consuming procedures. If sample preparation is the duty of another laboratory section, the analyst must be certain that sample preparation was correct; collect all information required to calculate the final results (volume or weight of the sample used for treatment, final volume after treatment, moisture of soil, dilution or concentration, etc.). Carefully select the accompanying pretreated QC check samples (duplicates, spiked samples, laboratory control samples, preparation blank, equipment blanks, trip blanks, etc.).
- When samples are stored in a refrigerator, remove the bottles prior to starting the analysis to allow samples to warm up to room temperature.
- Carefully check sample label for identification and scrupulously select samples according to proper bottle type and preservation.
- Collect and check stocks, standards, QC check standards, and respective preparation dates. When the expiration date indicates, discard these solutions and prepare new ones, and be sure that all related preparation log forms are completed. When reagents are stored in a refrigerator, allow them to warm up to room temperature before use.
- Standardize solution when applicable, and document on the designated form.
- Collect the appropriate work form, according to test and matrix, and prepare for starting.

- Collect the necessary glassware and check cleanliness and appropriate cleaning procedure. Improperly cleaned glassware will ruin the test. Mark glassware as needed and organize.
- Switch instrument on and let it warm up.
- Prepare calibration standards by selecting the proper range as indicated by the analytical method, and document each step properly.
- Prepare calibration curve and verify according to initial and continuing calibration criteria as detailed in Sections 6.6 and 13.6. Be sure that calibration curve and linear regression calculations are documented, and keep records (manual forms, strip charts, or tabular printouts).
- When a previously approved calibration curve is available, perform a continuing calibration check, and indicate acceptance or rejection of the calibration curve. Upon acceptance of the existing curve, sample analysis may begin. Upon rejection of the existing curve, discard the former calibration and prepare a fresh one.
- Measure the precision and accuracy of the analytical performance by required QC checks as described in the analytical methodology and laboratory SOP. When the QC data are not acceptable, take corrective action and solve the problem. Never report results with a doubtful QC check!
- If the analyte of interest shows a higher value than the concentration of the highest calibration standard, dilute the sample for an accurate reading. The dilution technique and proper calculation of the final value should be documented on the working paper or in the laboratory notebook.
- Turbidity, color, or other interferences should be corrected according to the analytical methodology.
- Run the analysis by following the approved method and incorporate all required QC checks.
- The analyst should be knowledgeable enough to recognize problems, initiate and conduct corrective actions, and keep all documentation related to the analysis clean and in order to be ready for inspection at any time.
- The analyst should be able to protect and defend all raw data as well as reported results.
- At the end of the analysis, switch the instrument off; collect analytical wastes in designated containers (if applicable); collect dirty glassware; and transfer stocks, standards, reagents, and samples to appropriate storage areas or to designated refrigerators.
- Collect all documentation, calculate results and QC checks, plot the accuracy and precision data on QC charts, and transfer figures to the tabulated summary log form. In the case of deviation from the confidence limits, the analyst should identify the cause of the failure and correct it, or report to the laboratory supervisor and QA (quality assurance) manager for further assistance. All sample data determined at the time the "out-of-control" condition occurred must be labeled as "suspect data" and reanalyzed after the problem is resolved.
- If all QC data agree with the analytical values, the analyst transfers the result to the analytical report summary log paper or enters it into the computer. Collect all documentation to be prepared for further questions or checking.

16.2 CALCULATIONS FOR FINAL VALUES

Raw data produced during the analytical process should be converted to appropriate measurement units in reportable, final results. By using the appropriate formula, the analyst calculates final values and registers them on the laboratory "result sheet" or enters them into the laboratory computer system, to be ready for further supervision and checks.

16.2.1 Dilution and Concentration

Dilution applied during the test should be recorded; the final result is obtained by multiplying the measured value by the *dilution factor*. The concentration technique used during sample preparation should be documented; the analytical value divided by the *concentration factor* produces the final result.

For instance, assume that a water sample was analyzed for Pb concentration, and that the appropriate reading should be diluted five times. If the result of the diluted sample was 0.2 mg/l, the final result will be $0.2 \times 5 = 1.0$ mg/l Pb in the sample. Next, assume that during the digestion technique, a water sample analyzed for chromium was concentrated by a factor of 10 (original 250-ml sample was cooked down to a 25-ml final volume) because of the low concentration of the metal in the sample. The reading of this concentrated sample was 0.06 mg/l, so the final result is $0.06/10 = 0.006$ mg/l or 6 μg/l.

16.2.2 Calculations for Solids, Moisture, and Ash

Complete and clear description of the reported analyte in appropriate measurement units is an important and critical part of generating a correct analytical result.

16.2.2.1 Solids

Determination of solids, moisture, or ash may be expressed as milligrams per liter or as percentage. Solids as total solids (TS), suspended solids (SS), and total dissolved solids (TDS) are reported in milligrams per liter; moisture and ash results are expressed in percentages. Respective formulas follow:

$$\text{mg/l solids (TS, TSS, TDS)} = [(A - B) \times 1000]/\text{ml sample} \tag{16.1}$$

$$\text{\% solids (TS, TSS, TDS)} = [(A - B) \times 100]/\text{ml sample} \tag{16.2}$$

$$\text{\% solids (TS, TSS, TDS)} = (\text{mg/l})/10{,}000 \tag{16.3}$$

where A = weight of the dish and residue dried at 105°C in grams (for TS), weight of dish and residue dried at 180°C in grams (for TDS), or weight of filter and residue dried at 105°C in grams (for TSS); B = weight of dish in grams.

16.2.2.2 Moisture

Moisture of any solid is determined by drying a known-quantity aliquot of sample at 103 to 105°C in a laboratory oven. After the dried sample is cooled in a desiccator, weigh it, and calculate its moisture percentage by using the following formula:

$$\text{\% moisture} = (\text{g of solid} - \text{g of dried solid}) \times 100 / \text{g of solid} \tag{16.4}$$

16.2.2.3 Ash

The ash content of any solid is determined by igniting a known-quantity aliquot of the well-mixed sample at 1000°C in a muffle furnace. By knowing the weight of the original sample and the weight of the remaining ash (ignited residue), and using the following formula, the percentage of ash content of the sample is determined and calculated:

$$\text{\% ash} = (\text{g of ash} \times 100) / \text{g of original sample} \tag{16.5}$$

16.2.2.4 Percentage Composition of Solid Sample

Determine the percentage of moisture at 105°C, as described above and using Formula 16.4. The residue from the moisture determination is ignited at 1000°C. The difference between the weight of the ignited residue (at 1000°C) and the dried solid (at 105°C) will give the volatile and organic compounds in the sample. It is expressed also as a percentage. The remaining residue is considered as ash (inorganic) and is calculated as a percentage.

The sum of the percentage of moisture, percentage of volatile, and percentage of ash values should be 100%. If the sum of the calculated values is not 100%, and the checked calculations are correct, the entire process should be repeated.

16.2.2.5 Calculating the Results of Solid Matrices

For solid matrices, the report is expressed as ppm (parts per million, or milligrams per kilogram) or ppb (parts per billion, or micrograms per kilogram). The report should state that the reported value is calculated on the "wet base" (also called "as is basis") or on the "dry base." See discussion and calculations of these parameters in Appendix I.

16.2.3 Conversion of Milligrams per Liter and Milliequivalents per Liter

16.2.3.1 Milligrams per Liter Converted to Milliequivalents per Liter

When the analyzed milligrams-per-liter value of the ion is converted to millequivalents per liter, the first milligram value is multiplied by the factor derived from the ratio of the valence and the atomic or formula weight:

$$\text{meq/l} = \text{mg/l} \times (v/\text{at. wt.}) \quad (16.6)$$

For example, Na has a valence of 1 and atomic weight of 22.9897; the factor is 1/22.9897 = 0.04340. For a sulfate (e.g., SO_4^{2-}), the valence is 2 and the atomic weight is 96.0636; therefore, the factor is 2/96.0636 = 0.0282.

16.2.3.2 Milliequivalents per Liter Converted to Milligrams per Liter

When the milliequivalents-per-liter value is converted to milligrams per liter, the milliequivalent value is multiplied by a factor derived from the ratio of the atomic or formula weight divided by the valence:

$$\text{mg/l} = \text{meq/l} \times (\text{at. wt.}/v) \quad (16.7)$$

Using the same chemicals as in Section 15.2.3.1, when milliequivalents for Na are converted to milligrams, the factor will be 22.9897/1 = 22.9897. For sulfates, the factor is 96.0636/2 = 48.03. Table 16.1 contains two-way conversion factors for milligrams and milliequivalents per liter.

16.2.4 Conversion of ppm (w/v) to mg/m³ and Vice Versa

The concentration of analytical values in air samples is generally expressed as milligrams per cubic meter. Conversion of parts per million to milligrams per cubic meter and vice versa at standard temperature and pressure (STP; 273 K and 1 atm or 760 torr) follow:

$$1 \text{ ppm} = (\text{mol wt}/22.4) = \text{mg/m}^3 \quad (16.8)$$

TABLE 16.1
Two-Way Conversion Factors for mg/l to meq/l and Vice Versa

Cations	Factor mg/l × = meq/l	Factor meq/l × = mg/l	Anions	Factor mg/l × = meq/l	Factor meq/l × = mg/l
Al^{3+}	0.1112	8.994	BO_2^-	0.02336	42.81
B^{3+}	0.2775	3.603	Br	0.01257	79.90
Ba^{2+}	0.01456	68.67	Cl^-	0.02821	35.45
Ca^{2+}	0.04990	20.04	CO_3^{2-}	0.03333	30.00
Cr^{3+}	0.05770	17.33	CrO_4^{2-}	0.01724	58.00
Cu^{2+}	0.03147	31.77	F^-	0.05264	19.0
Fe^{2+}	0.03581	27.92	HCO_3^-	0.01639	61.02
Fe^{3+}	0.05372	18.62	HPO_4^{3+}	0.02084	47.99
H^+	0.9922	1.008	H_2PO^{4-}	0.01031	96.99
K^+	0.02558	39.10	HS	0.03024	33.07
Li^+	0.1441	6.941	HSO_3^-	0.01234	81.07
Mg^{2+}	0.08229	12.15	HSO_4^-	0.01030	97.07
Mn^{2+}	0.03640	27.47	I^-	0.00788	126.9
Mn^{4+}	0.07281	13.73	NO_2^-	0.02174	46.01
Na^+	0.04350	22.29	NO_3^-	0.01613	62.0
NH_4^+	0.05544	18.04	OH^-	0.05880	17.01
Pb^{2+}	0.009653	103.6	PO_4^{3-}	0.03159	31.66
Sr^{2+}	0.02283	43.81	S^{2-}	0.05283	16.03
Zn^{2+}	0.03059	32.69	SO_4^{2-}	0.02082	48.03

Note: mg/l = milligrams/liter; meq/l = milliequivalent/liter; meq/l = mg/l × factor; factor = ionic charge/atomic or formula weight (Cl_f = 1/35.45 = 0.02821); mg/l = meq/l × factor; factor = atomic or formula weight/ionic charge (Cl_f = 35.45/1 = 35.45).

$$1 \text{ mg/m}^3 = (22.4/\text{mol wt}) = \text{ppm} \tag{16.9}$$

where

22.4 = molar volume at STP = 22.4 l.
mol wt = molecular weight.

For example, 1 ppm of cyclohexane at STP = 84/22.4 = 3.75 mg/m^3, and 2.93 mg/m^3 of benzene at STP = 2.93 (22.4/78) = 0.85 ppm.

16.2.4.1 Conversion of ppm to mg/m^3 at Nonstandard Temperature and Pressure

This conversion is done in two steps. First, calculate the molar volume and then calculate the reported milligrams per cubic meter.

16.2.4.2 Determination of Molar Volume at Given Temperature and Pressure Conditions

Molar volume can be calculated with the use of the combined gas law or the ideal gas law.

16.2.4.2.1 Calculation Using Combined Gas Law

1. Convert temperature to Kelvin (°C + 273 = K).
2. Convert pressure to atm (1 atm = 760 torr).
3. Calculate the molar volume using the combined gas law:

$$P_1V_1/T_1 = P_2V_2/T_2 \tag{16.10}$$

where

P_1 = initial pressure at STP = 1 atm.
V_1 = initial volume at STP = 22.4 l.
T_1 = initial temperature at STP = 0°C = 273 K.
P_2 = final pressure.
V_2 = final volume.
T_2 = final temperature.

16.2.4.2.2 Calculation Using Ideal Gas Law

1. Convert temperature to K (°C + 273 = K).
2. Convert pressure to atm if necessary (760 torr = 1 atm).
3. Calculate the molar volume by using the ideal gas law:

$$PV = nRT \text{ and } V = nRT/P \tag{16.11}$$

where

P = pressure.
V = volume.
n = number of moles of analyte.
R = ideal gas constant = 0.082 l atm/mol K.
T = temperature.

Assume that ppm and molar volume are known. Convert to milligrams per cubic meter by using the following formula:

$$\text{mg/m}^3 = \text{ppm} \times (\text{mol wt/molar volume}) \tag{16.12}$$

where ppm = mg/l and molar volume = l.

Assume that air sampling of SO_3 is performed at an altitude where the temperature is 7°C and the pressure is 725 torr. Calculate the molar volume of the compound:

T = 7°C = 7 + 273 = 280 K
P = 725 torr = 725/760 = 0.954
V = nRT/P
V = [(0.082) × (280)]/0.954
V = 24.6 l

The result of the analysis is 0.5 ppm of SO_3 in the sample. Convert the ppm value to milligrams per cubic meter:

$$\text{mg/m}^3 = (\text{mol wt/molar volume}) \times \text{ppm}$$
$$\text{mg/m}^3 = (80/24.6) \times 0.5 = 1.62$$

16.2.5 Significant Figures

The number of significant figures refers to the number of digits reported for the value of a measured or calculated quantity indicating the accuracy and precision of the value. The general rule is to report only figures that are justified by the accuracy of the analytical method.

The number of significant figures is said to be the number of digits remaining when the data are rounded. For example, when a measured value is 10.6 mg/l, the analyst should be quite certain of the 10, but may be uncertain as to whether the 0.6 should be 0.5 or 0.7, or even 0.4 or 0.8, because of the unavoidable uncertainty in the analytical procedure. The reported value of 10.6 mg/l has three significant figures. Rules for counting significant figures are summarized below.

16.2.5.1 Nonzero Integers

Always count as significant figures.

16.2.5.2 Leading Zeros

Leading zeros do not count as significant figures. The zeros simply indicate the position of the decimal point. For example, 0.00065 has only two significant figures.

16.2.5.3 Captive Zeros

Captive zeros are located between nonzero digits and always count as significant figures. For example, 1.034 has four significant figures.

16.2.5.4 Trailing Zeros

Trailing zeros are located at the right end of a number. They are significant only if the number contains a decimal point. For example, when reporting 4600 ppb, use 4.6 ppm instead, assuming the two significant figures are reliable.

16.2.5.5 Significant Figures in Multiplication and Division

The number of significant figures in the result is the same as the number in the least precise measurement used in the calculation. For example, in $1.6 \times 5.24 = 8.38$, round to 8.4 because 1.6 has two significant figures.

16.2.5.6 Significant Figures in Addition and Subtraction

The result should have the same number of decimal places as the least precise measurement used in the calculation. For example, in $15.62 + 12.5 + 20.4 = 48.52$, the correct result is 48.5 because 12.5 has only one decimal place.

16.2.6 Rounding Data

The following rules should be used in rounding data:

- If the digit to be removed is less than 5, the preceding digit stays the same. For example, 2.33 is rounded to 2.3.
- If the digit to be removed is greater than 5, the preceding digit is increased by 1. For example, 2.36 is rounded to 2.4.
- If the digit to be removed is 5, round off the preceding digit to the nearest even number. For example, 2.15 becomes 2.2 and 2.35 becomes 2.4.

16.2.7 Exponential Notation

Exponential notation (e.g., 1.15×10^3 or 11.5×10^2) is an acceptable way to express both the number and the significant figures. This form is not used in analytical reports, because it would be inconsistent with the normal expression of the results and might be confusing. The general rule is that the report should contain only figures that are justified by accuracy requirements of the analytical method.

16.3 RECORDS FOR RAW AND CALCULATED DATA

All documentation related to the raw data and the reported results should be in a bound form, easy to identify, and ready for inspection at all times.

16.3.1 Field and Laboratory Notebook

A bound notebook is preferred to a loose-leaf one. The size of the notebook should be comfortable to work with. Choose one that fits easily on the balance and working table, leaving space for glassware, reagents, and other materials necessary for the analysis. Large notebooks, especially when open, tend to get in the way and can cause spills and other problems. The first two pages should be reserved for an index, in which page numbers and titles are listed. If the pages of the notebook are not numbered, number each page in the upper right corner. Use the title heading to identify data on each page and date it. Write all data in ink to avoid smearing and erasures. All raw data and observations should be written in the notebook. Nothing should be erased; cross out errors and unnecessary information with a single line, followed by an explanatory notation, initials, and date.

16.3.2 Work Sheets

Work sheets are drawn up for every analysis. The analyst must complete all parts of the work sheet that contain the information relating to the sample (ID number, sample type, matrix, source, etc.), analytical method or method number with reference, MDL (method detection limit), instrumentation, analytical and QC raw data, calculated values, and proper measurement unit. The work sheet should contain complete and clear information on all elements necessary for validation of the analytical process. Tables 16.2, 16.3, and 16.4 provide general examples of forms used in recording various analytical performances on diverse matrices for metals analysis. Documentation forms vary from one institution or laboratory to another, according to requirements and preferences described and approved in respective QA/QC programs.

16.3.3 Other Documentation

Documentation should be stored in bound form, with the date started, date ended, ID number started and ended, document title, analytical group, and parameter prominent on covers. The storage area should be large enough to keep records on shelves or in storage cabinets and easy to find. Strip charts, AA calibration curves, and raw data collected should be stored in file boxes and identified as suggested above.

16.3.4 Documents To Be Saved

- Charts and other instrument-response readout records
- Calibration curves
- QC charts
- Target limits
- Method detection limits (MDLs)

TABLE 16.2
Working Paper for Spectrophotometric Analysis (Water Samples)

Parameter__________**Method No.**______________**Reference** ______________________________
Method detection limit (MDL) ____
Model of spectrophotometer__________________**Wavelength** ______________________________
Correlation coefficient of calibration curve ______________________________________
Analyzed by ______________________________________**Date** ______________________________

ID No.	Sample Identification	Sample type	Sample (ml)	Absorbance	Dilution	Result (mg/l)	Remarks
	Blank						
	Standard 1						
	2						
	3						
	4						
	CCS						
	CVS						
	Duplicate						
	Spike						
	Blank						

INFORMATION RELATED TO QC CHECKS
% Recovery of Continuing Calibration Standard (CCS)____________
True Value of Calibration Verification Standard (CVS) ____________
% Recovery of CVS ____________________**ID of Duplicate Sample**______________________________
ID No. of Spiked Sample ____________________**Sample ml Spiked**______________________________
Concentration of Spike Stock Solution__________**Added Spike Value**______________________________
Precision as Relative Percent Deviation (RPD) ________________
Accuracy as % Recovery (%*R*) ____________________________
% Recovery of Spike (%R_{sp})____________________________

Approval of Supervisor__________________________________**Date**______________________________

TABLE 16.3
Working Paper for Spectrophotometric Analysis (Solid Samples)

Parameter__________Method No. ______________Reference ________________________________
Method detection limit (MDL) ____
Model of spectrophotometer_________________Wavelength _____________________________
Correlation coefficient of calibration curve ____
Analyzed by ___________________________________Date ______________________________

ID No.	Sample Identification	g	Final (ml)	Tested (ml)	Abs	Dil	Res (mg/l)	Mois (%)	Rep. wet (mg/kg)	Rep. dry (mg/kg)
	Blank									
	Standard 1									
	2									
	3									
	4									
	CCS									
	CVS									
	Duplicate									
	Spike									
	Blank									

INFORMATION RELATED TO QC CHECKS
% Recovery of Continuing Calibration Standard (CCS)____________
True Value of Calibration Verification Standard (CVS)____________
% Recovery of CVS ____________________ID of Duplicate Sample______________________
ID No. of Spiked Sample ____________________Sample ml Spiked______________________
Concentration of Spike Stock Solution ____________________
ml or µl Spike Stock Added ________________Added Spike Value______________________
Precision as Relative Percent Deviation (RPD) ____________
% Recovery of spike (%R_{sp}) ______________________________

Approval of Supervisor ______________________________Date ______________________

TABLE 16.4
Working Paper for Atomic Absorption Spectroscopy

Parameter: ______________________ **Model:** ______________________
Method No.: ______________________ **Sensitivity Check:** ______________________
Method Ref.: ______________________ **Detection Limit:** ______________________
Correlation Coefficient: ______________________ **Analyzed by:** ______________________
Wavelength: ______________________ **Date:** ______________________

No.	Sample Identification	Matrix	Read abs. (mg/l)	Dil. or Conc.	Result (mg/l)	Sample		Result (mg/kg wet base)	Moisture (%)	Result (mg/kg dry base)	Remarks
						g	final ml				
	Blank										
	Standard low										
	mid										
	high										
	cont.										
	QC (CVS)										
	Sensitivity check										
	Prep. blank										
	Lab. control std. (LCS)										
	Sample 1										
	Sample 2										
	Sample 3										
	Sample 4										
	Sample 5										
	Sample 6										
	Sample 7										
	Sample 8										
	Sample 9										
	Sample 10										
	Duplicate spike										
	QC (CVS) std. cont.										
	Blank										

Note: Precision, RPD (relative percent deviation) = $(A - B)/(A + B) \times 200$. Accuracy, as % Recovery on QC = (measured value × 100)/(true value). % Recovery on spike = [(spiked sample value – sample value) × 100]/(true spike value). CVS = calibration verification standard.

- Summary log forms for precision and accuracy data
- Work sheets
- Records concerning receipt of stock solutions
- Preparation of calibration standards
- All notebooks, data forms, and log forms that pertain to laboratory operations
- Laboratory custody reports (holding times, sample transmittal forms, sample storage log, sample disposal log)
- Sample preparation log
- All calculations related to sample results and statistical calculations for QC limits
- Instrument maintenance and instrument performance check logs
- Copies of final reports
- Field records such as field notebooks, field test data, and field custody record

Records should be retained for a period of at least 3 years. Drinking water reports require a retention time of up to 10 years.

16.4 EVALUATION AND APPROVAL OF ANALYTICAL DATA

16.4.1 Checking Correctness of Analysis

To evaluate and approve reportable results, checks should be performed on sampling, sample handling and storage, methodology calibration, and quality control. If problems arise because of suspect data, initiation of corrective actions and providing the final approved results are the responsibility of the analyst, the laboratory supervisor, and the QC department. Analytical data for each parameter produced by the analyst are checked according to QC acceptance criteria, validated, corrected if necessary, and then converted to a reportable value. The approved and checked analytical results for each analyzed parameter are transferred to the sample report form. The sample report contains the results for each requested parameter analyzed for a particular sample.

16.4.2 Validation of QC Checks

As discussed previously, QC checks are performed on analytical measurement results. When an error occurs, it should be identified in detail, corrected, and replaced by a valid, reportable value. The reviewer should be focused on the acceptance criteria of the QC check, the probable source of the out-of-control item, and the steps taken in the specified corrective action. The QC check must include QC measures, such as blanks, calibration processes, QC check standards, QC check samples, certified reference standards, matrix and reagent water spikes, duplicates, split samples, standardization of titrant, instrument performance, proper storage of the reagents and standards, and quality checks of laboratory-pure water. No data should be released until statistically supported limits are reported. Data should be technically sound and defensible before reporting. The first requirement for reporting is that all pertinent documentation is available and referenced so that it can be consulted whenever a need arises.

16.4.2.1 Blanks

Analytical blanks measure the degree of contamination in a test and seriously affect the accuracy of low-level determinations. Blank types include method, reagent, calibration, instrument, preparation, blank, and trip, as described in Sections 13.5.3, 13.6.7, and 13.8.1. *Note:* Blank values should be less than the method detection limit (MDL).

16.4.2.1.1 Possible Causes of Unacceptable Blank Values

- Contaminated reagents
- Contaminated analyte-free water
- Contaminated glassware or sample containers
- Contaminated apparatus
- Contamination during sample collection and sample handling
- Contamination resulting from dirty laboratory environment or analyst's clothing

16.4.2.2 Calibration

Calibration in chemical measurements is the process by which the response of a measurement system is related to various concentrations of the analyte of interest. Generally, a measurement is a comparison process in which the unknown is compared with a known standard. (Calibration procedures are discussed in Sections 6.6 and 13.6.) The first criterion for correct calibration is the purity and accuracy of the standards. Standards may be prepared by the analyst or outside suppliers and should have an assigned expiration date indicating stable life expectancy. Standards should never be used beyond expiration dates.

16.4.2.2.1 Acceptance Criteria for Calibration

- The correlation coefficient should be at least 0.9950, but varies according to the analytical method and instrumentation.
- Deviation of the continuing calibration standard (CCS) from the original calibration standard should be ±10% for inorganic analyses (±15% in organic analyses).
- Criteria for the calibration verification standard (CVS) or QC check sample (reference standard) are generated in-house and cannot exceed the method range.

16.4.2.2.2 Probable Sources of Unacceptable Calibration

- Calculation error
- Incorrectly prepared stock and intermediate or calibration standards
- Outdated stocks and standards
- Faulty or expired QC check standard (CVS)
- Improperly stored stocks and standards
- Improperly selected and improperly used volumetric glassware
- Incorrect responses of the instrument
- Incorrectly cleaned glassware, containers, or dirty environment

Reanalyze all samples measured between the acceptable and unacceptable calibration check with the corrected calibration.

16.4.2.3 Duplicates and Split Samples

Duplicates and split samples are used to measure the precision of the analytical system. Duplicate types include field, laboratory, and matrix spike (Sections 13.5.3 and 13.8).

16.4.2.3.1 Accepted Limit or Target Limit

Calculated precision values vary by laboratory, parameter, and analytical method. When the measured and calculated precision values fall outside of this limit, corrective action is required to identify the problem.

16.4.2.3.2 Probable Sources and Correction of Unacceptable Precision Values

- *Sampling error for field duplicates*: Review sample collection and preservation protocol. Review sample preparation techniques and, if necessary, repeat the process for both duplicates and then reanalyze. If these activities do not resolve the problem, samples and duplicates should be redone.
- *Preparation error for laboratory duplicates*: A crowded work schedule in the laboratory may cause the use of unidentical samples for duplication. Check carefully and identify the sample for duplication prior to reanalyzing. Poor homogenization of the sample before duplication may also produce incompatible values.
- *Contamination error*: Different results for duplicate analyses are possibly caused by contamination originating from improperly washed or stored glassware and equipment. Check washing procedures and take all precautions to avoid contamination during duplication, sample pretreatment, and analysis.
- *Calculation error*: Recheck all calculations related to the data in the report.

When these steps do not bring the analysis back within the acceptable precision limits, then the entire analytical procedure must start again. It may be necessary to prepare new standards and calibrations. Other sample data generated with the same analytical run are questionable and must be reanalyzed after the control limits are justified.

16.4.2.4 Spikes

All daily spiking data should agree with the spike accuracy limit established for each parameter and analytical method.

16.4.2.4.1 Probable Sources of Unacceptable Spike Values

- Calculation error
- Error during spike preparation
- Improperly prepared and stored spike stock solution
- Expired spike stock solution
- Contamination during spiking
- Instrument malfunction

All samples associated with the unacceptable spiked sample must be reprocessed and analyzed again after correction.

16.4.2.5 Analytical Performance Checks

To validate the analytical system or measurement process, laboratories use *performance evaluation samples*. *Internal reference standards* are prepared by the laboratory. *External reference materials* are provided by an outside source and may originate from a technically acceptable and certified organization. Analytical values should be within the certified limits.

16.4.2.5.1 Causes of Performance Check Failure

- Calculation error
- Improper sample preparation
- Improper analysis; if samples were concentrated or diluted to the proper range of the analysis, carefully check these techniques and the accompanying calculations

- Contamination error; after correction, reprocess and reanalyze all samples that were measured at the same time as the failed reference standard

16.4.2.6 Quality of Laboratory Pure Water

Acceptance criteria for the quality of laboratory pure water are discussed in Section 13.4. If the measured values deviate from this standard, corrective action is needed immediately. Change the cartridges and call the contracted company for regular maintenance of the system.

16.4.3 Documentation of Out-of-Control Conditions

All out-of-control conditions must be documented immediately by personnel at all levels of responsibility. Documentation by the analyst is done by plotting the out-of-control data on the QC charts and labeling the sample results pertaining to that test parameter as "suspect data."

The laboratory supervisor will document all "out-of-control" conditions by issuing a noncompliance report to the quality assurance (QA) officer. The report must include the out-of-control test parameter, date of occurrence, ID number, copy of the failed QC data, description of the analytical problem, and the corrective action. The QA officer will issue a written report to the laboratory manager defining any QC noncompliance occurrence and specifying suspect sample data. The report must include the out-of-control parameter, date, time, ID numbers of samples containing suspect data, description of the QC problem, and the corrective action. A typical noncompliance report form is shown in Table 16.5.

TABLE 16.5
Noncompliance Report Form

Test Out of Control	Date of Problem	Name of Analyst	Suspect Samples ID	QC True Value	QC Measured Value	(%*R*)	(%*R*) Limit

Description of Analytical Problem

Corrective Action

Laboratory supervisor ______________________ Date ______________

17 Reporting Analytical Data

Reports are the written records of analytical work. The content of analytical reports should include all necessary information presented in a transparent and well-organized fashion. The designer of the final report must keep in mind that the reader frequently does not understand the results generated by instruments and laboratory protocols. Brief interpretation of the reported values is essential.

Chapter 16 discusses how analytical data are calculated and checked. Never issue a report until all checks have been completed and approved.

17.1 REQUIRED DOCUMENTATION

All documentation used to approve and defend reported data must be collected and should be available and referenced so that it can be consulted whenever necessary. The content of these documents should be detailed and clear enough to explain to the interested party how the final reported values were generated.

17.1.1 Documentation Required To Approve and Defend Reported Data

17.1.1.2 Documentation for Sample Collection and Identification

- Sampling plan and sample collection methods
- Field QC
- Sample identification and chain of custody
- Field notebook and documentation of field tests

17.1.1.3 Documentation of Analytical Performance

- Analytical method used and method detection limit (MDL)
- Instrumentation (manufacturer, model, performance checks, maintenance log)
- Calibration data (initial and continuing)
- Detailed analytical work (work sheets, standards and reagent preparation, calculations)

17.1.1.4 QA/QC Documentation and Data

- Analysis of blanks
- Precision and accuracy data
- QC charts
- Acceptable ranges for precision and accuracy per parameter
- Source of QC check standards
- Preparation of spike stock solution and how spikes were prepared

17.1.1.5 Checks and Validation of Analytical Data

- Documentation of how analytical data were checked and validated
- Corrective actions (when applicable)
- Date and signature of approval of reportable data for each parameter tested
- Date and signature of approval for final analytical report

17.2 SIGNIFICANT FIGURES IN ANALYTICAL REPORTS

Numerical data are often obtained with more digits than are justified for required accuracy and precision. Report only figures that are justified by the accuracy of the analytical method. Do not use the common practice of making all numbers in one column have the same number of significant figures to the right of the decimal point. The reported numbers pertain to different parameters and they are analyzed by different methods, so the significant figures will also be different. Significant figures are discussed in Section 15.5.

- If an analytical result is 16.6 mg/l, and the analyst is certain of the 16 but not certain of the 0.6, the rounded-off result reported should be 17.
- When an analytical result is 16.61 and is generated by a method that justifies these significant figures, the reported number should be 16.61.
- If a calculated value is 2346 mg/l, but the analyst is not certain about the last two numbers, the rounded-off number reported should be 2350.
- If a number is written as 5.000, it is understood that the zeros are significant, or the number would have to be rounded off to 5.00, 5.0, or 5, whichever is appropriate.
- If a result is 360 mg/l, there should be certainty that the zero is significant and cannot be deleted.

17.3 UNITS USED TO EXPRESS ANALYTICAL RESULTS

Measurement units used to express analytical results depend on the analytical method used, the concentration of the analyte, and the matrices of the sample analyzed. The most common unit used to express analytical results in environmental samples is *ppm*. This unit is equal to milligram per liter (mg/l) or milligram per kilogram (mg/kg).

When the concentration is less than 0.1 ppm, it is more convenient to express the result as *ppb*, microgram per liter (μg/l), or microgram per kilogram (μg/kg).

If the concentration is greater than 10,000 ppm, the result is expressed as a percent (%). One percent (1%) is equal to 10,000 ppm when the specific gravity is 1.0.

If the result is issued in ppm or percent by weight for solid samples or liquids with high specific gravity, a correction is necessary as follows:

$$\text{mg/kg} = (\text{mg/l})/\text{specific gravity} \tag{17.1}$$

$$\%\text{ by weight} = (\text{mg/l})/(10{,}000 \times \text{specific gravity}) \tag{17.2}$$

Analytical results for solid samples are reported in milligrams per kilogram (mg/kg) or micrograms per kilogram (μg/kg), or, as stated above, as percentages (%), according to the concentration value. The report must also indicate that the result is on an "as-is basis" (also called "wet base") of the solid, when the solid is not corrected to a dry, moisture-free solid. When results are reported on the "dry base," the value is corrected to dry, moisture-free solid. Calculation of these units is discussed in Section 16.2.2 and Appendix I.

Calculations related to water treatment require analytical results in milliequivalents per liter (meq/l). Conversion of milligrams per liter to millequivalents per liter are discussed in Section 16.2.3.

17.4 CONFIDENCE INTERVAL

Reporting the calculated confidence interval (CI) of a measurement helps the reader to estimate the reliability of the result. The calculated confidence interval is rounded to two significant figures in reports. Usually the 95% confidence limit is used in the calculation. Calculation of CIs is discussed in Section 13.10.2.3.

17.5 REPORT FORMAT

A simple tabular form is acceptable for regular, routine reporting. It should be clear, easy to follow, and contain all necessary information to evaluate the analytical report. For special, nonroutine purposes, reports are more detailed, and a comprehensive QA/QC report should be attached to the analytical report.

Title: The title should be brief but descriptive and identify the goal of the analytical work.

Who requested the analysis: Identify the organization or person for whom the work was done, including name, organization, address, phone number, work order, and so on.

Report number: Provide the laboratory identification number (ID number, such as calendar year/sequence number).

Date: Provide date the report was completed.

Objective: Include a short statement about the reason the work was done.

Sample identification: Provide a physical description of the sample, sampling area, and all information related to the sample that may impact the data. If possible, include a photograph of the sampling area.

Sampling details: This information includes sampling procedures, sample type, sample preservation, name of the sample collector, date and time of the collection, chain of custody, sample field custody and transportation.

Analyzed parameters with method numbers and method references: References should be specific and provide all necessary information (reference number, revision date, etc.). Modifications of the original method should be stated.

Method detection limit (MDL) or practical quantitation limit (PQL): Provide the specified MDL or PQL for each analysis.

Numerical values and units: Numerical values are reported with correct significant figures and corresponding unit.

References: Provide information related to previous analytical work, references to other reports on the same sample location.

Precision and accuracy data: State the precision and accuracy data with the acceptance limit for each analytical performance.

Discussion: This secton includes interpretation of the results, recommendation of additional work or corrections, and any special observations related to the sample or the analytical report that serves the objective of the analysis.

Signatures: Include signatures and titles of all persons responsible for the data and the report.

Distribution list: Provide the full distribution list of the report.

Attachment: Attach the QC report, which contains the basis of calibration, standardization, statement, and where the documentation is found. Precision and accuracy data with corresponding acceptance limits should be included for each analytical performance.

The pages of the report should be properly numbered. Readers must be certain that they have the full report for review (e.g., use page 1 of 5, page 2 of 5, etc.).

18 Selected Methods for Determination of Metals in Environmental Samples

18.1 METHODOLOGY

Methods are developed to analyze diverse media for specific parameters. Each method is approved by the Environmental Protection Agency (EPA), which specifies the procedures, instrument calibration, sample preparation, analytical procedures, and quality control requirements for the analytical work. EPA methods are differentiated according to the media (matrix) of the sample analyzed. Each laboratory has a written guidebook that contains specific procedures used, known as standard operating procedures (SOPs). SOPs should be constantly revised to include new methodologies and procedural changes. The SOPs are an important tool for the quality assurance/quality control (QA/QC) operation of the laboratory.

18.1.1 EPA-Approved Methods and References for Analyzing Water Samples

18.1.1.1 Methods and References for Analyzing Drinking Water

Methods for Chemical Analysis of Water and Wastes (EPA 600/4–79–020, revised March 1983)
Methods for Determining Organic Compounds in Drinking Water (EPA 600/4–88–039, December 1988)
Standard Methods for the Examination of Water and Wastewater (APHA-AWWA-WPCF, 19th ed., 1998) (an updated edition is issued every 5 years)
Manual for Certification of Laboratories Analyzing Drinking Water (EPA 570/9–90/008, April 1990)
CFR Part 141, Subpart C and Subpart E (monitoring and analytical requirements)
EPA 500 series (should be used for organic analyses of drinking waters and raw source waters)

18.1.1.2 Methods and References for Analyzing Surface Waters and Wastewater Effluents

Methods for Chemical Analysis of Water and Wastes (EPA 600–4–79–020, revised in March 1983)
Test Methods for Evaluating Solid Waste (EPA SW-846, 3rd ed., 1986; rev. ed., December 1987)
40 CFR, Part 136 (Tables IA, IB, IC, ID, and IE, July 1989)

18.1.1.3 Methods and References for Analyzing Water Sources (Surface and Groundwater) Pursuant to 40 CFR Part 261 (RCRA)

Test Methods for Evaluating Solid Waste (EPA SW-846, 3rd ed., 1986; rev. ed., December 1987) *40 CFR, Part 261* (Methods, Appendix III, 1989)

USEPA Contract Laboratory Program Statement of Work for Inorganic Analyses (EPA SOW ILMO3.0, March 1990)

USEPA Contract Laboratory Program Statement of Work for Organic Analyses (EPA SOW OLMO3.1, August 1994)

18.1.1.4 Methods and References for Microbiological and Biological Tests of Water Samples

Microbiological Methods for Monitoring the Environment (EPA 600/8–78–017, 1987)

40 CFR, Part 141 (Subpart C, monitoring and analytical requirements, July 1989)

40 CFR, Part 136 (Table IA, July 1989)

Methods for Measuring the Acute Toxicity of Effluent to Freshwater and Marine Organisms (EPA 600/4–85–013, 3rd ed., 1985)

Short-Term Methods for Estimating the Chronic Toxicity of Effluent and Receiving Waters to Freshwater Organisms (EPA 600/4–89–1990)

Short-Term Methods for Estimating the Chronic Toxicity of Effluent and Receiving Waters to Marine and Estuarine Organisms (EPA 600/4–87–028, 1988)

18.1.2 EPA-Approved Methods and References for Analyzing Sediments and Residuals

18.1.2.1 Methods and References for Analyzing Soils, Sediments, Domestic and Industrial Sludges, Solid and Hazardous Wastes

Test Methods for Evaluating Solid Waste (EPA SW-846, 3rd ed., 1986; rev. ed., December 1987)

40 CFR, Part 261 (Appendix III, July 1989)

Procedures for Handling and Chemical Analysis of Sediments and Water Samples (EPA/Corps of Engineers, CE-81–1, 1981)

USEPA Contract Laboratory Program Statement of Work for Inorganic Analysis (EPA SOW ILMO3.0, March 1990)

USEPA Contract Laboratory Program Statement of Work for Organic Analysis (EPA SOW OLMO3.1, August1994)

POTW Sludge Sampling and Analysis Guidance Document (EPA Permits Division, August 1989)

18.1.3 Approved Modification of EPA Methods

18.1.3.1 EPA Method 300.0

This method may be used for the analysis of specified ions in ground water and surface water, except for fluoride. It is currently approved for drinking water analysis.

18.1.3.2 EPA Methods 601, 602, 624, and 625

Capillary columns may be used instead of the specified packed columns if the laboratory meets the pertinent accuracy and precision criteria and detection limit with this modification.

18.1.3.3 EPA Methods 601 and 602

The photoionization detector and electrolytic conductivity detector may be used in a series if the laboratory can meet the performance criteria.

18.1.3.4 EPA Methods 602, 8020, 8021

These methods may include analysis of xylene and methyl-tert-butyl-ether (MTBT).

18.1.3.5 EPA Methods 610, 625, 8100, 8310, 8250, 8270

These methods may include analysis of methylnaphthalenes.

18.1.3.6 EPA Method 5030/8010

This method must be modified to analyze EDB in soils. An electron-capture detector instead of an electrolytic conductivity detector must be used.

18.1.4 EPA Contract Laboratory Protocol (CLP)

This protocol was developed for the Superfund program. CLP specifies a set of methods based on the existing methodology for organic and inorganic parameters, but which are modified to incorporate certain *quality control, calibration,* and *deliverable requirements*. The data package includes a full reporting of quality control procedures and data, making it particularly useful if litigation is a possibility. The results of the analyses are provided in many different formats, ranging from a sample report only to a full-documentation data package.

The CLP, as stated in the EPA statement of work (SOW), has a high level of quality assurance requirements. The deliverable requirements include quality control summaries (method blank, initial calibration verification, duplicate analysis, and matrix spike/matrix spike duplicates) and quality control data, as well as data on a diskette. Consequently, CLP has become a commonly requested methodology and has the effect of separating larger laboratories — which have the equipment, certifications, and trained personnel capable of producing data according to this protocol — from the thousands of smaller environmental laboratories which do not.

Because EPA methods, as now written, are not interchangeable, it is very difficult for an analytical laboratory to accommodate all quality control criteria for all methods. Thus, the EPA's current intent is to create a unified method to minimize the requirement differences.

18.1.5 Determination of Selected Metals in Environmental Samples

Table 18.1 summarizes the methods, method numbers, and references used for determination of metals in environmental samples.

18.2 ALUMINUM

Aluminum (Al) is the third most abundant element of the Earth's crust, occurring in mineral rocks and clays. Soluble, insoluble, and colloidal aluminum may appear in treated water or wastewater as

TABLE 18.1
Methods for Determination of Metals

Parameter	FL	GR	Other	Method No.	Ref.	Method No.	Ref.
Aluminum	+	+	—	202.1&2	R-1	7020	R-3
Antimony	+	+	—	204.1&2	R-1	7040	R-3
Arsenic	–	+	—	206.2	R-1	7060	R-3
Barium	+	+	—	208.1&2	R-1	7080	R-3
Beryllium	+	+	—	210.1&2	R-1	7090	R-3
Boron	–	–	Curcumin	4500-BB	R-2	—	—
Boron	–	–	Carmine	4500-BC	R-2	—	—
Cadmium	+	+	—	213.1&2	R-1	7130	R-3
Calcium	+	–	—	215.1	R-1	7140	R-3
Calcium	–	–	EDTA titrimetric	215.2	R-1	—	—
Chromium	+	+	—	218.1&2	R-1	7190	R-3
Chromium^{6+}	–	–	Colorimetric	3500CrD	R-2	7196	R-3
Cobalt	+	+	—	219.1&2	R-1	7200	R-3
Copper	+	+	—	220.1&2	R-1	7210	R-3
Iron	+	+	—	236.1&2	R-1	7380	R-3
Lead	+	+	—	239.1&2	R-1	7420	R-3
Magnesium	+	+	—	242.1&2	R-1	7450	R-3
Manganese	+	+	—	243.1&2	R-1	7460	R-3
Mercury	–	–	Cold vapor	245.1.	R-1	7470, 7471	R-3
Molybdenum	+	+	—	246.1&2	R-1	7480	R-3
Nickel	+	+	—	249.1&2	R-1	7520	R-3
Potassium	+	–	—	258.1	R-1	7610	R-3
Selenium	–	+	—	270.2	R-1	7740	R-3
Silver	+	+	—	272.1&2	R-1	7760	R-3
Sodium	+	–	—	273.1	R-1	7770	R-3
Thallium	+	+	—	279.1&2	R-1	7840	R-3
Tin	+	+	—	282.1&2	R-1	7870	R-3
Titanium	+	+	—	283.1&2	R-1	—	—
Vanadium	+	+	—	286.1&2	R-1	7910	R-3
Zinc	+	+	—	289.1&2	R-1	7950	R-3

Note: Metals analysis by inductively coupled plasma (ICP) method is widely used according to method 6010, with reference to R-3. Fl = flame atomic absorption technique; Gr = graphite furnace atomic absorption technique; R-1 = methods for Chemical Analysis of Water and Wastes (EPA-600/4-79-020, Revised March 1983); R-2 = *Standard Methods for the Examination of Water and Wastewater* (AWWA, 18th ed., 1992); R-3 = *Test Methods for Evaluating Solid Wastes* (EPA SW-846 EPA SW-846, 3rd ed., 1986).

a residual of coagulation with aluminum-containing material. Filtered water from a modern, rapid-sand filtration plant should have an aluminum concentration less than 50 µg/l.

Selection of method: The FAAS, GrAAS, and ICP methods are preferred. For discussion of instrumentation and analysis procedures, see Chapters 8, 9, and 12, respectively.

18.2.1 Flame Atomic Absorption Spectroscopy (FAAS)

Aluminum may be as much as 15% ionized in a nitrous oxide/acetylene flame. Use an ionization suppressor of 1000 µg/ml K as KCl (dissolve 95 g of KCl and dilute to 1000 ml). The calibration standards should contain the same type of acid in the same concentration as in the sample (usually 5 ml of acid per 100 ml), and 2 ml/100 ml of KCl solution as suppressor (see above).

18.2.1.1 Instrument Parameters

- *Instrument*: Aluminum hollow cathode lamp
- *Fuel*: Acetylene
- *Oxidant*: Nitrous oxide
- *Type of flame*: Rich fuel
- *Background correction*: Not required

18.2.1.2 Performance Characteristics

- *Optimum concentration range*: 5 to 50 mg/l
- *Detection limit*: 0.1 mg/l
- *Sensitivity*: 1 mg/l
- *Wavelength*: 309.3 nm

18.2.2 Graphite Furnace Atomic Absorption Spectrometry (GrAAS)

Background correction may be required if the sample contains highly dissolved solids. Chloride ion and nitrogen used as a purge gas reportedly suppress the aluminum signal; therefore, the use of halide acids and nitrogen as a purge gas should be avoided.

18.2.2.1 Instrument Parameters

- *Drying time and temperature*: 30 sec at 125°C
- *Ashing time and temperature*: 30 sec at 1300°C
- *Atomizing time and temperature*: 10 sec at 2700°C
- *Purge gas*: Argon
- *Wavelength*: 309.3 nm

Other operating parameters should be set as specified by the instrument manufacturer.

18.2.2.2 Performance Characteristics

- *Optimum concentration range*: 20 to 200 mg/l
- *Detection limit*: 3 mg//l

18.3 ANTIMONY

The level of antimony (Sb) present in natural waters is usually less than 10 μg/l and may be present in higher concentrations in hot springs or waters draining mineralized areas. Antimony is a regulated contaminant under various federal and state programs.

Selection of method: The GrAAS method (Chapter 8) is the method of choice because of its sensitivity. Alternatively, use the FAAS method (Chapter 9) or the ICP method (Chapter 12) when high sensitivity is not required.

18.3.1 Flame Atomic Absorption Spectroscopy (FAAS)

In the presence of lead (1000 mg/l), spectral interference may occur at the 217.6-nm resonance line. In this case, the 231.1-nm antimony line should be used.

18.3.1.1 Instrument Parameters

- *Instrument*: Antimony hollow cathode lamp
- *Wavelength*: 217.6 nm
- *Fuel*: Acetylene
- *Oxidant*: Air
- *Type of flame*: Lean fuel

18.3.1.2 Performance Characteristics

- *Optimum concentration range*: 1 to 40 mg/l
- *Sensitivity*: 0.5 mg/l
- *Detection limit*: 0.2 mg/l

18.3.2 Graphite Furnace Atomic Absorption Spectrometry (GrAAS)

High Pb concentration may cause a measurable spectral interference on the 217.6 nm-line. In this case, a secondary wavelength or Zeeman background correction should be used. See Chapter 9 for general discussion of the furnace technique. A soft-digestion procedure is the only recommended one for Sb, as discussed in Sections 15.2.2 and 15.8. The addition of HCl to the digestate prevents furnace analysis of many metals.

18.3.2.1 Instrument Parameters

- *Drying time and temperature*: 30 sec at 125°C
- *Ashing time and temperature*: 30 sec at 800°C
- *Atomizing time and temperature*: 10 sec at 2700°C
- *Purge gas*: Argon or nitrogen
- *Wavelength*: 217.6 nm (primary); 231.1 nm (alternate)
- *Background correction*: Required

Other operating parameters should be set as specified by the instrument manufacturer.

18.3.2.2 Performance Characteristics

- *Optimum concentration range*: 20 to 300 mg/l
- *Detection limit*: 3 mg/l

The above concentration values and instrument conditions are based on the use of a 20-μl injection, continuous-flow purge gas and nonpyrolytic graphite. See instrument manufacturer's operations manual for information.

18.4 ARSENIC

Severe poisoning can arise from the ingestion of arsenic trioxide (As_3O_2) in amounts as small as 100 mg; chronic effects may result of the accumulation of arsenic compounds in the body at low intake levels. Carcinogenic properties are also known. The toxicity of arsenic depends on its chemical form. The As concentration in potable waters is usually less than 10 μg/l, but values as high as 100 μg/l have been reported. Aqueous arsenic may result from mineral dissolution, industrial discharges, or the application of herbicides.

Selection of methods: The hydride-generation atomic absorption method (Chapter 11) is the method of choice, although the GrAAS (Chapter 9) is simpler.

18.4.1 Gaseous Hydride Atomic Absorption Method

This method is applicable for sample matrices that do not contain high concentrations of Cr, Cu, Hg, Ni, Ag, Co, and Mo. Instrumentation and analytical procedures are discussed in Chapter 11. The typical detection limit for this method is 0.002 mg/l.

18.4.2 Graphite Furnace Atomic Absorption Spectrometry (GrAAS)

Following the appropriate dissolution (acid digestion) of the sample, a representative aliquot of the digestate is spiked with nickel nitrate solution and is placed manually or by means of an automatic sampler into a graphite furnace. See Chapter 9 for details of the GrAAS technique.

18.4.2.1 Instrument Parameters

- *Drying time and temperature*: 30 sec at 125°C
- *Ashing time and temperature*: 30 sec at 1100°C
- *Atomizing time and temperature*: 10 sec at 2700°C
- *Purge gas*: Argon
- *Wavelength*: 193.7 nm

Other operating parameters should be set as specified by the instrument manufacturer.

18.4.2.2 Performance Characteristics

- *Optimum concentration range*: 5 to 100 mg/l
- *Detection limit*: 1 mg/l

18.4.2.3 Interferences

Elemental As and many of its compounds are volatile; therefore, samples may be subject to losses of As during sample preparation. Spike samples and standard reference materials should be processed to determine if the chosen dissolution method is appropriate.

Caution must be employed during the selection of temperature and times for the dry and char cycles. A nickel nitrate solution must be added to all digestates prior to analysis to minimize volatilization losses during drying and ashing.

Arsenic analysis may be subject to severe nonspecific absorption and light scattering caused by matrix components during atomization. Aluminum is a severe positive interferant in the analysis of arsenic. Zeeman background correction is very useful in this situation.

If the analyte is not completely volatilized and removed from the furnace during atomization, memory effects will occur. If this situation is detected by means of blank burns, the tube should be cleaned by operating the furnace at full power at regular intervals during the analysis.

18.4.2.4 Reagents

- Concentrated HNO_3
- Hydrogen peroxide, H_2O_2 (30%)
- As stock solution, 1000 mg/l (commercially available or prepared according to recipe in Appendix H)
- Nickel nitrate, 5% (dissolve 24.780 g of $Ni(NO_3)_2.6H_2O$ in reagent-grade water and dilute to 100 ml)
- Nickel nitrate, 1% (dilute 20 ml of the 5% nickel nitrate solution to 100 ml with reagent-grade water)

18.4.2.5 Procedure

1. Prepare samples for the analysis as described in Sections 15.6.2 and 15.6.3.
2. Pipet 5 ml of digested solution into a 10-ml volumetric flask, add 1 ml of the 1% nickel nitrate solution, and dilute to 10 ml with reagent-grade water. The sample is ready for injection into the furnace.
3. The 193.7-nm wavelength line is recommended.
4. A background correction system is required. For other spectrophotometric parameters, follow the manufacturer's instructions.
5. Furnace parameters suggested by the manufacturer should be employed as guidelines. Because temperature-sensing mechanisms and temperature controllers can vary among instruments or with time, the validity of the furnace parameters must be periodically confirmed by systematically altering the furnace parameters while analyzing a standard. In this manner, losses of analyte due to overly high temperature settings or losses in sensitivity due to less-than-optimum settings can be maintained. Similar verification of furnace parameters may be required for complex sample matrices.
6. Calibration curves must be composed of a minimum of a blank and three standards. A calibration curve should be made for every hour of continuous sample analysis.
7. Inject a measured microliter aliquot of sample into the furnace and atomize. If the concentration found is greater than the highest standard, the sample should be diluted in the same acid matrix and reanalyzed. The use of multiple injections can improve accuracy and help detect furnace pipeting errors.
8. Run a check standard after every ten injections of samples. Standards are run in part to monitor the life and performance of the graphite tube. Lack of reproducibility or significant change in the signal for the standard indicates that the graphite tube should be replaced.
9. Employ a minimum of one blank with a sample batch to verify any contamination.
10. The standard addition method (Section 7.7.1.1.1) should be employed for the analysis of all EPTOX extracts.
11. QC requirements are listed in Chapter 13.

18.5 BARIUM

Barium (Ba) stimulates the heart muscle. However, a barium dose of 550 to 600 mg is considered fatal to human beings. Despite its relative abundance in nature (16th in order of rank), barium occurs only in trace amounts in water (0.7 to 900 μg/l, with a mean of 49 μg/l). Higher concentrations in drinking water often signal undesirable industrial waste pollution.

Selection of method: Preferably, analyze via the FAAS (Chapter 8), GrAAS (Chapter 9), or ICP (Chapter 12) method.

18.5.1 Flame Atomic Absorption Spectroscopy (FAAS)

The FAAS technique is described in Chapter 8. A high, hollow, cathode-current setting and a narrow spectral band pass must be used, because both barium and calcium emit strongly at barium's analytical wavelength. Barium undergoes significant ionization in the nitrous oxide/acetylene flame, resulting in a significant decrease in sensitivity. All samples and standards must contain 2 ml of potassium chloride (KCl) ionization suppressant per 100 ml of sample. (Dissolve 95 g of KCl in reagent-grade water and dilute to 1 liter.)

Prepare calibration standards via dilutions of the stock solution at the time of analysis. The calibration standards should be prepared to contain the same type and concentration of acid as the samples to be analyzed after digesting. All calibration standards should contain 2 ml of the KCl (ionization suppressant) solution.

18.5.1.1 Instrument Parameters

- *Instrument*: Barium hollow cathode lamp
- *Wavelength*: 553.6 nm
- *Fuel*: Acetylene
- *Oxidant*: Nitrous oxide
- *Type of flame*: Rich fuel
- *Background correction*: Not required

18.5.1.2 Performance Characteristics

- *Optimum concentration range*: 1 to 20 mg/l
- *Sensitivity*: 0.4 mg/l
- *Detection limit*: 0.1 mg/l

18.5.2 Graphite Furnace Atomic Absorption Spectrometry (GrAAS)

The use of halide acid should be avoided. Because of possible chemical interaction, nitrogen should not be used as a purge gas.

18.5.2.1 Instrument Parameters

- *Drying time and temperature*: 30 sec at 125°C
- *Ashing time and temperature*: 30 sec at 1200°C
- *Atomizing time and temperature*: 10 sec at 2800°C
- *Purge gas*: Argon
- *Wavelength*: 553.6 nm

Other operating parameters should be set as specified by the instrument manufacturer.

18.5.2.2 Performance Characteristics

- *Optimum concentration range*: 10 to 200 mg/l
- *Detection limit*: 2 mg/l

18.6 BERYLLIUM

Beryllium (Be) and its compounds are very poisonous and in high concentrations can cause death. Inhalation of beryllium dust can cause a serious disease called berylliosis. Beryllium disease also can cause dermatitis, conjunctivitis, acute pneumonitis, and chronic pulmonary berylliosis. Beryllium is used in atomic reactors, aircraft, rockets, and missile fuels. Entry into water can result from the discharges of these industries. The usual range of beryllium in drinking waters is 0.01 to 0.7 μg/l.

Selection of methods: FAAS, GrAAS, and ICP methods may be used (see Chapters 8, 9, and 12, respectively).

18.6.1 Flame Atomic Absorption Spectroscopy (FAAS)

Background correction may be required. Concentration of aluminum greater than 500 ppm may suppress beryllium absorbance. The addition of 0.1% fluoride has been found effective in eliminating this interference. High concentrations of magnesium and silicon cause similar problems and require the use of the standard additions method.

18.6.1.1 Instrument Parameters

- *Instrument*: Beryllium hollow cathode lamp
- *Wavelength*: 234.9 nm
- *Fuel*: Acetylene
- *Oxidant*: Nitrous oxide
- *Type of flame*: Rich fuel
- *Background correction*: Required

18.6.1.2 Performance Characteristics

- *Optimum concentration range*: 0.05 to 2 mg/l
- *Sensitivity*: 0.025 mg/l
- *Detection limit*: 0.005 mg/l

For concentrations below 0.02 mg/l, the furnace procedure is recommended.

18.6.2 Graphite Furnace Atomic Absorption Spectrometry (GrAAS)

Long residence time and high concentrations of the atomized sample in the optical path of the graphite furnace can result in severe physical and chemical interference. Furnace parameters must be optimized to minimize these effects. In addition to the normal interferences experienced during graphite furnace analysis, beryllium analysis is subject to severe nonspecific absorption and light scattering during atomization. Simultaneous background correction is required to avoid erroneous high results.

18.6.2.1 Instrument Parameters

- *Drying time and temperature*: 30 sec at 125°C
- *Ashing time and temperature*: 30 sec at 1000°C
- *Atomizing time and temperature*: 10 sec at 2800°C
- *Purge gas*: Argon
- *Wavelength*: 234.9 nm
- *Background correction*: Required

Other operating parameters should be set as specified by the instrument manufacturer.

The above concentration values and instrument conditions are for a Perkin Elmer HGA-2100, based on the use of a 20-µl injection, continuous-flow purge gas, and nonpyrolytic graphite. Smaller sizes of furnace devices or those employing faster rates of atomization can be operated using lower atomization temperatures for shorter time periods than the recommended settings above.

18.6.2.2 Performance Characteristics

- *Optimum concentration range*: 1 to 30 mg/l
- *Detection limit*: 0.2 mg/l

18.7 BISMUTH

Bismuth is extremely insoluble in natural waters and is generally present only in trace amounts (less than 10 μg/l). It may be present in higher concentrations in waters draining mineralized areas.

18.7.1 Flame Atomic Absorption Spectroscopy (FAAS)

See Chapter 8.

18.8 CADMIUM

Cadmium (Cd) is highly toxic and has been implicated in some cases of poisoning through food. A cadmium concentration of 200 μg/l is toxic for certain fish. Cadmium may enter water as a result of industrial discharges or the deterioration of galvanized pipes.

Selection of methods: The GrAAS method (Chapter 9) is preferred. The FAAS (Chapter 8) and ICP (Chapter 12) methods provide acceptable precision and bias with higher concentration limits.

18.8.1 Flame Atomic Absorption Spectroscopy (FAAS)

Nonspecific absorption and light scattering can be significant at the analyte wavelength. Background correction is required.

18.8.1.1 Instrument Parameters

- *Instrument*: Cadmium hollow cathode lamp
- *Wavelength*: 228.8 nm
- *Fuel*: Acetylene
- *Oxidant*: Air
- *Type of flame*: Oxidizing (lean fuel)
- *Background correction*: Required

18.8.1.2 Performance Characteristics

- *Optimum concentration range*: 0.05 to 2 mg/l
- *Sensitivity*: 0.025 mg/l
- *Detection limit*: 0.005 mg/l

For concentrations of cadmium below 0.02 mg/l, the furnace procedure is recommended.

18.8.2 Graphite Furnace Atomic Absorption Spectrometry (GrAAS)

In addition to the normal interferences experienced during graphite furnace analysis, cadmium analysis may be subject to severe nonspecific absorption and light scattering caused by matrix components during atomization. Simultaneous background correction is required to avoid erroneous high results.

Excess chloride may cause premature volatilization of cadmium. Ammonium phosphate used as a matrix modifier minimizes this loss.

Calibration standards should be prepared at the time of analysis. To each of the 100-ml standards and the sample, add 2.0 ml of 40% ammonium phosphate solution (40 g $(NH_4)_2HPO_4$ per 100 ml of reagent-grade water). The calibration standards should be prepared to contain 0.5% (v/v) HNO_3.

Many plastic pipet tips (yellow) contain cadmium. Use “cadmium-free” tips.

18.8.2.1 Instrument Parameters

- *Drying time and temperature*: 30 sec at 125°C
- *Ashing time and temperature*: 30 sec at 500°C
- *Atomizing time and temperature*: 10 sec at 1900°C
- *Purge gas*: Argon
- *Wavelength*: 228.8 nm
- *Background correction*: Required

Other operating parameters should be set as specified by the instrument manufacturer.

18.8.2.2 Performance Characteristics

- *Optimum concentration range*: 0.5 to 10 mg/l
- *Detection limit*: 0.1 mg/l

The above concentration values and instrument conditions are for a Perkin Elmer HGA-2100, based on the use of a 20-μl injection, continuous-flow purge gas and nonpyrolytic graphite. Smaller sizes of furnace devices or those employing faster rates of atomization can be operated using lower atomization temperatures for shorter time periods than the above-recommended settings.

18.9 CALCIUM

The presence of calcium (Ca, fifth among the elements in order of abundance) in water supplies results from the passage of water through or over deposits of limestone, dolomite, gypsum, and gypsiferous shale. Cadmium content may range from zero to several hundred milligrams per liter. Small concentrations of calcium carbonate combat corrosion of metal pipes by laying down a protective coating. Appreciable quantities of calcium salts, on the other hand, precipitate on heating to form harmful scale in boilers, pipes, and cooking utensils. Calcium contributes to the total hardness of water. Chemical softening treatment, reverse osmosis, electrodialysis, or ion exchange is used to reduce calcium and associated hardness.

Selection of method: FAAS (Chapter 8) and ICP (Chapter 12) methods are accurate means of determining calcium. The EDTA (ethylene diamine tetraacetic acid) disodium salt titration method provides good results for control and routine applications.

18.9.1 Flame Atomic Absorption Spectroscopy (FAAS)

All elements forming stable oxyanions will complex calcium and interfere unless lanthanum is added. The addition of lanthanum to prepared samples rarely presents a problem because virtually all environmental samples contain sufficient calcium to require dilution to obtain results in the method's linear range. Phosphates, sulfates, and aluminum, as well as high concentrations of magnesium, sodium, and potassium are interferants.

Calibration standards should be prepared at the time of the analysis and should contain the same type of acid and at the same concentrations as the preserved samples. Add 1 ml of lanthanum chloride solution (carefully dissolve 29 g of La_2O_3 in 250 ml of concentrated HCl and dilute to 500 ml with reagent-grade water) per 10 ml of standards and samples.

18.9.1.1 Instrument Parameters

- *Instrument*: Calcium hollow cathode lamp
- *Wavelength*: 422.7 nm
- *Fuel*: Acetylene

- *Oxidant*: Nitrous oxide
- *Type of flame*: Stoichiometric
- *Background correction*: Not required

18.9.1.2 Performance Characteristics

- *Optimum concentration range*: 0.2 to 7 mg/l
- *Sensitivity*: 0.08 mg/l
- *Detection limit*: 0.01 mg/l

18.9.2 Determination of Hardness by EDTA Titrimetric Method

The EDTA disodium salt forms a chelated soluble complex when added to a solution of certain metal cations. If a small amount of dye such as Eriochrom black T is added to an aqueous solution containing calcium and magnesium ions at a pH of 10, the solution becomes wine red. If EDTA disodium salt is added as a titrant, the calcium and magnesium will be complexed, and the solution turns from red wine to blue, marking the endpoint of the titration. The sharpness of the endpoint increases with increasing pH. Magnesium ions must be present for a satisfactory endpoint. To ensure the presence of Mg ions, a small amount of complexometrically neutral magnesium salt of EDTA is added to the buffer.

18.9.2.1 Apparatus and Materials

- Buret, 25 ml or 50 ml
- Volumetric pipets, 5 ml, 10 ml, 25 ml
- Graduated cylinders, 100 ml
- Mohr pipets, 5 ml
- Erlenmeyer flasks, 250 ml
- Volumetric flasks, various sizes
- Disposable transfer pipets
- Magnetic stirrer
- Teflon magnetic stirring bars
- pH paper, full range

18.9.2.2 Reagents

18.9.2.2.1 Buffer Solution

1. Solution 1:
   a. Weigh 1.179 g of EDTA disodium salt (Na_2EDTA) and transfer to a 150-ml beaker.
   b. Weigh 0.780 g of magnesium sulfate heptahydrate ($MgSO_4.7HO$) or 0.644 g of magnesium chloride hexahydrate ($MgCl_2.6H_2O$) and transfer to the same beaker.
   c. Add deionized (DI) water to the beaker until the volume is about 100 ml and mix until solids are dissolved.
2. Solution 2:
   a. Weigh 16.9 g of ammonium chloride (NH_4Cl), transfer into a 250-ml volumetric flask, and add 143 ml concentrated ammonia solution (NH_4OH).
   b. Transfer solution 1 from the beaker into solution 2 in the 250-ml volumetric flask. Rinse the beaker well with DI water and add the rinsate to the volumetric flask. Fill to 250 ml with DI water. Stopper the volumetric flask and mix well. Store buffer solution in polyethylene bottle and tighten stopper.

The buffer solution can be used for about 1 month. Discard the buffer when 1 or 2 ml are added to the sample and it fails to produce a pH of 10.0 at the titration endpoint.

18.9.2.2.2 Eriochrom Black T Indicator

Weigh 0.5 g of indicator and 100 g of NaCl into a porcelain mortar, and mix well. Alternatively, a coffee grinder may be used for complete mixing.

18.9.2.2.3 0.02N EDTA Titrant

Dissolve 3.723 g of EDTA disodium salt in about 700 ml of DI water in a 1-liter volumetric flask and dilute to the mark with DI water. Standardize against standard 0.02N $CaCO_3$ solution.

18.9.2.2.4 Calcium Carbonate Standard Solution

Weigh 1.0000 g of anhydrous $CaCO_3$ (primary standard) and transfer to a 500-ml Erlenmeyer flask. Place a funnel in the flask neck and add drops of 1+1 HCl solution until all $CaCO_3$ is completely dissolved. Add 200 ml of distilled water and boil for a few minutes to expel CO_2. Cool at room temperature. Add a few drops of methyl red indicator while stirring. Adjust the color, while stirring, to an intermediate orange color with 3N NH_4OH or 1+1 HCl. Transfer quantitatively into a volumetric flask and dilute to 1 liter with DI water. Store in a polyethylene bottle.

$$1 \text{ ml} = 1 \text{ mg } CaCO_3$$

18.9.2.2.5 Reference Stock Solution, 33,333 mg/l Total Hardness, as $CaCO_3$

Transfer 12.4860 g of anhydrous, primary-standard calcium carbonate ($CaCO_3$) into a 1-liter volumetric flask. Add about 200 ml of DI water and slowly add concentrated hydrochloric acid (HCl) until calcium carbonate is completely dissolved. Transfer 19.5847 g of anhydrous magnesium chloride ($MgCl_2$) into the 1-liter volumetric flask containing the calcium carbonate solution. Mix well for complete dissolution and fill up to the 1-liter volume. Mix well again. The solution contains 33,333 mg/l total hardness as $CaCO_3$.

18.9.2.2.6 Reference (Independent) Standard, 166 mg/l Total Hardness as $CaCO_3$

Pipet volumetrically 5 ml of the reference stock solution into a 1-liter volumetric flask and dilute to the required volume with DI water. This solution is the actual working reference with a value of 166 mg/l total hardness as $CaCO_3$.

18.9.2.2.7 Standardization of EDTA Titrant with $CaCO_3$ Standard Solution

1. Pipet 10 ml of $CaCO_3$ standard solution into a 100-ml Erlenmeyer flask.
2. Using a Mohr pipet, add 5 ml of buffer solution and one scoopful of Eriochrom black T indicator. Mix well. Solution should be wine red.
3. Rinse the buret three times with the EDTA disodium salt titrant.
4. Fill buret with the EDTA titrant.
5. Remove any air bubbles from the buret and bring level of titrant to 0.00 ml.
6. Titrate the contents of the Erlenmeyer flask with EDTA solution until red tint disappears. The color will turn purple. Continue titration slowly until the solution turns blue, which is the endpoint. Record the volume of EDTA used.
7. Perform this titrant check two more times.
8. Calculate the normality of the EDTA as follows:

$$\text{Normality}_{\text{EDTA}} = (0.02 \times S)/V(21.7)$$

where

0.02 = prepared normality of EDTA.
S = volume of titrated $CaCO_3$ solution (ml).
V = volume of EDTA used for titration.

Determine the normality with three parallel titrations. The exact normality is calculated by averaging the three results.

18.9.2.3 Procedure

1. Measure a 100-ml sample or portion diluted to 100 ml into a 250-ml Erlenmeyer flask.
2. Add 5 ml of buffer solution and a scoopful of Eriochrom black T indicator and mix. Solution should be wine red.
3. Rinse the buret with standardized EDTA titrant three times.
4. Fill buret with standardized EDTA titrant.
5. Remove air bubbles from the buret and check 0.00 level.
6. Titrate sample until red tint disappears. The color should turn pink. Continue titration slowly until the solution turns blue.
7. If the volume of the titrant used is over 25 ml, repeat titration by using a smaller sample size or appropriate dilution.

18.9.2.4 Calculation

$$\text{mg/l hardness as } CaCO_3 = (V - B) \times N \times 50 \times 1000/SV \qquad (18.1)$$

where

V = volume of titrant used for sample (ml).
B = volume of titrant used for blank (ml).
N = the determined normality of EDTA.
50 = equivalent weight of $CaCO_3$ (100/2).
SV = sample volume (ml).

Use appropriate dilution factor as necessary.

18.9.2.4.1 Total Hardness Calculation

$$\text{mg/l hardness as } CaCO_3 = 2.497 \text{ (Ca mg/l)} + 4.118 \text{ (Mg mg/l)}$$

or

$$\text{mg/l hardness as } CaCO_3 = [(\text{Ca, mg/l})/0.4] + [(\text{Mg, mg/l})/0.24]$$

For example, calcium and magnesium have been determined by the atomic absorption technique with the following results:

$$\text{Ca} = 16 \text{ mg/l}$$
$$\text{Mg} = 9.6 \text{ mg/l}$$

Calculated total hardness value as $CaCO_3$ is:

$$(2.497 \times 16) + (4.118 \times 9.6) = 79.5 \text{ mg/l}$$

or

$$(16/0.4) + (9.6/0.24) = 80 \text{ mg/l}$$

18.9.3 Calcium Determination by EDTA Titrimetric Method

When EDTA is added to water containing both calcium and magnesium, it combines first with calcium. Calcium can be determined directly with EDTA when the pH is made sufficiently high so that the magnesium is largely precipitated as the hydroxide and an indicator is used that combines with calcium only.

18.9.3.1 Apparatus and Materials

Apparatus and materials are the same as those listed for total hardness determination (Section 18.9.2.1).

18.9.3.2 Reagents

18.9.3.2.1 Sodium Hydroxide, NaOH, 1N

Place a 2-liter beaker or Erlenmeyer flask on a magnetic stirrer under a laboratory hood. Add about 500 ml of DI water and a magnetic stirring bar and add 40 g of NaOH slowly to the water while stirring.

Caution: This reaction liberates heat! After complete dissolution, transfer into a 1-liter volumetric flask and fill up to the mark with DI water. Mix well. Store in polyethylene bottle.

18.9.3.2.2 Murexide (Ammonium Purpurate) Indicator

A ground mixture of dye powder and sodium chloride provides a stable form of the indicator. Weigh 0.200 g of murexide (ammonium purpurate) and 100 g NaCl, and grind the mixture to 40 to 50 mesh in a porcelain mortar or in a coffee grinder used for this purpose.

18.9.3.2.3 Standard EDTA Titrant, 0.02N

Prepare as described in Section 18.9.2.2.3.

$$1 \text{ ml} = 400.8\ \mu\text{g Ca}$$

18.9.3.2.4 Reference Stock Solution, 12,500 mg/l Ca as $CaCO_3$

Prepare as described in Section 18.9.2.2.5 with a value of 12,5018 mg/l of Ca as $CaCO_3$.

18.9.3.2.5 Reference Standard Solution, 62.5 mg/l

Pipet 5 ml of reference stock solution (Section 18.9.3.2.4) into a 1-liter volumetric flask and dilute to the required volume with DI water.

18.9.3.3 Procedure

1. Measure 100-ml sample or smaller portion diluted to 100 ml.
2. Add 2 ml of 1N NaOH solution or a volume sufficient to produce a pH of 12 to 13. Stir.
3. Add a scoopful of indicator. The color of the sample becomes pink.
4. Titrate with standardized EDTA solution until the pink color changes to purple, which is the endpoint. Titrate immediately after adding indicator because the solution is unstable under alkaline conditions.
5. Check endpoint by adding one to two drops of titrant in excess to make certain that no further color change occurs. Facilitate endpoint recognition by preparing a color-comparison

blank containing 2 ml of 1N NaOH and a scoopful of indicator powder and sufficient EDTA titrant (0.05 to 0.10 ml) to produce an unchanging color.

18.9.3.4 Calculation

$$\text{Calcium as CaCO}_3 = (\text{ml} \times N \times 50 \times 1000)/\text{ml sample} \tag{18.2}$$

where

ml = ml of ETA standard used for titration.
N = exact normality of the EDTA titrant.
50 = equivalent weight of $CaCO_3$.

$$\text{Calcium as Ca (mg/l)} = \text{Ca as CaCO}_3\text{ (mg/l)} \times 0.4 \tag{18.3}$$

Magnesium may be estimated based on the difference between total hardness and calcium as $CaCO_3$:

$$\text{Mg as CaCO}_3\text{ (mg/l)} = \text{total hardness as CaCO}_3\text{ (mg/l)} - \text{calcium as CaCO}_3\text{ (mg/l)}$$

$$\text{Magnesium as Mg (mg/l)} = \text{magnesium as CaCO}_3\text{ (mg/l)} \times 0.24 \tag{18.4}$$

18.10 CHROMIUM

Chromium salts are used extensively in industrial processes and may enter a water supply through waste discharge. Chromate compounds frequently are added to cooling water for corrosion control. Chromium may exist in water supplies in both the hexavalent and the trivalent states, although the trivalent form rarely occurs in potable water.

Selection of method: Use the colorimetric method for the determination of hexavalent chromium in natural or treated water intended to be potable. Use the GrAAS method for determination of low levels of total chromium (less than 50 mg/l) in water and wastewater. Use the FAAS or ICP method to measure concentrations up to the milligram per liter level.

18.10.1 Flame Atomic Absorption spectroscopy (FAAS)

If the sample contains a higher level of alkali metal content than the standards, ionization interference may cause problems. To avoid this interference, add potassium-chloride, ionization-suppressant solution to standards and samples.

Background correction may be required because nonspecific absorption and scattering can be significant at the analytical wavelength. Background correction with certain instruments may be difficult at this wavelength due to low-intensity output from hydrogen or deuterium lamps. Consult the instrument manufacturer's literature for details.

18.10.1.1 Instrument Parameters

- *Instrument*: Chromium hollow cathode lamp
- *Wavelength*: 357.9 nm
- *Fuel*: Acetylene
- *Oxidant*: Nitrous oxide
- *Type of flame*: Rich fuel
- *Background correction*: Not required

18.10.1.2 Performance Characteristics

- *Optimum concentration range*: 0.5 to 10 mg/l
- *Sensitivity*: 0.25 mg/l
- *Detection limit*: 0.05 mg/l

For concentration of chromium below 0.2 mg/l, the furnace procedure is recommended.

18.10.2 Graphite Furnace Atomic Absorption Spectrometry (GrAAS)

Low concentrations of calcium and/or phosphate may cause interferences; at concentrations above 200 mg/l, calcium's effect is constant and eliminates the effect of phosphate. Calcium nitrate is therefore added to ensure a known constant effect. Nitrogen should not be used as the purge gas because of possible CN band interference.

Background correction may be required because nonspecific absorption and scattering can be significant at the analytical wavelength. Background correction with certain instruments may be difficult at this wavelength due to low-intensity output from hydrogen or deuterium lamps. Consult the instrument manufacturer's literature for details.

Prepare calibration standards at the time of analysis. These standards should be prepared to contain 0.5% (v/v) HNO_3, 1 ml of 30% H_2O_2, and 1 ml of calcium nitrate solution (dissolve 11.8 g of calcium nitrate ($Ca(NO_3)_2.4H_2O$), and dilute to 1 liter with reagent-grade water).

18.10.2.1 Instrument Parameters

- *Drying time and temperature*: 30 sec at 125°C
- *Ashing time and temperature*: 30 sec at 1000°C
- *Atomizing time and temperature*: 10 sec at 2700°C
- *Purge gas*: Argon (N should not be used!)
- *Wavelength*: 357.9 nm
- *Background correction*: Not required

Other operating parameters should be set as specified by the instrument manufacturer.

The above concentration values and instrument conditions are for a Perkin Elmer HGA-2100, based on the use of a 20-µl injection, continuous-flow purge gas, and nonpyrolytic graphite. Smaller sizes of furnace devices or those employing faster rates of atomization can be operated using lower atomization temperatures for shorter time periods than the recommended settings above.

18.10.2.2 Performance Characteristics

- *Optimum concentration range*: 5 to 100 mg/l
- *Detection limit*: 1 mg/l

18.11 HEXAVALENT CHROMIUM

18.11.1 Chelation/Extraction Method

This method is suitable for determining the concentration of dissolved hexavalent chromium, Cr(VI) in EP toxicity-characteristic extracts, groundwaters, and domestic and industrial wastes provided that no interfering substances are present. The method is based on the chelation of hexavalent chromium with *ammonium pyrrolidine dithiocarbamate* (APDC) and extraction with *methyl isobutyl ketone* (MIBK). The extract is aspirated into the flame of an atomic absorption spectrophotometer.

High concentrations of other metals may interfere. Because the stability of Cr(VI) is not completely understood, the chelation and extraction should be carried out as soon as possible. To retard the chemical activity of hexavalent chromium, samples and should be stored at 4°C until analysis.

18.11.1.1 Reagents

18.11.1.1.1 Ammonium Pyrrolidine Dithiocarbamate (APDC) Solution

Dissolve 1.0 g of APDC and dilute to 100 ml with reagent-grade water. Prepare fresh daily.

18.11.1.1.2 Bromphenol Blue Indicator Solution

Dissolve 0.1 g of Bromphenol blue in 100 ml of 50% ethanol.

18.11.1.1.3 Potassium Dichromate Standard Solution I

Dissolve 0.2829 g of pure dried $K_2Cr_2O_7$ and dilute to 1000 ml with reagent-grade water.

$$1 \text{ ml} = 100\ \mu\text{g Cr}$$

18.11.1.1.4 Potassium Dichromate Standard Solution II

Dilute 100 ml of potassium chromium standard I (Section 18.11.1.1.3) to 1 liter with reagent-grade water.

$$1 \text{ ml} = 10\ \mu\text{g Cr}$$

18.11.1.1.5 Potassium Dichromate Standard Solution III

Dilute 10 ml of potassium dichromate standard II (Section 18.11.1.1.4) to 1 liter with reagent-grade water.

$$1 \text{ ml} = 0.10\ \mu\text{g Cr}$$

18.11.1.1.6 Methyl Isobutyl Ketone (MIBK)

Avoid material that comes into contact with metal or metal-lined caps.

18.11.1.1.7 Sodium Hydroxide 1M Solution

Dissolve 40 g NaOH in reagent-grade water. *Caution:* Do not forget that the reaction of NaOH and water liberates extreme heat! Make the dissolution slowly under a chemical hood. Cool and dilute to 1 liter with reagent-grade water.

18.11.1.1.8 Sulfuric Acid, 0.12M

Slowly add 6.5 ml of spectrograde-quality H_2SO_4 to reagent-grade water and dilute to 1 liter.

18.11.1.2 Procedure

1. Pipet a volume of sample containing less than 2.5 μg chromium (maximum 100 ml) into a 200-ml volumetric flask and adjust the volume to approximately 100 ml.
2. Prepare a blank and sufficient standards, and adjust the volume of each to approximately 100 ml.
3. Add two drops of Bromphenol blue indicator solution (Section 18.11.1.1.2).
4. Adjust the pH by the addition of 1M NaOH (Section 18.11.1.1.7) by drops until a blue color persists.
5. Add 0.12M H_2SO_4 (Section 18.11.1.1.8) dropwise until the blue color just disappears in both the standards and sample. Then add 2.0 ml of 0.12M H_2SO_4 in excess. At this point, pH should be 2.4.
6. Add 5.0 ml of APDC solution (Section 18.11.1.1.1) and mix. The pH should then be approximately 2.8.

7. Add 10.0 ml of MIBK (Section 18.11.1.1.6) and shake vigorously for 3 min.
8. Allow the layers to separate and add reagent-grade water until the ketone layer is completely in the neck of the flask.
9. Aspirate the ketone layer and record the scale reading for each sample and standard against the blank. Repeat and average the duplicate results.
10. Determine Cr concentration in milligrams per liter.

A working curve must be prepared with each set of samples.

18.11.1.3 Verification

- For every sample matrix analyzed, verification is required to ensure that neither a reducing condition nor chemical interference is affecting chelation. To verify the absence of interference, the spike recovery must be between 85 and 115%.
- If addition of the spike extends the concentration beyond the calibration curve, the analysis solution should be diluted with blank solution and the calculated results adjusted accordingly.
- If the result of verification indicates a suppressive interference, the sample should be diluted and reanalyzed. If the interference persists after sample dilution, an alternative method should be used.
- Acidic extracts that yield recoveries of less than 85% should be retested to determine if the low spike recovery is due to the presence of residual reducing agent. This determination should be performed by first making an aliquot of the extract alkaline (8.0–8.5 pH) using 1N NaOH and then respiking and analyzing. If a spike recovery of 85 to 115% is obtained in the alkaline aliquot of an acidic extract that initially was found to contain less than 5 mg/l Cr(VI), the analytical method has been verified.

18.11.1.4 Quality Control

Calibration curves must be composed of a minimum of a blank and three standards. A calibration curve should be made for every hour of continuous sample analysis. Employ a minimum of one blank per sample batch to determine whether contamination or memory effects are occurring.

Verify calibration with an independently prepared check standard every 15 samples. Run one spike duplicate sample for every ten samples. A duplicate sample is a sample brought through the entire sample preparation and analytical process. The standard additions method should be used for the analysis of all EP extracts and when a new sample matrix is being analyzed.

18.11.2 Colorimetric Method

Dissolved hexavalent chromium, in the absence of interfering amounts of substances such as molybdenum, vanadium, and mercury, may be determined colorimetrically by reaction with *diphenylcarbazide* in acid solution. A red-violet color of unknown composition is produced. The reaction is very sensitive — the absorbancy index per gram atom of chromium is about 40,000 at 540 nm. Addition of an excess of diphenylcarbazide yields the red-violet product, and its absorbance is measured photometrically at 540 nm.

Iron in concentrations greater than 1.0 mg/l may produce a yellow color, but the ferric iron color is not strong and difficulty is not normally encountered if the absorbance is measured photometrically at the appropriate wavelength.

18.11.2.1 Reagents

18.11.2.1.1 Potassium Dichromate Stock Solution

Dissolve 141.4 mg of dried $K_2Cr_2O_7$ in reagent-grade water and dilute to 1 l.

1 ml = 50 μg Cr

18.11.2.1.2 Potassium Dichromate Standard Solution

Dilute 10 ml of potassium dichromate stock solution (Section 18.11.2.1.1) to 100 ml.

1 ml = 5 μg Cr (5 mg/l)

18.11.2.1.3 H_2SO_4 10% (v/v)

Dilute 10 ml of spectrograde-quality H_2SO_4 to 100 ml with reagent-grade water.

18.11.2.1.4 Diphenylcarbazide Solution

Dissolve 250 mg 1,5-diphenylcarbazide in 50 ml of acetone. Store in brown bottle. Discard when the solution becomes discolored.

18.11.2.1.5 Acetone (Analytical Reagent Grade)

Avoid material that comes in containers with metal or metal-lined caps.

18.11.2.2 Procedure

1. Collect samples as outlined in Section 14.5.4.
2. Transfer 95-ml sample to a 100-ml volumetric flask.
3. Add 2.0 ml of diphenylcarbazide solution (Section 18.11.2.1.4) and mix.
4. Add H_2SO_4 solution (Section 18.11.2.1.3) to obtain a pH of 2±0.5, dilute to 100 ml with reagent-grade water, and let stand 10 min for full color development.
5. Measure absorbance at 540 nm against blank.
6. An aliquot of the sample containing all reagents except diphenylcarbazide should be prepared and used to correct the sample.
7. From the corrected absorbance, determine the milligrams per liter of chromium present by reference to the calibration curve for turbidity.
8. Prepare calibration curve to compensate for the possible slight losses of chromium during digestion or other operations, and treat the chromium standards by the same procedure as the sample. Prepare calibration standards from the potassium dichromate standard solution (Section 18.11.2.1.2) in a concentration of 0.5 to 5.0 mg/l Cr(VI).

18.11.2.3 Quality Control

See chelation/extraction procedure in Section 18.11.1.

18.12 COBALT

Cobalt (Co) normally occurs at levels of less than 10 μg/l in natural waters. Wastewaters may contain higher concentrations.

Selection of method: Use the FAAS (Chapter 8), GrAAS (Chapter 9), or ICP (Chapter 12) method.

18.12.1 Graphite Furnace Atomic Absorption Spectrometry (GrAAS)

18.12.1.1 Instrument Parameters

- *Drying time and temperature*: 30 sec at 125°C
- *Ashing time and temperature*: 30 sec at 900°C
- *Atomizing time and temperature*: 10 sec at 2700°C
- *Purge gas*: Argon
- *Wavelength*: 240.7 nm
- *Background correction*: Required

Other operating parameters should be set as specified by the instrument manufacturer.

The above concentration values and instrument conditions are for a Perkin Elmer HGA-2100, based on the use of a 20-μl injection, continuous-flow purge gas, and nonpyrolytic graphite. Smaller sizes of furnace devices or those employing faster rates of atomization can be operated using lower atomization temperatures for shorter time periods than the recommended settings above.

18.12.1.2 Performance Characteristics

- *Optimum concentration range*: 5 to 100 mg/l
- *Detection limit*: 1 mg/l

18.13 COPPER

Copper (Cu) salts are used in water supply systems to control biological growth in reservoirs and distribution pipes and to catalyze the oxidation of manganese. Corrosion of copper-containing alloys in pipe fittings may introduce measurable amounts of copper into the water in a pipe system. Copper is essential to humans; the adult daily requirement has been estimated at 2.0 mg.

Selection of method: FAAS, GrAAS, and ICP are recommended (see Chapters 8, 9, and 12, respectively), because of their freedom from interferences.

18.13.1 Flame Atomic Absorption Spectroscopy (FAAS)

Background correction may be required because nonspecific absorption and scattering can be significant at the analytical wavelength. Background correction with certain instruments may be difficult at this wavelength due to low-intensity output from hydrogen or deuterium lamps. Consult the instrument manufacturer's literature for details.

18.13.1.1 Instrument Parameters

- *Instrument*: Copper hollow cathode lamp
- *Wavelength*: 324.7 nm
- *Fuel*: Acetylene
- *Oxidant*: Air
- *Type of flame*: Oxidizing (lean fuel)
- *Background correction*: Recommended, if possible

18.13.1.2 Performance Characteristics

- *Optimum concentration range*: 0.2 to 5 mg/l
- *Sensitivity*: 0.1 mg/l
- *Detection limit*: 0.02 mg/l

18.14 IRON

Iron (Fe) in water can cause staining of laundry and porcelain. Under reducing conditions, iron exists in the ferrous state. In the absence of complex-forming ions, ferric iron is not significantly soluble unless the pH is very low. On exposure to air or addition of oxidants, ferrous iron is oxidized to the ferric state and may hydrolyze to form insoluble, hydrated ferric oxide. In water samples, iron may occur in true solution, in a colloidal state that may be peptized by organic matter, in inorganic or organic iron complexes, or in suspended particles. It may be ferrous or ferric, or suspended or dissolved. Silt and clay in suspension may contain acid-soluble iron. Oxide particles are sometimes collected with a water sample as a result of flaking of rust from pipes. Iron from a metal cap used to close the sample bottle may contaminate the sample.

Selection of method: Sensitivity and detection limits for the FAAS procedure, the ICP method are similar and generally adequate for analysis of natural or treated waters.

18.14.1 Flame Atomic Absorption Spectroscopy (FAAS)

Iron is a universal contaminant, and great care should be taken to avoid contamination.

18.14.1.1 Instrument Parameters

- *Instrument*: Iron hollow cathode lamp
- *Wavelength*: 248.3 nm (primary); 248.7, 248.8, 271.9, 302.1, 252.7, or 372.0 nm (alternates)
- *Fuel*: Acetylene
- *Oxidant*: Air
- *Type of flame*: Oxidizing (lean fuel)
- *Background correction*: Required

18.14.1.2 Performance Characteristics

- *Optimum concentration range*: 0.3 to 5 mg/l
- *Sensitivity*: 0.12 mg/l
- *Detection limit*: 0.03 mg/l

18.14.2 Graphite Furnace Atomic Absorption Spectrometry (GrAAS)

Background correction is recommended. Nitrogen may also be used as the purge gas.

18.14.2.1 Instrument Parameters

- *Drying time and temperature*: 30 sec at 125°C
- *Ashing time and temperature*: 30 sec at 1000°C
- *Atomizing time and temperature*: 10 sec at 2700°C
- *Purge gas*: Argon
- *Wavelength*: 248.3 nm

Other operating parameters should be set as specified by the instrument manufacturer.

18.14.2.2 Performance Characteristics

- *Optimum concentration range*: 5 to 100 mg/l
- *Detection limit*: 1 mg/l

18.15 LEAD

Lead (Pb) is a serious cumulative body poison. Natural waters seldom contain more than 5 μg/l, although much higher values have been reported. Lead in a water supply may come from industrial, mining, and smelter discharges or from the dissolution of old lead plumbing. Tap waters that are soft, acid, and not suitably treated may contain lead resulting from an attack on lead service pipes or solder pipe joints.

Selection of method: FAAS has a relatively high detection limit and requires an extraction procedure for the low concentrations common in potable water. The GrAAS method is much more sensitive for low concentrations and thus does not require extraction. The ICP method has a sensitivity similar to that of FAAS method.

18.15.1 Flame Atomic Absorption Spectroscopy (FAAS)

18.15.1.1 Instrument Parameters

- *Instrument*: Lead hollow cathode lamp
- *Wavelength*: 283.3 nm (primary); 217.0 nm (alternate)
- *Fuel*: Acetylene
- *Oxidant*: Air
- *Type of flame*: Oxidizing (lean fuel)
- *Background correction*: Required

18.15.1.2 Performance Characteristics

- *Optimum concentration range*: 1 to 20 mg/l
- *Sensitivity*: 0.5 mg/l
- *Detection limit*: 0.1 mg/l

For concentrations of lead below 0.2 mg/l, the furnace technique is recommended.

18.15.2 Graphite Furnace Atomic Absorption Spectrometry (GrAAS)

If poor recoveries are obtained, a matrix modifier may be necessary. Add 10 μl of phosphoric acid to 1 ml of prepared sample in the furnace sampler cup and mix well.

18.15.2.1 Instrument Parameters

- *Drying time and temperature*: 30 sec at 125°C
- *Ashing time and temperature*: 30 sec at 500°C
- *Atomizing time and temperature*: 10 sec at 2700°C
- *Purge gas*: Argon
- *Wavelength*: 283.3 nm
- *Background correction*: Required

Other operating parameters should be set as specified by the instrument manufacturer.

The above concentration values and instrument conditions are for a Perkin Elmer HGA-2100, based on the use of a 20-μl injection, continuous-flow purge gas, and nonpyrolytic graphite. Smaller sizes of furnace devices or those employing faster rates of atomization can be operated using lower atomization temperature for shorter time periods than the recommended settings above.

18.15.2.2 Performance Characteristics

- *Optimum concentration range*: 5 to 100 μg/l
- *Detection limit*: 1 μg/l

18.16 LITHIUM

A minor constituent of minerals, lithium (Li) is present in fresh waters in concentrations below 0.2 mg/l. Brines and thermal waters may contain higher lithium levels. The use of lithium or its salts in dehumidifying units, medical waters, and metallurgical processes and the manufacture of some types of glass and storage batteries may contribute to its presence in wastes. Lithium hypochlorite is available commercially as a source of chlorine and is used in swimming pools.

Selection of methods: FAAS (Chapter 8) and ICP (Chapter 12) methods are preferred.

18.17 MAGNESIUM

All elements forming stable oxyanions will complex magnesium and interfere unless lanthanum is added. Addition of lanthanum to prepared samples rarely presents a problem because virtually all environmental samples contain sufficient magnesium to require dilution to be in the linear range of the analytical method.

Calibration standards should be prepared by using the same type of acid and at the same concentration as will result in the sample to be analyzed after processing, including 1 ml of lanthanum solution per 10 ml of solution. To prepare lanthanum chloride solution, dissolve 29 g of La_2O_3 in 250 ml of concentrated HCl. *Caution:* Reaction is violent! Dilute to 500 ml with reagent-grade water.

Selection of methods: Direct determination can be made with the FAAS (Chapter 8) and ICP (Chapter 12) methods.

18.17.1 Flame Atomic Absorption Spectroscopy (FAAS)

18.17.1.1 Instrument Parameters

- *Instrument*: Magnesium hollow cathode lamp
- *Wavelength*: 285.2 nm
- *Fuel*: Acetylene
- *Oxidant*: Air
- *Type of flame*: Oxidizing (lean fuel)
- *Background correction*: Required

18.17.1.2 Performance Characteristics

- *Optimum concentration range*: 0.02 to 0.05 mg/l
- *Sensitivity*: 0.007 mg/l
- *Detection limit*: 0.001 mg/l

18.18 MANGANESE

Although manganese (Mn) in ground water is usually present in the soluble divalent ionic form because of the absence of oxygen, part of or all manganese at a water treatment plant is in higher valence states. Excess manganese in higher oxidation states must be detected with great sensitivity to control treatment and prevent discharge into a distribution system. Although rarely present in excess of 1 mg/l, manganese causes tenacious stains in laundry and plumbing fixtures. The low manganese limits imposed on an acceptable water stem from these, rather than toxicological, considerations. Special means of removal are necessary, such as chemical precipitation, pH adjustment, aeration, and use of special ion-exchange materials. Manganese occurs in domestic wastewaters, industrial effluent, and receiving streams.

Selection of method: FAAS, GrAAS, and ICP methods permit direct determination with acceptable sensitivity.

18.18.1 Flame Atomic Absorption Spectroscopy (FAAS)

18.18.1.1 Instrument Parameters

- *Instrument*: Manganese hollow cathode lamp
- *Wavelength*: 279.5 nm (primary); 403.1 nm (alternate)
- *Fuel*: Acetylene
- *Oxidant*: Air
- *Type of flame*: Slightly oxidizing (slightly lean fuel, stoichiometric)
- *Background correction*: Required

18.18.1.2 Performance Characteristics

- *Optimum concentration range*: 0.1 to 3 mg/l
- *Sensitivity*: 0.05 mg/l
- *Detection limit*: 0.01 mg/l

18.18.2 Graphite Furnace Atomic Absorption Spectrometry (GrAAS)

The use of background correction is recommended. Nitrogen may also be used as purge gas.

18.18.2.1 Instrument Parameters

- *Drying time and temperature*: 30 sec at 125°C
- *Ashing time and temperature*: 30 sec at 1000°C
- *Atomizing time and temperature*: 10 sec at 2700°C
- *Purge gas*: Argon
- *Wavelength*: 279.5 nm

Other operating parameters should be set as specified by the instrument manufacturer.

18.18.2.2 Performance Characteristics

- *Optimum concentration range*: 1 to 30 mg/l
- *Detection limit*: 0.2 mg/l

18.19 MERCURY

Organic and inorganic mercury (Hg) salts are very toxic and their presence in the environment, especially in water, should be monitored.

Selection of method: The cold-vapor atomic absorption method should be used for all types of samples.

18.19.1 Cold-Vapor Atomic Absorption Technique

For both liquid and solid samples, see Section 10.3.5.

18.20 MOLYBDENUM

Molybdenum (Mo) occurs in trace levels (<10 μg/l) in natural waters. In waters draining mineralized areas or in wastewaters from processes using Mo, concentrations maybe much higher.

Selection of methods: The FAAS, GrAAS, or ICP method can be used, as described in Chapters 8, 9, and 12, respectively.

18.20.1 Flame Atomic Absorption Spectroscopy (FAAS)

Interferences in an air–acetylene flame from Ca, Sr, SO4, and Fe are severe. These interferences are greatly reduced in the nitrous oxide flame and by addition of 1 mg/l of aluminum to samples and standards. The calibration standards should be prepared using the same type of acid and at the same concentration as will result in the sample to be analyzed after processing. To each 100 ml of sample and standard alike, add 2 ml of aluminum nitrate solution. (To prepare aluminum nitrate solution, dissolve 139 g of $Al(NO_3)_3.9H_2O$ in 150 ml of reagent-grade water, and heat to effect solution. Allow to cool and make up to 200 ml.)

18.20.1.1 Instrument Parameters

- *Instrument*: Molybdenum hollow cathode lamp
- *Wavelength*: 313.3 nm
- *Fuel*: Acetylene
- *Oxidant*: Nitrous oxide
- *Type of flame*: Rich fuel
- *Background correction*: Required

18.20.1.2 Performance Characteristics

- *Optimum concentration range*: 1 to 40 mg/l
- *Sensitivity*: 0.4 mg/l
- *Detection limit*: 0.1 mg/l

For concentrations of molybdenum below 0.2 mg/l, the furnace technique is recommended.

18.20.2 Graphite Furnace Atomic Absorption Spectrometry (GrAAS)

Molybdenum is prone to carbide formation. Use a pyrolytically coated graphite tube. Memory effects are possible, and cleaning of the furnace may be required after analysis of highly concentrated samples or standards.

18.20.2.1 Instrument Parameters

- *Drying time and temperature*: 30 sec at 125°C
- *Ashing time and temperature*: 30 sec at 1400°C
- *Atomizing time and temperature*: 5 sec at 2800°C
- *Purge gas*: Argon (N should not be used!)
- *Wavelength*: 313.3 nm
- *Background correction*: Required

Other operating parameters should be set as specified by the instrument manufacturer.

The above concentration values and instrument conditions are for a Perkin Elmer HGA-2100, based on the use of a 20-µl injection, continuous-flow purge gas.

18.20.2.2 Performance Characteristics

- *Optimum concentration range*: 3 to 60 mg/l
- *Detection limit*: 1 mg/l

18.21 NICKEL

Selection of method: The FAAS and ICP methods are preferred for all types of samples.

18.21.1 Flame Atomic Absorption Spectroscopy (FAAS)

High concentrations of Fe, Co, or Cr may interfere, requiring either matrix matching or use of a nitrous oxide–acetylene flame. A nonresonance line of Ni at 232.14 nm causes nonlinear calibration curves at moderate to high Ni concentrations, requiring sample dilution or use of the 352.4 nm line.

18.21.1.1 Instrument Parameters

- *Instrument*: Nickel hollow cathode lamp
- *Wavelength*: 232.0 nm (primary); 352.4 nm (alternate)
- *Fuel*: Acetylene
- *Oxidant*: Air
- *Type of flame*: Oxidizing (lean fuel)
- *Background correction*: Required

18.21.1.2 Performance Characteristics

- *Optimum concentration range*: 0.3 to 5 mg/l
- *Sensitivity*: 0.15 mg/l
- *Detection limit*: 0.04 mg/l

18.21.2 Graphite Furnace Atomic Absorption Spectrometry (GrAAS)

Background correction is recommended. Nitrogen may also be used as the purge gas.

18.21.2.1 Instrument Parameters

- *Drying time and temperature*: 30 sec at 125°C
- *Ashing time and temperature*: 30 sec at 800°C
- *Atomizing time and temperature*: 10 sec at 2700°C
- *Purge gas*: Argon
- *Wavelength*: 232.0 nm

Other operating parameters should be set as specified by the instrument manufacturer.

18.21.2.2 Performance Characteristics

- *Optimum concentration range*: 5 to 50 µg/l
- *Detection limit*: 1 µg/l

18.22 POTASSIUM

Potassium (K) ranks seventh among the elements in order of abundance, yet its concentration in most drinking waters seldom reaches 20 mg/l. Occasionally, brines may contain more than 100 mg/l.

Selection of method: The FAAS and ICP methods are acceptable.

18.22.1 Flame Atomic Absorption Spectroscopy (FAAS)

In air–acetylene or other high-temperature flames (>2800°C), potassium can experience partial ionization, which indirectly affects absorption sensitivity. The presence of other alkali salts in the sample can reduce ionization and thereby enhance analytical results. The ionization-suppressive effect of sodium is small if the ratio of Na to K is under 10. Enhancement due to sodium can be stabilized by adding excess sodium (1000 μg/ml) to both sample and standard solutions. If more stringent control of ionization is required, the addition of cesium should be considered. Reagent blanks should be analyzed to correct for potassium impurities in the buffer stock.

18.22.1.1 Instrument Parameters

- *Instrument*: Potassium hollow cathode lamp
- *Wavelength*: 766.5 nm
- *Fuel*: Acetylene
- *Oxidant*: Air
- *Type of flame*: Slightly oxidizing (lean fuel)
- *Background correction*: Not required

18.22.1.2 Performance Characteristics

- *Optimum concentration range*: 0.02 mg/l
- *Sensitivity*: 0.04 mg/l
- *Detection limit*: 0.01 mg/l

18.23 SELENIUM

Selenium (Se) is an essential trace nutrient, and selenium-deficiency diseases are well known in veterinary medicine. Above trace levels, ingested selenium is toxic to animals and may be toxic to humans. The selenium concentration in most drinking waters and natural waters is less than 10 μg/l. However, the pore water in seleniferous soils in semiarid areas may contain up to hundreds of micrograms of selenium per liter. Certain plants that grow in such areas accumulate large concentrations of selenium and may poison livestock that graze on them. Water drained from such soil may cause severe environmental pollution and wildlife toxicity. Soluble selenium can leach from coal ash at electric power plants that burn seleniferous coal. Selenium derives from microbial degradation of seleniferous organic matter. Nonvolatile organic selenium compounds may be released in water by microbial processes.

Selection of method: Selenium can be determined by the hydride-generation atomic absorption spectrometric or GrAAS methods. For higher concentrations of selenium, the ICP method may be used.

18.23.1 Graphite Furnace Atomic Absorption Spectrometry (GrAAS)

Prior to analysis, samples must be prepared in order to convert organic forms of selenium to inorganic forms, minimize organic interferences, and convert samples to suitable solutions for analysis. Elemental selenium and many of its compounds are volatile; therefore, samples may be subject to losses of selenium during sample preparation. Spike samples and relevant standard reference materials should be processed to determine if the chosen dissolution method is appropriate.

Caution must be taken during the selection of temperature and times for the dry and char (ash) cycles. A nickel-nitrate solution must be added to all digestates prior to analysis to minimize volatilization losses during drying and ashing.

Selenium analysis can suffer from severe nonspecific absorption and light scattering caused by matrix components during atomization. Selenium analysis is particularly susceptible to these problems because of its low analytical wavelength. Simultaneous background correction is required to avoid erroneous high results. High iron levels can produce overcorrection in a deuterium background. Zeeman background correction can be useful in this situation.

If the analyte is not completely volatilized and removed from the furnace during atomization, memory effects will occur. If this situation is detected, the tube should be clean by operating the furnace at full power at regular intervals in the analytical scales.

Selenium analysis suffers interference from chlorides (>800 mg/l) and sulfate (>200 mg/l). The addition of nickel nitrate such that the final concentration is 1% nickel will reduce this interference.

18.23.1.1 Reagents

18.23.1.1.1 Concentrated HNO_3

The reagent is commercially available.

18.23.1.1.2 30% H_2O_2

The reagent is commercially available.

18.23.1.1.3 Nickel-Nitrate Solution 5%

Dissolve 24.780 g of $Ni(NO_3)_2.6H_2O$ in reagent-grade water and dilute to 100 ml.

18.23.1.1.4 Nickel Nitrate Solution 1%

Dissolve 20 ml of the 5% nickel-nitrate solution (Section 18.23.1.1.3) to 100 ml with reagent-grade water.

18.23.1.1.5 Selenium Calibration Standards

Withdraw an appropriate aliquot of the stock solution (1000 mg/l) and add 1 ml of concentrated HNO_3, 2 ml of 30% H_2O_2, and 2 ml of 5% nickel-nitrate solution. Dilute to 100 ml with reagent-grade water.

18.23.1.2 Procedure

1. Transfer 100 ml of well-mixed sample to a 250-ml Griffin beaker; add 2 ml of 30% H_2O_2 and sufficient concentrated HNO_3 to result in an acid concentration of 1% (v/v).
2. Heat for 1 h at 95°C or until the volume is slightly less than 50 ml.
3. Cool and bring back to 50 ml with reagent-grade water.
4. Pipet 5 ml of this digested solution to a 10-ml volumetric flask, add 1 ml of the 1% nickel-nitrate solution, and dilute to 10 ml with reagent-grade water.
5. The sample is ready now for injection into the furnace. Furnace parameters suggested by the manufacturer should be employed as guidelines.
6. Inject a measured microgram-per-liter aliquot of sample into the furnace and atomize. The use of multiple injections can improve accuracy. Run a check standard after every ten samples. Standards are run in part to monitor the life and performance of the graphite tube. Lack of reproducibility or significant change in the signal for the standard indicates that the tube should be replaced.
7. Duplicates, spiked samples, and check standards should be analyzed every 20 samples.

18.23.1.3 Instrument Parameters

- *Drying time and temperature*: 30 sec at 125°C
- *Ashing time and temperature*: 30 sec at 1200°C
- *Atomizing time and temperature*: 10 sec at 2700°C
- *Purge gas*: Argon
- *Wavelength*: 196.0 nm

Other operating parameters should be set as specified by the instrument manufacturer.

18.23.1.4 Performance Characteristics

- *Optimum concentration range*: 5 to 100 µg/l
- *Detection limit*: 2 µg/l

18.23.2 Atomic Absorption Gaseous Hydride Technique

See Chapter 11 for instrumentation and procedure details.

18.24 SILVER

The silver (Ag) concentration in drinking waters varies from 0 to 2 µg/l, with a mean of 0.13 µg/l. Silver can cause argyria, a blue-gray discoloration of the skin and eyes. Concentrations in the range of 0.4 to 1 mg/l have caused pathological changes in the kidneys, liver, and spleen of rats. Toxic effects in freshwater fish have been observed at concentrations as low as 0.17 µg/l. Relatively small quantities of silver are bactericidal or bacteriostatic and are sometimes used in disinfecting swimming pool waters.

Selection of method: The FAAS and ICP methods are preferred. The GrAAS method is the most sensitive.

18.24.1 Flame Atomic Absorption Spectroscopy (FAAS)

Background correction is required because nonspecific absorption and light scattering may occur at the analytical wavelength. Silver nitrate solutions are light sensitive and tend to plate out on container walls. Silver standards should be stored in brown bottles. Silver chloride is insoluble; therefore, HCl should be avoided.

18.24.1.1 Instrument Parameters

- *Instrument*: Silver hollow cathode lamp
- *Wavelength*: 328.1 nm
- *Fuel*: Acetylene
- *Oxidant*: Air
- *Type of flame*: Oxidizing

18.24.1.2 Performance Characteristics

- *Optimum concentration range*: 0.1 to 4 mg/l
- *Detection limit*: 0.01 mg/l

18.24.2 Graphite Furnace Atomic Absorption Spectrometry (GrAAS)

Background correction may be required if the sample contains highly dissolved solids. The use of halide acids should be avoided.

18.24.2.1 Instrument Parameters

- *Drying time and temperature*: 30 sec at 125°C
- *Ashing time and temperature*: 30 sec at 400°C
- *Atomizing time and temperature*: 10 sec at 2700°C
- *Purge gas*: Argon
- *Wavelength*: 328.1 nm

Other operating parameters should be set as specified by the instrument manufacturer.

18.24.2.2 Performance Characteristics

- *Optimum concentration range*: 1 to 25 μg/l
- *Detection limit*: 0.2 μg/l

18.25 SODIUM

Sodium (Na) ranks sixth among the elements in order of abundance and is present in most natural waters. Levels may vary from less than 1 mg/l to more than 500 mg/l. Relatively high levels may be found in brines and hard waters softened by the sodium exchange process. The ratio of sodium to total cations is important in agriculture and human pathology. A limiting concentration of 2 to 3 mg/l is recommended in feed waters destined for high-pressure boilers.

Selection of method: The FAAS or ICP method is recommended.

18.25.1 Flame Atomic Absorption Spectroscopy (FAAS)

Ionization interferences can affect analysis for sodium; therefore, samples and standards must be matrix matched or an ionization suppressant employed. Sodium is a universal contaminant, and great care should be taken to avoid contamination.

18.25.1.1 Instrument Parameters

- *Instrument*: Sodium hollow cathode lamp
- *Wavelength*: 589.6 nm
- *Fuel*: Acetylene
- *Oxidant*: Air
- *Type of flame*: Oxidizing (lean fuel)
- *Background correction*: Not required

18.25.1.2 Performance Characteristics

- *Optimum concentration range*: 0.03 to 1 mg/l
- *Sensitivity*: 0.015 mg/l
- *Detection limit*: 0.002 mg/l

18.26 THALLIUM

Thallium (Ta) occurs normally at trace levels (<10 μg/l) in natural waters. Because thallium is toxic, it has been placed on several regulatory lists.

Selection of method: Use the FAAS method (Chapter 8) or the ICP method (Chapter 12).

18.26.1 Flame Atomic Absorption Spectroscopy (FAAS)

Background correction is required. HCl should not be used.

18.26.1.1 Instrument Parameters

- *Instrument*: Thallium hollow cathode lamp
- *Wavelength*: 276.8 nm
- *Fuel*: Acetylene
- *Oxidant*: Air

- *Type of flame*: Oxidizing (lean fuel)
- *Background correction*: Required

18.26.1.2 Performance Characteristics

- *Optimum concentration range*: 1 to 20 mg/l
- *Sensitivity*: 0.5 mg/l
- *Detection limit*: 0.1 mg/l

18.26.2 Graphite Furnace Atomic Absorption Spectrometry (GrAAS)

The use of background correction is recommended. Nitrogen may also be used as the purge gas.

18.26.2.1 Instrument Parameters

- *Drying time and temperature*: 30 sec at 125°C
- *Ashing time and temperature*: 30 sec at 400°C
- *Atomizing time and temperature*: 10 sec at 2400°C
- *Purge gas*: Argon
- *Wavelength*: 276.8 nm

Other operation parameters should be set as specified by the instrument manufacturer.

18.26.2.2 Performance Characteristics

- *Optimum concentration range*: 5 to 100 μg/l
- *Detection limit*: 1 μg/l

18.27 TIN

Tin (Sn) is normally soluble only at trace levels in natural waters (<100 μg/l), except in process waste waters or mineral waters.

Selection of method: Use either the FAAS (Chapter 8) or GrAAS (Chapter 12) method.

18.27.1 Flame Atomic Absorption Spectroscopy (FAAS)

18.27.1.1 Instrument Parameters

- *Instrument*: Tin hollow cathode lamp
- *Wavelength*: 286.3 nm
- *Fuel*: Acetylene
- *Oxidant*: Nitrous oxide
- *Type of flame*: Rich fuel
- *Background correction*: Not required

18.27.1.2 Performance Characteristics

- *Optimum concentration range*: 10 to 300 mg/l
- *Sensitivity*: 4 mg/l
- *Detection limit*: 0.8 mg/l

18.27.2 Graphite Furnace Atomic Absorption Spectrometry (GrAAS)

The use of background correction is recommended. Nitrogen may also be used as a purge gas. Tin analysis is sensitive to chloride concentration. If chloride concentration presents a matrix problem or causes loss prior to atomization, add an excess of 5 mg of ammonium nitrate to the furnace and ash using a ramp as necessary or with incremental steps until the recommended ashing temperature is reached. Extended ashing times have been reported to improve precision.

18.27.2.1 Instrument Parameters

- *Drying time and temperature*: 30 sec at 125°C
- *Ashing time and temperature*: 30 sec at 600°C
- *Atomizing time and temperature*: 10 sec at 2700°C
- *Purge gas*: Argon
- *Wavelength*: 224.6 nm

Other operating parameters should be set as specified by the instrument manufacturer.

18.27.2.2 Performance Characteristics

- *Optimum concentration range*: 20 to 300 µg/l
- *Detection limit*: 1 µg/l

18.28 VANADIUM

Vanadium (V) may play a beneficial role in the prevention of heart disease. The mean concentration found in U.S. drinking waters is 6 µg/l.

Selection of method: The FAAS (Chapter 8) and ICP (Chapter 12) methods are recommended.

18.28.1 Flame Atomic Absorption Spectroscopy (FAAS)

High concentrations of aluminum, titanium, or the presence of bismuth, chromium, cobalt, iron, acetic acid, phosphoric acid, surfactants, detergents, or alkali metals, may interfere. The interference can be controlled by adding 1000 mg/l aluminum to samples and standards.

The calibration standards should be prepared using the same type of acid and at the same concentration as will result in the sample to be analyzed after processing. In addition, 2 ml of the aluminum nitrate solution should be added to each 100 ml of standards and samples. (To prepare the aluminum nitrate solution, dissolve 139 g $Al(NO_3)_3.9H_2O$ in 150 ml of reagent-grade water; heat to complete dissolution. Allow to cool and dilute to 200 ml with reagent-grade water.)

18.28.1.1 Instrument Parameters

- *Instrument*: Vanadium hollow cathode lamp
- *Wavelength*: 318.4 nm
- *Fuel*: Acetylene
- *Oxidant*: Nitrous oxide
- *Type of flame*: Rich fuel
- *Background correction*: Required

18.28.1.2 Performance Characteristics

- *Optimum concentration range*: 2 to 100 mg/l
- *Sensitivity*: 0.8 mg/l
- *Detection limit*: 0.2 mg/l

For concentrations of vanadium below 0.5 mg/l, the furnace technique is recommended.

18.28.2 Graphite Furnace Atomic Absorption Spectrometry (GrAAS)

Vanadium is refractory and prone to form carbides. Consequently, memory effects are common, and care should be taken to clean the furnace before and after analysis. Nitrogen should not be used as a purge gas.

18.28.2.1 Instrument Parameters

- *Drying time and temperature*: 30 sec at 125°C
- *Ashing time and temperature*: 30 sec at 1400°C
- *Atomizing time and temperature*: 15 sec at 2800°C
- *Purge gas*: Argon (N should not be used)
- *Wavelength*: 318.4 nm
- *Background correction*: Required

Other operating parameters should be set as specified by the instrument manufacturer.

The above concentration values and instrument conditions are for a Perkin Elmer HGA-2100, based on the use of a 20-µl injection, continuous-flow purge gas, and nonpyrolytic graphite. Smaller sizes of furnace devices or those employing faster rates of atomization can be operated using lower atomization temperatures for shorter time periods than the recommended settings above.

18.28.2.2 Performance Characteristics

- *Optimum concentration range*: 10 to 200 µg/l
- *Detection limit*: 4 µg/l

18.29 ZINC

Zinc (Zn) is an essential and beneficial element in human growth. Concentrations above 5 mg/l can cause a bitter astringent taste and an opalescence in alkaline waters. Zinc concentrations in U.S. drinking waters range from 0.06 to 7.0 mg/l with a mean of 1.33 mg/l. Zinc most commonly enters the domestic water supply from deterioration of galvanized iron and dezincification of brass. In such cases, lead and cadmium may also be present because they are impurities of the zinc used in galvanizing. Zinc in water also may result from industrial waste pollution.

Selection of method: The FAAS (Chapter 8) and ICP (Chapter 12) methods are preferable.

18.29.1 Flame Atomic Absorption Spectroscopy (FAAS)

High levels of silicon, copper, or phosphate may interfere. Addition of strontium (1500 mg/l) removes the copper and phosphate interference.

18.29.1.1 Instrument Parameters

- *Instrument*: Zinc hollow cathode lamp
- *Wavelength*: 213.9 nm
- *Fuel*: Acetylene
- *Oxidant*: Air
- *Type of flame*: Oxidizing (lean fuel)
- *Background correction*: Required

18.29.1.2 Performance Characteristics

- *Optimum concentration range*: 0.05 to 1 mg/l
- *Sensitivity*: 0.02 mg/l
- *Detection limit*: 0.005 mg/l

18.29.2 Graphite Furnace Atomic Absorption spectrometry (GrAAS)

18.29.2.1 Instrument Parameters

- *Drying time and temperature*: 30 sec at 125°C
- *Ashing time and temperature*: 30 sec at 400°C
- *Atomizing time and temperature*: 10 sec at 2500°C
- *Purge gas*: Argon
- *Wavelength*: 213.9 nm

Other operating parameters should be set as specified by the instrument manufacturer.

18.29.2.2 Performance Characteristics

- *Optimum concentration range*: 0.2 to 4 μg/l
- *Detection limit*: 0.05 μg/l

19 Laboratory Safety Rules

Before beginning any type of laboratory work, it is important to understand the potential hazards in the laboratory, to be familiar with the precautions and rules, and to recognize and avoid the causes of those hazards. According to the Occupational Safety and Health Act (OSHA), "[E]ach employer has the general duty to furnish each of the employees a workplace free from recognized hazards causing or likely to cause death or serious harm." *Comprehensive safety training is essential for all laboratory workers.*

19.1 LABORATORY HAZARDS

Laboratory hazards can be categorized as chemical hazards, fire hazards, and careless habits.

19.1.1 Chemical Hazards

Virtually all chemicals are toxic to some extent, and care should be taken in handling them. Chemical hazards may be minimized by the following precautions.

19.1.1.1 Cleanliness

Cleanliness in the laboratory is essential:

- *Wash hands* periodically and immediately after contact with chemicals and just before leaving the laboratory.
- *Never drink from laboratory glassware.*
- *Keep work areas clean.* Clean working areas before and after work.
- *Use clean laboratory coats and aprons.* These garments are designed to protect the body from chemical spills. Dirty clothing can be a source of health hazards and contamination.

19.1.1.2 Eye Protection

The eyes are especially susceptible to injury from chemicals. Breakage of glass containers of acid, bases, and other chemicals and out-of-control chemical reactions are the principal hazards. Safety glasses, goggles, or face shields should be worn during laboratory work. In the event of chemical spray in eyes, immediately flood the eyes with water using a specially designed eye-wash fountain or quick flushing with water from the nearest tap, and seek medical attention as soon as possible.

19.1.1.3 Skin Contact with Certain Chemicals

Chemical burns can result from contact with strong acids or bases. Certain chemicals are *absorbed* through the skin. Because many chemicals absorb rapidly through the skin, prompt clean-up is important. Remove contaminated clothing immediately and flush affected areas with a large quantity of water. Medical attention may be necessary, depending on the amount of chemical involved.

19.1.1.4 Body Protection

- *Use laboratory coats or aprons.* Laboratory coats are made from materials that provide protection against acids and bases. Laboratory aprons are not affected by ordinary corrosive fluids or other chemicals.
- *Never wear open-toe shoes or sandals.* This type of footwear offers little or no protection against chemical spills or broken glass.
- *Secure ties or scarves with fasteners.*
- *Put long hair up and out of the way.*
- *When handling corrosive chemicals, use protective gloves.* Protective gloves are selected according to need. Asbestos gloves protect against heat, but they are not advisable for handling corrosive chemicals (acids or bases), because asbestos absorbs the substance and increases contact time and area. When working with hot objects or organic solvents, do not use rubber or plastic gloves, because they may soften and dissolve.

19.1.1.5 Ingestion of Toxic Chemicals

Do not consume or store food or beverages in the laboratory. Food is easily contaminated, such as by traces of chemicals on hands. To avoid any possibility of ingesting chemical solutions while using a pipet, use a pipeting bulb and not the mouth.

19.1.1.6 Inhalation of Volatile Liquids and Gases

The presence of these substances in the air (even in low concentrations) is hazardous. Acute exposure to extremely high concentrations in vapors (above the maximum allowable concentration) can cause unconsciousness and even death, if the person is not removed from the area and if medical attention is delayed. The exposure to solvent and chemical vapors can be avoided by working with such chemicals under chemical hoods and wearing protective respiratory devices. Good ventilation is essential to a safe laboratory.

19.1.1.6.1 Toxicity of Metallic Elements

Metals with a specific gravity of greater than 5 are called heavy metals. In the metallic state they are harmless, but in the vapor state these elements and their soluble compounds are toxic. The most common heavy metals are antimony (Sb), arsenic (As), cadmium (Cd), chromium (Cr), lead (Pb), mercury (Hg), nickel (Ni), silver (Ag), and thallium (Tl).

19.1.1.6.2 Chemical Dust

Fine-powder chemicals can be inhaled as dust; therefore, these chemicals should also be handled under a laboratory hood.

19.1.1.7 Chemical Spills

19.1.1.7.1 Solid, Dry Substances

Spills of chemicals in this form can be swept together, brushed into a dustpan or cardboard receptacle, and then deposited in an appropriate waste container.

19.1.1.7.2 Acid Spills

Clean up acid spills by using the appropriate spill kit and following the instructions. The material in such kits neutralizes and absorbs the acid for easy clean-up. Afterwards the area should be washed with water. Alternatively, use soda ash (Na_2CO_3) or sodium bicarbonate ($NaHCO_3$) solution for neutralization, and then flush the area with water.

Caution: When water is poured on spills of concentrated sulfuric acid (H_2SO_4), tremendous heat is released (exothermic reaction) and the acid splatters. Deluge with water to dilute the acid to minimize heat generation and splattering.

19.1.1.7.3 Alkaline Spills

Alkaline spills are treated similarly to acid spills; use an alkaline spill kit. Alternatively, use a weak acid solution, such as diluted acetic acid, for neutralization. The area should then be flushed with water to a floor drain. If a mop and bucket are used, flush by replacing water frequently.

Caution: Alkali solutions make the floor slippery!

Clean sand can also be used to clean up alkaline spills. Throw sand over the spill and sweep up. The wet sand is then discarded.

19.1.1.7.4 Volatile Solvent Spills

Volatile solvents evaporate very rapidly because of the extremely large surface area. This kind of the spill can create a fire hazard if the solvent is flammable and will invariably cause highly dangerous concentrations of fumes in the laboratory. When inhaled, these fumes cause serious injuries. They may also become explosive upon mixing with air.

To clean up a small spill, wipe up the liquid with absorbent cloths or towels and discard them in an appropriate waste receptacle. If a large amount of solvent is involved in the spill, use a mop and pail. Squeeze out the mop in the pail and continue as needed.

19.1.1.7.5 Oily Substance Spills

This type of spill should be cleaned up with an appropriate nonflammable volatile solvent. Pour solvent on an absorbent cloth, and wipe up the spilled substance. Rinse the cloth in a pail of solvent to remove all spilled material, because oily floors are slippery and dangerous. Finally, thoroughly scrub with detergent and water to remove oily residue.

19.1.1.7.6 Mercury Spills

Spills are one of the most common sources of mercury vapor in laboratory air. In a spill, mercury may be distributed over a wide area, exposing a large surface area of the metal, and droplets become trapped in crevices. Unless the laboratory has adequate ventilation, mercury vapor concentration (accumulated over time) may exceed the recommended limit. Vibration increases mercury vaporization.

Caution: Surfaces that appear to be free of mercury will harbor microscopic droplets.

To clean up mercury spills, push droplets together to form pools, and then use a suction device to pick up the mercury. If there are cervices or cracks in the floor that can trap small droplets of mercury that cannot be picked up, seal over the cracks with a thick covering of floor wax or an aerosol hair spray. The covering will dramatically reduce vaporization. Sulfur powder can also be used to fix mercury. Mercury spill kits are also available for proper mercury clean-up.

19.1.2 Fire Hazards

Fire in a chemical laboratory can be dangerous and devastating. In case of fire, stay calm and think! Sources of fires include electrical equipment, friction, mechanical sparks, flames, hot surfaces, and flammable organic compounds. Accidental ignition of volatile organic solvents is perhaps the most common source of laboratory fires. To avoid accidental spills and reduce fire hazards, *keep volatile solvents in small containers and never work with a volatile solvent around an open flame.* The sooner you respond to put out a fire, the easier it is to control.

19.1.2.1 Fire Classifications

The appropriate response to a fire depends on the type of material being consumed. The use of the wrong type of firefighting equipment may increase the intensity of the fire. Fire classifications are described below.

19.1.2.1.1 Class A Fires

These fires are caused by the burning of paper, wood, and textiles. Almost any type of extinguisher is satisfactory.

19.1.2.1.2 Class B Fires

This type of fire is caused by the burning of oil, grease, organic solvents, and paint. Use a dry-chemical, liquid, CO_2, or foam extinguisher.

19.1.2.1.3 Class C Fires

This type consists of electrical fires in equipment. Do not use a water or foam extinguisher, because you may become a part of the electrical circuit and be electrocuted! Use a CO_2 or dry-chemical fire extinguisher only.

19.1.2.1.4 Class D Fires

Class "D" fires are caused by sodium, potassium, magnesium, lithium, and all metal hydrides. Use a dry, soda-ash fire extinguisher, sodium chloride, or dry sand.

19.1.2.2 Fire-Fighting Responses

19.1.2.2.1 Fire in Clothing

Wrap the person in a fire blanket or heavy towels. Use an emergency shower.

19.1.2.2.2 General Fires

Select the proper fire extinguisher according to the type of fire. First, cool the area around the fire with the extinguisher to prevent the fire from spreading. Next, use the extinguisher at the core area of the fire. Finally, extinguish scattered remnants of the fire.

19.1.2.2.3 Electrical Fires

First, disconnect the apparatus by pulling the safety switch to avoid the possibility of being electrocuted. Then, use class C (CO_2 or dry chemical) extinguisher.

19.1.2.2.4 Poisonous Gas Fires

Use an appropriate respirator, and select the proper fire extinguisher. If the fire gets beyond the control of the available fire extinguisher, get out of the room immediately. Close the door to prevent drafts and gas spread. Always be certain that no one is left behind.

In case of fire, immediately notify the local fire department!

19.1.3 Carelessness

Most laboratory accidents are caused by impulsive acts that later seem thoughtless, careless, and even reckless. Thus, always think about the possible consequences of your actions before you act.

19.1.3.1 Hazards from Falling Objects

Falling objects can cause serious injuries. Do not place heavy objects on high shelves! If a heavy object must be placed on a shelf, secure it with a belt or chain. Be careful when moving heavy instruments and other heavy objects; use a laboratory cart whenever possible.

19.1.3.2 Hazards from Falling

Never climb on drums, cartons, or boxes to reach objects located on high shelves. You may be severely injured, and the injury can be compounded by breakage of glassware or chemical splash. Always use a safety stepladder; special locking devices ensure that the rubber-tipped legs do not move.

19.1.3.3 Transporting Large Bottles

Moving large bottles and carboys is a dangerous operation because of the potential for bottle breakage and liquid spillage. Always use safety carts and safety bottle carriers when transporting large bottles of chemicals. Safety bottle carriers prevent shock and breakage.

19.2 SAFE HANDLING OF COMPRESSED GASES

Cylinders of compressed gas can be dangerous because gases are contained under very high pressure. Always follow safety precautions when handling such cylinders.

19.2.1 General Precautions When Working with Compressed Gases

19.2.1.1 General Precautions

- Close off main cylinder valve when not in use.
- Close needle valve or auxiliary cut-off valve in the line and the cylinder. Do not rely solely on the cylinder valve.
- Replace cylinders within reasonable time periods. Corrosive gas cylinders should be replaced every 3 months or less.
- Always use gases in areas where adequate ventilation is provided.
- Keep cylinders in outside storage, or use manifolds that pipe low-pressure gas into buildings.
- Use the smallest cylinder that is practical for the purpose.

19.2.1.2 Safety Rules for Using Compressed Gases

- *Cylinder contents must be properly identified*: Do not use cylinders without written content identification. Do not rely on color codes for identification. Do not destroy identification tags or labels.
- *Protect cylinder valves*. Use only cylinders equipped with protective valve caps. Leave caps in place until ready to use the gas.
- *Store properly*. Provide specifically assigned locations for cylinder storage, preferably in a dry, fire-resistant, and well-ventilated area away from sources of ignition or heat. Outdoor storage areas should have proper drainage and be protected from direct sunlight. Secure cylinders by chains or other means to prevent accidental tipping or falling..
- *Transport correctly*. Transport cylinders by means of a suitable hand truck. Do not roll cylinders on the ground!
- *Do not drop*. Never drop cylinders or permit them to strike each other.
- *Return in condition received.* Close valve, and replace cylinder-valve protective cap and dust cap. Mark or label cylinder "EMPTY" or "MT."
- *Prevent confusing empties with full cylinders*: Store empty cylinders in an area separate from full cylinders. Connecting an empty cylinder to a pressurized system could cause contamination or violent reaction in the cylinder.

19.2.2 Hazardous Properties of Compressed Gases

The properties of a compressed gas must be well known and understood before the gas is put to use. Hazards include flammability, toxicity, and corrosivity.

19.3 STOCKROOM SAFETY RULES

The laboratory stockroom should be adequate and efficiently planned for safe operation.

19.3.1 Safety Checklist for Storage Rooms: Room Characteristics and Organization

- Wide aisles, adequate lighting, and no blind alleys; the entire complex should be orderly and clean
- Adequate ventilation and emergency exhaust system
- Well-marked exits, including emergency exits
- Adequate fire-protection and firefighting equipment
- Heavy items stored near the floor
- Proper storage for glass apparatus and tubing (never projecting beyond shelf limits)
- Fragile and bulky equipment secured to shelving.
- Shelving fitted with ledges to prevent items from sliding or rolling off
- Appropriate grouping and separation of liquids and hazardous chemicals
- No waste accumulation of any kind
- Safety ladders available; all laboratory personnel should be encouraged to use safety ladders, because they prevent accidents and save time and effort
- No excessive heat, because of fire hazard
- Regular housekeeping activities aimed at maintaining safe storage practices

19.3.2 Chemical Storage

Chemicals are manufactured in varying degrees of purity. Carefully select the grade of the chemical that meets the need of the work to be done. Always recheck the label of the chemical that you are using! The use of a wrong chemical can cause an explosion or ruin the analytical work. Carefully check the information on the chemical container, including name, formula, formula weight, percent impurities, analytical grade, health hazards, and safety codes.

19.3.2.1 Acids

Acids should be stored in original containers in cabinets labeled "Acids" and grouped by safety color codes. Bottles with impact-resistant plastic coatings are preferred.

19.3.2.2 Flammable Solvents

Store these chemicals in original containers, in cabinets labeled "Flammable." Large quantities should be stored in metal safety cans outside of the laboratory in an area marked "Flammable Storage Area."

19.3.2.3 Solvents

Solvents should be stored in original containers in a separate cabinet labeled "Solvents" and in a well-ventilated area.

19.3.2.4 Chemicals Used in Volatile Organic (VOC) Analysis

These chemicals should be stored in original containers in a separate, appropriately labeled cabinet and in a well-ventilated area. No other chemicals should be stored along with them.

19.3.2.5 Storage Organization

Chemicals should be stored in alphabetical order in the storage room, with records of date of arrival and date of opening affixed to each container. Store phenol and hydrogen peroxide in a refrigerator labeled with "Chemical Storage." The *LabGuard Safety Label System* on chemical bottles assists in the proper storage of chemicals. Each chemical used in the laboratory should be accompanied by a *Material Safety Data Sheet* (MSDS). MSDSs contain ingredients, physical and chemical characteristics of the substance, physical hazards, reactivity and health hazards involved, and safe handling and safety precautions. In addition, control measures to reduce harmful exposures are also listed in every MSDS.

19.4 SUMMARY OF LABORATORY SAFETY RULES

1. Safety glasses/corrective glasses should be worn at all times in the laboratory. Visitors to the laboratory must be appropriately warned and safety glasses made available to them.
2. Participation in practical jokes or "horseplay" in the laboratory is not permitted.
3. Each laboratory worker is expected to cooperate in keeping his or her working area in a neat and orderly condition and to cooperate with others in keeping the entire laboratory neat and orderly. *A clean laboratory is a safe laboratory.*
4. Proper techniques should be utilized when lifting, pushing, pulling, or carrying materials to prevent injuries.
5. All laboratory personnel must know the location of fire extinguishers, safety showers, eyewash stations, and spill kits.
6. All laboratory workers must know how and when to use the equipment listed in item 5.
7. Eating, drinking, and smoking in the laboratory are never allowed. Never use laboratory containers (beakers or flasks) for drinking.
8. No food or beverages intended for human consumption are stored in refrigerators in the laboratory.
9. MSDSs must be attached to all chemicals used in the laboratory.
10. All chemicals should be clearly labeled. Do not use material from unlabeled containers. Ensure that chemicals are clearly identified before using them.
11. In the event of chemical spraying in the eyes, use the eyewash station and report the incident to the laboratory supervisor.
12. Respirators must be used when working with hot acids or solvents that are handled when not under a fume hood.
13. Pouring of volatile liquids should be done only in a well-ventilated hood remote from sources of ignition.
14. Only minimum amounts of flammable liquids that are necessary for running a test should be kept on workbenches.
15. Heavy reagent containers, such as 5-gallon containers, must not be carried or placed on a shelf by one person working alone.
16. Face shields, rubber gloves, and protective rubber aprons should be used when preparing, transporting, or pouring corrosive chemicals, such as concentrated acids and bases.
17. When diluting acid with water, always add the acid to the water, stirring constantly. Never

add water to the acid, as this produces a violent reaction.

18. When drawing liquid into a pipet, always use a suction bulb. Mouth pipeting is never allowed.
19. Pouring mercury into a sink or drain is strictly prohibited. Mercury will remain in the trap and continue to vaporize and contaminate the air.
20. In the event of an acid spill on a person, flush thoroughly with water *immediately.*

Caution: Acid–water mixtures produce heat. Removal of clothing from the affected area while flushing may be important so as not to trap hot acid–water mixtures against the skin. Acids or acid–water mixtures can cause very serious burns if left in contact with skin for even a very short period of time.

21. Weak acids should be used to neutralize base spills, and weak bases should be used to neutralize acid spills. Such solutions should be available in the laboratory in case of emergency. Acid and base spill kits are also available.
22. Unsupervised or unauthorized work in the laboratory is not permitted.
23. Never wear open-toed shoes or sandals because they offer little or no protection against chemical spills and broken glassware.
24. Keep ties and scarves secured with fasteners. Do not wear medallions, pendants, or other hanging objects.
25. Tie long hair up and out of the way.
26. Asbestos gloves should be worn when handling or working with hot materials.
27. Gloves should be worn when exerting pressure is necessary to open jars, bottles, or other containers.
28. A face shield should be worn when handling a receptacle containing more than 1 liter of acid, alkali, or corrosive liquid.
29. Chemicals should never be transported, transferred, poured, or otherwise handled at a height above one's head.
30. Any injury, regardless of how superficial, should be reported to the laboratory supervisor (or instructor in a school laboratory), and appropriate first-aid action taken.
31. A leakage check should be made on all gas lines and connections whenever a line is broken and reconnected.
32. Immediately report to the laboratory supervisor any failure of exhaust fans to evacuate vapors completely, defective electrical equipments, faulty or empty fire extinguishers, and worn or defective rubber gas-burner hoses or other gas hazards.
33. Use a stepladder provided for this purpose when reaching into high shelving.
34. Never leave operations involving explosives or flammable mixtures unattended.
35. When transporting a large quantity of bottles, do so with a basket or receptacle designed for this purpose.
36. Do not use damaged glassware.
37. Do not place glassware close to the edge of the laboratory bench; a passerby may knock it off.
38. Wear goggles or a face shield when working with a glass apparatus that is under pressure or vacuum.
39. When making a vacuum distillation, use a shield to guard to protect against explosion and fire hazard.
40. Clean up broken glass immediately and place it in containers provided for broken glass. Never dispose of broken glassware in a regular garbage container!

Appendix A: Operation of Mass Spectrophotometer

MASS SPECTROSCOPY

Mass spectroscopy is a technique used to determine relative atomic masses and the relative abundance of isotopes, in chemical composition analysis and the study of ion reactions. In a mass spectrometer, a sample (usually gaseous) is ionized and the positive ions produced are accelerated into a high-vacuum region containing electric and magnetic fields. These fields deflect and focus the ions onto a detector. The fields can be varied in a controlled way so that ions of different types hit the detector.

OPERATION OF MASS SPECTROPHOTOMETER

1. All the air is pumped out of the instrument.
2. The sample (gaseous vapor of liquid or solid) is fed into the *ionization chamber* of the spectrophotometer.
3. The sample is then exposed to a beam of rapidly moving electrons. When an accelerated electron collides with an atom and knocks another electron out of it, the atom becomes a positively charged ion.
4. The positive ions are accelerated out of the chamber by a strong *electric field*. Speeds attained by the ions depend on their masses, with light ions reaching higher speeds than heavy ones.
5. When the accelerated ions pass through a *magnetic field* generated by an electromagnet, their paths are bent to an extent dependent on speed and hence on mass.
6. A signal is produced when the strength of the magnetic field is just enough to bend the beam of ions so that they arrive at the *detector*.
7. The mass of the ion formed is then calculated based on the accelerating voltage and strength of the magnetic field used to produce the *signal*.

The process of sample inlet system → Ionization chamber → Mass analyzer → Detector → Signal A is shown in Figures A.1 and A.2, respectively.

MASS SPECTRUM

The mass spectrum obtained in the spectrophotometer signal consists of a series of peaks of variable intensity to which mass/charge (m/e) values can be assigned. The mass spectrum is a plot of the detector signal against the magnetic field. The positions of the peaks are used to calculate the mass of accelerated ions, and the relative heights of the peaks indicate the proportions of ions of various types. For organic molecules, the mass spectrum consists of a series of peaks, one corresponding to the parent ion and the others to fragment ions produced in the ionization process. Molecule composition can be identified by characteristic patterns of lines.

Gas Chromatography–Mass Spectroscopy (GC–MS)

Gas chromatography is used to separate a mixture into its components, which are then directly injected into a mass spectrometer. The combined technique is known as gas chromatography–mass spectroscopy (GC–MS).

FRACTIONAL ABUNDANCE OF ISOTOPES

In 1913, J.J. Thomson determined that the mass of neon is 20 amu, but he also found a less abundant mass of 22 amu, which he thought was a contaminant. Later, with the benefit of improved equipment, F.W. Aston showed that most elements are mixtures of isotopes. Isotopes are one or more atoms of the same element that have the same number of protons in their nucleus but different numbers of neutrons.

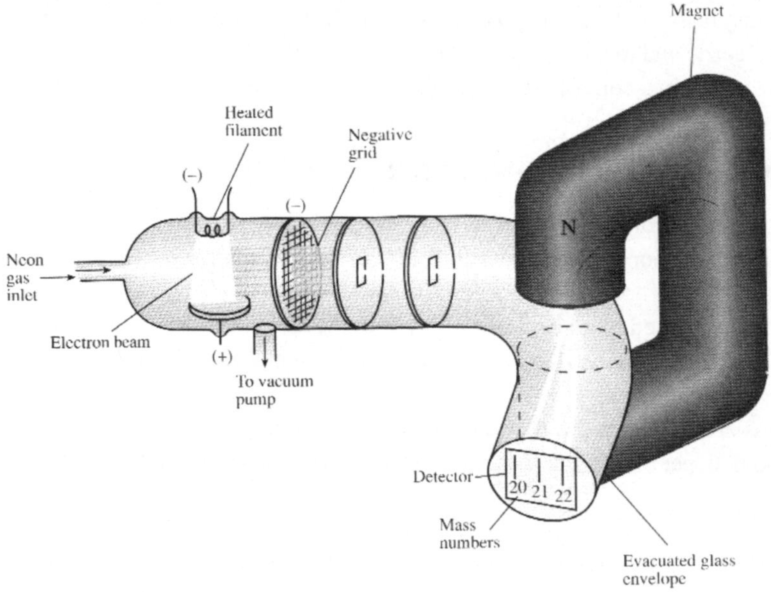

FIGURE A.1 Diagram of a simple mass spectrophotometer showing separation of neon isotopes.

FIGURE A.2 Mass spectrophotometer. This instrument measures the mass of atoms and molecules.

For instance, hydrogen isotopes include hydrogen (1 proton, no neutrons), deuterium (1 proton, 1 neutron), and tritium (1 proton, 2 neutrons). Most elements in nature consist of a mixture of isotopes.

The mass spectrum provides all information necessary to calculate atomic weight: the mass of each isotope and relative numbers, or fractional abundance of the isotopes. The *fractional abundance* of an isotope is the fraction of the total number of atoms composed of a particular isotope. The atomic weight of an element is calculated by multiplying each isotopic mass by its fractional abundance and summing the values.

For example, positively charged neon atoms split into three beams corresponding to the three isotopes of neon. Each atom has a charge of +1 but has a mass number of 20, 21, or 22. Neon isotopes and respective atomic mass units follow: neon 20, 19.992 amu; neon 21, 20.994 amu; and neon 22, 21.991 amu. Figure A.1 shows the mass spectrum of neon. The fractional abundance of the neon isotopes in naturally occurring neon follow: neon 20, 0.9051; neon 21, 0.0027; and neon 22, 0.0922. To calculate the atomic weight of an element, multiply each isotope mass by its fractional abundance and sum all values, as follows:

$$\text{Isotope mass} \times \text{Fractional abundance} = \text{Isotope atomic weight}$$

Neon 20: $19.992 \times 0.9051 = 18.0950$

Neon 21: $20.994 \times 0.0027 = 0.0567$

Neon 22: $21.991 \times 0.0922 = 2.0276$

$$\text{Element atomic weight} = \Sigma \text{ isotope atomic weight}$$

Neon atomic weight: $18.0950 + 0.0567 + 2.0276 = 20.1793$

Appendix B: Silicon Chips

METALLIC CONDUCTION

In a metallic solid, cations lie in a regular array and are surrounded by a sea of electrons, as illustrated in Figure 1.3. This structure gives unique properties to metals. One of the most striking properties of a metal is its ability to conduct an electric current. The general term for this property is *electronic conduction*, and the specific term as applied to metals is *metallic conduction*. The ability of a substance to conduct electricity is measured by its *resistance* — the lower the resistance, the better it conducts.

INSULATORS

Insulator substances do not conduct electricity. Insulators include gases, most ionic solids, most network solids, almost all organic compounds, and all molecular and covalent liquids and solids.

METALLIC CONDUCTORS

A metallic conductor is an electronic conductor with a resistance that increases as temperature increases. All metals are metallic conductors.

SEMICONDUCTORS

A semiconductor is an electronic conductor with a resistance that decreases as the temperature increases. Semiconductor properties are a feature of metalloid elements such as silicon and germanium.

SUPERCONDUCTORS

A superconductor is an electronic conductor that conducts electricity with zero resistivity. Most superconductors are metals (e.g., lead) or compounds cooled to near-absolute zero. A superconductor substance has zero resistance below a certain temperature.

SEMICONDUCTING ELEMENTS

Semiconducting elements exhibit very low electrical conductivity at room temperature when pure; electrical conductivity increases with temperature or with the addition of a certain element. The process of adding small quantities of other elements to a semiconducting element to increase its conductivity is called *doping*. See Figure B.1 for a schematic drawing of silicon semiconductor crystal layers.

N-TYPE SEMICONDUCTORS

In an n-type semiconductor, a minute amount of a group VA (15) element, such as arsenic (As), is added to very pure silicon (Si). The As increases the number of electrons in the solid: Each Si atom (Group IVA, 14) has four valence electrons, whereas each As atom has five. The additional electrons

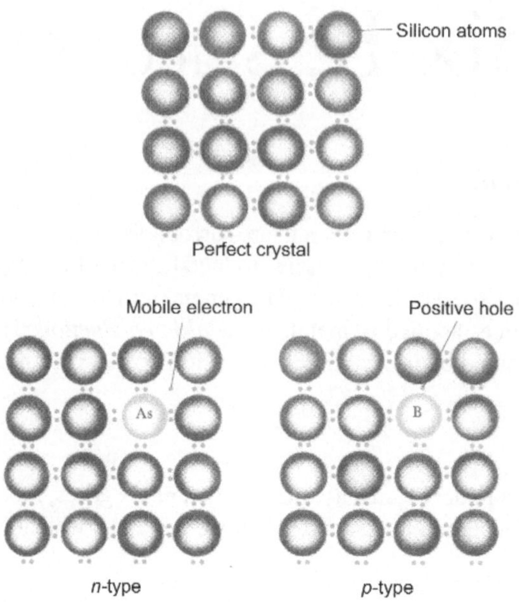

FIGURE B.1 Schematic drawing of silicon semiconductor crystal layers. (From *World of Chemistry*, 1st ed., by M.D. Joesten, D.O. Johnston, J.T. Netterville, J.L. Wood © 1990. Reprinted with permission of Brooks/Cole, an imprint of the Wadsworth Group, a division of Thomson Learning. Fax 800 730-2215.)

enter the upper, normally empty conduction band of silicon and allow the solid to conduct. This type of material is called an n-type semiconductor because it contains excess negatively charged electrons.

p-Type Semiconductors

In a p-type semiconductor, Si (Group IVA, 14) is doped with an element from group IIIA (13), such as boron (B). In this case, B has fewer valence electrons than Si, so the valence band is not completely full. The band now has "holes." Because the valence band is no longer full, it has turned into a conduction band and thus a current can flow. This type of material is called a p-type semiconductor because the absence of negatively charged electrons is equivalent to the presence of a positive charge.

TRANSISTORS AND OTHER ELECTRONIC DEVICES

Transistors

One of the most significant discoveries of the twentieth century was how electrical characteristics of semiconductors can be modified by the controlled introduction of carefully selected impurities. This led to the development of transistors, which have made possible all the electronic devices we now take for granted, such as portable televisions, compact disc players, radios, calculators, and microcomputers.

Various types of transistors (devices for controlling electrical signals) can be made by combining p- and n-type semiconductors. Transistors can be formed directly on the surface of a silicon chip, which has made possible the microcircuits used in computers and calculators. Some of the latest computer chips contain microscopic electrical circuits integrated with as many as a million transistors per centimeter of surface area.

Chips

The chip, a nickname for the integrated circuit, is a small slice of silicon that contains an intricate pattern of electronic switches (transistors) joined by "wires" etched from a thin film of metal. Some chips, known as *memory chips*, store information, while others combine memory with logic functions to produce *computer* or *microprocessor chips*. Chip applications are almost infinite. A microprocessor chip, for example, can provide a machine with decision-making ability, memory for instructions, and self-adjusting controls. In everyday life, we see many examples of chip applications: digital watches; microwave oven controls; hand calculators; electronic cash registers for calculating total bills, posting sales, and updating inventories; and computers in a variety of sizes and capacity.

Solar Batteries

The solar cell directly converts solar energy into electron flow. A *silicon solar battery* (also called a *solar cell*) is composed of a silicon wafer doped with arsenic (an n-type semiconductor) over which is placed a thin layer of silicon doped with boron (p-type semiconductor). The makeup of a silicon solar battery is illustrated in Figure B.1.

In the absence of light, equilibrium exists between electrons and holes at the interface between the two layers, which is called a p–n junction. Some electrons from the n-type layer diffuse into the holes in the p-layer and are trapped. This leaves some positive holes in the n-layer. Equilibrium is achieved when the positive holes in the n-layer prevent further movement of electrons into the p-layer. When light falls on the surface of the cell, the equilibrium is upset. Energy is absorbed, which permits electrons that were trapped in the p-layer to return to the n-layer. As electrons move across the p–n junction into the n-layer through the wire, they pass through the electrical circuit and enter the p-layer. Thus, an electric current flows when light falls on the cell and the external circuit is completed. The electric current can be used to run a motor or charge a battery, among many other tasks.

Appendix C: Lasers

A laser is a source of an intense, highly directed beam of monochromatic light. The word "laser" is an acronym for *light amplification by stimulated emission of radiation*. In a laser, electrons are raised to a higher energy state by the absorption of energy in one form or another. If conditions are right, the number of excited atoms exceeds the number in the ground state, and a *population inversion* exists. Not all substances can function as lasers. The laser process begins when one excited atom emits a photon, which strikes another excited atom that is stimulated to emit a photon. These emissions initiate further emissions and so on, until a cascade of photons is produced. In this way, the intensity of the original one-photon emission is amplified enormously.

Lasers can be solid, liquid, or gas devices. Population inversion is achieved by optical pumping with flashlights or with other lasers. It can also be achieved by such methods as chemical reactions and discharges in gases.

RUBY LASERS

The ruby laser was one of the earliest. A very bright flashlight, similar to the kind used in the electronic flash in the modern camera, wraps around a ruby rod and provides the energy to pump the laser into an excited state. The laser beam then emerges from the ruby through the partially reflecting end as seen in Figure C.1. Ruby is comprised of aluminum oxide containing a small concentration of chromium (III) ions (Cr^{3+}) in place of some aluminum ions. The electron transitions in a ruby laser are those of Cr^{+3} ions in solid Al_2O_3. Most of the Cr^{3+} ions are initially in the lowest energy level (level 1). If you shine light of wavelength 545 nm on a ruby crystal, the light is absorbed and Cr^{3+} ions undergo transitions from level 1 to level 3. A few of these ions in level 3 emit photons and return to level 1, but most of them undergo radiationless transitions to level 2. In these transitions, the ions lose energy as heat to the ruby crystal, rather than emit photons. However, this *spontaneous emission* of Cr^{3+} is relatively slow. If you flash a ruby rod with a bright light at 545 nm, most of the Cr^{3+} ions end up in level 2 for perhaps a fraction of a millisecond. This buildup of many excited species is crucial to the operation of a laser. If these excited ions can be triggered to emit simultaneously, an intense emission will be obtained. The process of *simultaneous emission* is ideal for this triggering. When a photon corresponding to 694 nm encounters a Cr^{3+} ion in level 2, it stimulates the ion to undergo the transition from level 2 to level 1. The ion emits a photon corresponding to exactly the same wavelength as the original photon. In the place of just one photon, there are now two photons, the original one and the one obtained by stimulated emission. The net effect is to increase the intensity of the light at this wavelength. Thus, a weak light at 694 nm can be amplified by stimulated emission of the excited ruby. A sketch of the ruby laser is shown in Figure C.1.

GAS LASERS

One of the most powerful and efficient gas lasers uses CO_2 mixed with He and N_2. It produces laser light with a wavelength in the infrared region of the spectrum.

APPLICATIONS

The light from a laser has some unique properties. Laser light is *coherent.* This means that the waves forming the beam are all in phase; that is, the waves' maxima and minima occur at the same points in space and time. The property of coherence of a laser beam is used in compact disc (CD) audio players.

Other properties of laser light are used in diverse applications. The ability of a laser to *focus intense light* on a spot is used in the surgical correction of a detached retina in the eye. In effect, the laser beam is used to "spot weld" on the retina.

The *intensity* of the laser beam is used in laser printers. These printers follow the principle of photocopiers but use a computer to direct the laser light in a pattern of dots to form an image.

In chemical research, laser beams provide intense monochromatic light to locate energy levels in molecules, study the products of very fast chemical reactions, and analyze samples for small amounts of particular substances.

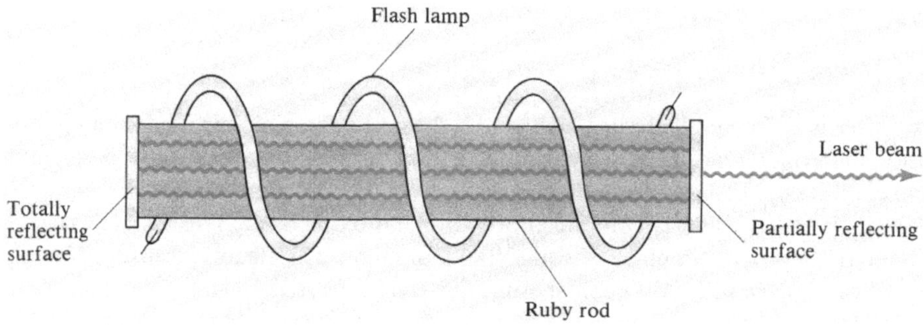

FIGURE C.1 Ruby laser.

Appendix D: Metals and Plants

Laszlo Gy. Szabo

Plants require several mineral substances. The uptake and assimilation of these substances are just as important as those of carbon, hydrogen, oxygen, nitrogen, phosphorus, or sulfur. Because the role of metals cannot be understood without an appreciation of the six "biogen" elements, they will also be discussed indirectly in this short review.

Plants take up metals necessary for metabolism as well as several metals that are not necessary (or at least the role of these metals in plant metabolism is not yet understood). These "unnecessary" elements (mainly heavy metals) and excess quantities of micronutrients may not be absorbed; if absorbed, these substances are accumulated or excreted (and thus rarely cause toxic symptoms in plants). However, plants are quite diverse in this respect; some taxa are sensitive to these unnecessary elements and others are tolerant.

Humans ingest toxic substances (e.g., lead or cadmium) in plant foods, both directly by eating contaminated plants and indirectly by eating the products of animals fed contaminated plants. The question of whether the toxic substance is found in plants (in the form of molecules within plant cells or excreted and thus neutralized from a plant physiological point of view) or in dust on the surface of the plant (epidermis, areoles, adsorbed to trichomes, etc.), is perhaps secondary.

Optimal concentrations of metals that are essential for plants depend on the plant genotype. The optimum amounts vary, not only by taxa but also by cultivar. Deficiency symptoms can often be recognized via simple visual inspection, but chemical analysis is usually necessary for precise identification. Adsorption of excessive quantities causes metabolic disorders. The relative proportions of certain metals must also be optimal. Nutritive disorders are characterized by changes in element composition, but a significant and sometimes conspicuous change in element proportions may also occur in plants damaged by pathogens or parasites. These changes can be measured especially well in the case of metals. Much current research on plant physiology is focused on the synergy and antagonism of metals. Metals in plant physiology are categorized according to relative quantities used by plants and effects on plants.

Macronutrients (%, g/100 g): Potassium, calcium, and magnesium always taken up by plants together with the nonmetal macronutrients nitrogen, phosphorus, and sulfur (usually in the form of anions) (In the case of nitrogen fixing, bacteria have a significant role.)

Micronutrients (mg/g): Iron, manganese, zinc, copper, cobalt, molybdenum, selenium, sodium, silicon (Chlorine is the only halogen considered to be essential for photosynthesis of higher plants.)

Uncertain role: Vanadium, chromium, nickel, strontium, and aluminum

Mostly toxic: Arsenic, cadmium, and lead

MOST IMPORTANT METALS

The most important metals in plant physiology are briefly described below.

Potassium

Form of uptake: ion
Role: enzyme activation, photosynthesis, respiration, osmotic potential (especially the stomatal opening mechanism), turgor, maintenance
Deficiency symptom: spotted lower leaves, necroses with browning, intercostal wilting, root mucosity

Calcium

Form of uptake: ion
Role: cell membranes, enzyme activation (calmoduline), polysaccharide (Ca-pectate), inclusion formation (Ca-oxalate, Ca-sulphate), gravitropism, cell-cycle control, senescence (calcification)
Deficiency symptom: decay of apical buds, root mucosity

Magnesium

Form of uptake: ion
Role: chlorophyll, ATP, cAMP, enzyme activation, DNA synthesis, RNA synthesis
Deficiency symptom: chlorosis, intercostal necrosis (midrib remaining green), root mucosity

Iron

Form of uptake: ion (II, III). Deficiency can be caused by excess phosphate, bicarbonate, Cu, Zn, Co, Cd, Mn, or Ni. Chelate-forming siderophores (iminocarbonic acid polymers) bind Fe(III), being reduced to Fe(II) in root tissue, which is transported and utilized in this form.
Role: chlorophyll synthesis, redox processes in photosynthesis and respiration (cytochromes, Fe-S proteins, ferredoxin), nitrate and nitrogen reduction, cell division (phytopherritins)
Deficiency symptom: intercostal chlorosis in younger, then older leaves and later senescence
Accumulation: in older leaves

Copper

Form of uptake: ion (I, II)
Role: redox processes, photosynthetic electron transport (plastocyanine), respiration (cytochrome oxidase), metalloenzymes (e.g., aminooxidase, superoxide dismutases), nitrogen fixation and nitrogen reduction, resistance to fungal diseases
Deficiency symptom: young leaves are dark green and spiraled, later necrosis

Zinc

Form of uptake: ion
Role: enzyme activation (e.g., peptidase, proteinase, phosphohydrolase, superoxide dismutase, dehydrogenase, carboanhydrase), auxin biosynthesis, growth, seed formation
Deficiency symptom: small leaves, rosette formation, withering along leaf veins

Manganese

Form of uptake: ion
Role: chlorophyll biosynthesis, enzyme activation (e.g., pyruvate carboxylase, superoxide dismutase)

Deficiency symptom: uneven withering of young leaves, necroses (vein remaining green)

Molybdenum

Form of uptake: molybdenate anion (II)
Role: nitrate reduction (nitrate reductase), nitrogen fixation (nitrogenase), chlorophyll biosynthesis
Deficiency symptom: intercostal chlorosis in older leaves, wrapped leaf lamina

Selenium

Form of uptake: selenate anion (II)
Role: antioxidant systems (glutathione peroxidase). Se analogs of S-containing amino acids (selenomethionine, selenocysteine) in nontolerant plants take part in enzyme synthesis, thereby producing toxic symptoms. Tolerant plants are able to distinguish between Se and S; the nonprotein-forming amino acids are stored in the vacuole and do not take part in metabolism.

Sodium

Form of uptake: ion
Role: osmotic substance in the form of NaCl may be important in low concentrations; toxic in high concentrations, causing potassium loss and membrane depolarization and calcium loss of plasmalemma

Cobalt

Form of uptake: ion
Role: nitrogen fixation, growth of nitrogen-fixing plants

Silicon

Form of uptake: silicate anion (II), silicic acid
Role: incrustating in cell wall, strengthens (e.g., by forming polysaccharide esters with ortosilicic acid and iso-polyacids)

Aluminum

Form of uptake: ion
Role: not clear, growth stimulator in tolerant plants

Arsenic

Form of uptake: arsenite (III), arsenate (III) anion
Role: toxic. Arsenite is more phytotoxic than arsenate accumulating in roots and older leaves (low concentrations of phosphate (III) remove arsenate or arsenite from soil particles, thus increasing their uptake; in higher concentrations, however, the effect is the opposite, displacing them from root surface)

Cadmium

Form of uptake: ion

Role: toxic, although being bound to phytochelatins disturbs enzyme activity if free; binding to the living parts of root zone damages the root; growth inhibitor causing chlorosis in leaves

LEAD

Form of uptake: ion
Role: toxic; enzyme inhibitor that causes chlorosis and red necroses in leaves, roots blacken

The most important micronutrients are the redox metals (Fe and Cu), which have an indispensable role in photosynthetic and mitochondrial electron transport as electron carriers (cytochromes, iron-sulfur proteins, ferredoxin, and plastocyanin).

METALLOENZYMES

Metal-containing enzymes (metalloenzymes) are also essential in plants. A few examples are listed below.

Zinc Metalloenzymes

Carbonic anhydrase (1 Zn)
Carboxypeptidase A (1 Zn)
Alcohol dehydrogenase (2 Zn or 4 Zn)
Superoxide dismutase (2 Zn + 2 Cu)

Manganese Metalloenzymes

Pyruvate carboxylase (3–4 Mn)
Superoxide dismutase (2 Mn)

Copper Metalloenzymes

Superoxide dismutase (2 Cu + 2 Zn)
Cytochrome oxidase (2 Cu + 2 hemes)
Amine oxidase (3 Cu)
Ascorbic acid oxidase (4 Cu)

Iron Metalloenzymes

NADH dehydrogenase (4 Fe)
Succinate dehydrogenase (8 Fe)
Aldehyde oxidase (8 Fe + 2 Mo)
Sulphite oxidase (2 hemes + 2 Mo)
Cytochrome oxidase (2 hemes + 2 Cu)

Iron is also present in cytochrome P-450, which is important in detoxification and hydroxylation.

METAL UPTAKE SYSTEMS

Metal uptake levels are also determined by the nutrient available in the smallest amount in the soil or other nutritive medium (Liebig's "law of minimum"). This fact must be taken into consideration when preparing nutritive solutions and cultures with a fluid or solid medium, as well as in agricultural practices (Salisbury and Ross, 1992; Lea and Leegood, 1993).

Metals dissolved in water and available for uptake get into the vascular tissue system of the root as ions or complexes via root hairs. The type of movement of substances dissolved in water can be classified as short, medium length, or long distance.

Short-distance metal transport (similar to the transport of water and nonmetallic ions or small molecules of monosaccharides and disaccharides) takes place via membranes at the intracellular level. However, the volume increase of the plant cell is restricted by the cell wall, which ensures a flexible but solid structure. Nonosmotic uptake of substances sized under approximately 10 nm through the cell wall is possible; thus, the cell wall almost fulfills the role of a filter. Larger particles can get only to the border of the plasmalemma; meanwhile, they are mostly adsorbed to the macromolecules of the cell wall. The cell wall consists mainly of more or less lignified cellulose and hemicellulose macromolecules, but special fibrillar glycoproteins (extensine, expansine) ensure its flexibility. The cell wall is relatively poor in enzymes; usually only hydrolases and anionic peroxidases are found here. Substances getting through the plasmalemma, including metal ions, get into the inner part of the cell (cytosol) via osmotic forces. The plasmalemma is basically a barrier consisting of a semipermeable membrane. The negative water potential maintained by dissolved salts (mainly cations and anions dissolved in water) includes the swelling of the cytoplasm, the character and intensity of genotype-dependent biosynthesis change, cell metabolism, and the beginning or continuity of growth or division. Transport of materials takes place through monolayer membranes (e.g., plasmalemma at the border of the cytoplasm, tonoplast inside the cell, and at the border of the vacuole) and bilayer membranes (e.g., in the case of chloroplast and mitochondrium). Controlled ion transport through these membranes is possible with specific carrier proteins and channel proteins.

Thus, ion transport is controlled by the plasmalemma and the tonoplast, allowing evolution of the membrane potential, function of the proton pump through the proton-transporting ATP-ases, and finally, ATP-synthesis connected to the proton gradient and electron transport. The proton pump regulates the chemical reaction of the cytosol, and its balance ensures cation antiport and anion symport (Tanner and Caspari, 1996).

At a medium-length distance, ions move more easily via the apoplast than via the symplast (e.g., through plasmodesmata). Transfer cells can be found around vascular bundles, mediating material flow from the cortex parenchyma toward the tracheas. Metals may form complexes with organic substances that came into being by a shorter or longer biosynthetic pathway. For example, the phosphate ester (polyphosphate) of meso-inositol (cyclic alcohol) is able to bind a large amount of iron, calcium, or other metal temporarily and thus store it for a period of intensive metabolism (e.g., germination). Another example is the most common plant pigment, anthocyanin, as a polyphenol. Sugar molecules (polyphenol glycosides) forming polyethers (glycosides) with phenolic hydroxy groups allow many types of chelate formation. Metallo-anthocyanins, for instance, remain quite stable; they can prevent for a long period the quick transformation or decomposition of cell-protecting molecules (recently freed radical scavengers) and pigment-coloring (insect attracting) molecules. Furthermore, due to their strongly hydrophilic nature, they bind large quantities of water molecules (hydration), making the water potential of the cell more negative, thereby increasing the stress-enduring ability of cells or tissues (drought, frost, etc.).

Transport in the xylem (tracheas, tracheids) and in the phloem (sieve tubes and companion cells, at the gymnosperms sieve cells and albuminous cells) is called long-distance transport. The composition of xylem and phloem sap is not the same. Generally, phloem sap is richer than xylem sap, not only in sucrose and amino acids synthesized by the plant but to a smaller extent also in transported metals. Transport of dissolved substances in the phloem can be observed even when water movement is not detectable.

The flow of aqueous solutions in the sieve tube occurs through the pores of the cribrum via plasmodesmata. The osmotic pressure of the sieve-tube sap depends on the amount of substances dissolved in it. A positive pressure gradient emerges in the phloem, ensuring material flow from the apex toward the base. Xylem and phloem transport occurs in the opposite direction in leaves, roots, and stem, but in the same direction in fruits. The latter may be responsible for significant metal accumulation in seeds and fruits.

METAL ACCUMULATION

Supplying micronutrients (essential metals) continues to be an important issue in plant cultivation. For optimal dosage, the needs of the plant species (and the cultivar) must be known, as well as soil composition and ecostability characteristics of the grown cultivar. The timing and method of micronutrient supply are also of great importance for plants. Leaf sprays are frequently applied, as aqueous solutions can be absorbed by leaf tissues, as well as through stomata and an epidermis that is thinner than other parts of the plant. The optimum can be determined only by careful chemical analysis of plant and soil, but the quality of irrigation water must also be taken into account.

Metal accumulation is of great interest, especially from the viewpoint of environment protection. Ions with an inhibitory character (heavy metals, such as lead and cadmium) can be bound with chelate-forming substances (e.g., EDTA). The chelate-forming substances in living organisms are usually called metallothioneins. These detoxifying polypeptides are rich in sulfur (hence their name). Metallothioneins consisting of approximately 60 amino acids have been detected in fungi and vertebrate animals.

The form of accumulation in plants containing certain mineral substances in high amounts cannot be easily determined. In general, high concentrations are found dissolved in the vacuoles or bound to peptides in the cytoplasm. Characteristics of accumulation sites can only be determined by special, selective extraction techniques and analytical separation methods. At present few data are available that can be used for comparison, and competent conclusions are also rare (Tolgyesi, 1969) reported on selected plant species in Hungary that accumulate extremely high amounts of metals. Mean concentrations of selected metals by plant species reported by Tolgyesi are listed below(sample number in parentheses).

Calcium, g/kg

Althaea officinalis, 22.5 (5)
Cirsium canum, 20.4 (10)
Cornus sanguinea, 21.5 (10)
Echium vulgare, 20.2 (5)
Euonymus europaeus, 21.6 (9)
Fagus sylvatica, 35.0 (3)
Ononis hircina, 18.6 (9)
Onosma arenaria, 72.0 (1)
Salix caprea, 22.0 (2)
Urtica dioica, 31.8 (9)

Iron, mg/kg

Alnus glutinosa, 632 (5)
Calamintha clinopodium, 1130 (1)
Cirsium canum, 603 (10)
Eupatorium cannabinum, 440 (6)
Inula britannica, 495 (7)
Linaria vulgaris, 453 (11)
Matricaria recutita, 496 (8)

Ononis hircina, 607 (9)
Pulmonaria officinalis, 436 (6)
Schoenoplectus tabernaemontani, 630 (6)
Tussilago farfara, 500 (6)
Viburnum lantana, 450 (4)

Manganese, mg/kg

Abies alba leaf, 2200 (5)
Alnus glutinosa leaf, 208 (5)
Bolboschoenus maritimus, 235 (5)
Carex elata, 257, (8)
Carex pilosa, 261 (6)
Carpinus betulus, 540 (12)
Castanea sativa leaf, 850 (4)
Fagus silvatica leaf, 434 (6)
Glyceria maxima, 206 (5)
Larix decidua leaf, 1050 (5)
Luzula albida, 830 (5)
Picea abies leaf, 295 (4)
Quercus cerris leaf, 612 (6)
Quercus petraea leaf, 564 (6)
Quercus pubescens leaf, 706 (3)
Quercus robur leaf, 354 (7)
Salix alba leaf, 200 (13)
Salix caprea leaf, 285 (3)
Vaccinium myrtillus, 930 (4)

Zinc, mg/kg

Aristolochia clematitis, 50 (13)
Carex pilosa, 49 (6)
Datura stramonium, 47 (5)
Lactuca serriola, 47 (9)
Lepidium draba, 64 (5)
Picea abies leaf, 66 (4)
Populus alba leaf, 105 (6)
Populus italica leaf, 101 (5)
Populus nigra leaf, 79 (5)
Salix alba leaf, 83 (13)
Salix caprea leaf, 100 (3)

Copper, mg/kg

Alnus glutinosa leaf, 14.0 (5)
Aristolochia clematitis, 24.4 (13)
Artemisia vulgaris, 17.0 (21)
Bidens tripartita, 24.2 (6)
Caltha palustris, 14.4 (6)

Datura stramonium, 15.4 (5)
Erigeron canadensis, 15.0 (7)
Ononis hircina, 18.1 (9)
Papaver rhoeas, 20.9 (5)
Solanum dulcamara, 17.1 (5)
Solanum nigrum, 15.4 (4)
Symphytum officinale, 14.8 (13)
Taraxacum officinale, 14.6 (5)

High levels of accumulation are primarily characteristics of the inherited physiological/biochemical habits of taxa. However, it must be emphasized that the element composition of character species in a natural plant association may vary within specific ranges under diverse edaphic and ecological conditions (Szabo et al., 1985).

On the basis of several studies, Tolgyesi (1969) confirmed that inorganic chemotaxonomy is justified because certain taxonomic categories can be characterized by a specific ability to accumulate inorganic elements. For example, plants belonging to the Boraginaceae and Betulaceae families are especially rich in Ca; Lamiaceae, in Fe; Fagaceae and Betulaceae, in Mn; Solanaceae, Laminaceae, Boraginaceae, and Asteraceae, in Cu; and Salicaceae, in Zn.

The sensitivity of most nonaccumulating, nontolerant plant species indicates the toxicity of metals. Effects can manifest as membrane damage, inhibition of enzymes, induction of enzymes, defense mechanisms against metal phytotoxicity, and interaction of metals with nucleic acids (Farago, 1994).

Metal inhibition is reported for many enzymes. The high affinity of metals for sulfhydryl groups is suggested as one of the main mechanisms of enzyme inhibition, including metal inhibition of enzymes related to photosynthesis (δ-ALA-dehydratase, protochlorophyllide reductase); inhibition of photosynthetic electron transport and photophosphorylation; photosynthetic carbon dioxide fixation; carbonic anhydrase; and superoxide dismutase.

Besides irreversible biochemical changes, enzyme induction effects are also known. These secondary indirect effects of metals are considered to play an important role in the stress metabolism induced by toxic metal concentrations. Peroxidase induction has been observed in leaves and roots of various plant species after application of toxic amounts of cadmium, zinc, copper, nickel, and mercury. Similarly, the activity of catalase, esterases, and superoxide dismutase also increases due to the effect of heavy metals.

Recent studies of special metal-binding proteins that are synthesized because of the effect of heavy-metal stress are of special importance. These chelate-forming proteins are called phytochelatins. Grill et al. (1989) were the first to report that *Acer platanoides* react in a highly sensitive but specific way to heavy metals. The trees were planted in the zinc-rich soil of a closed mine. These plants synthesized the phytochelatin suitable for binding Zn, whereas in the absence of Zn, synthesis did not even begin.

Thus, in plant cells chelate-forming proteins are produced via the inductive effect of heavy metal ions. To a certain extent, binding the ions maintains the normal metabolism of the cell. Phytochelatins can be derived from glutathione (glutamyl-cysteinyl-glycine). Glutathione added to the medium of the root culture of *Rubia tinctorum* increased the amount of cadmium-induced phytochelatins (Kubota et al., 1995). Synthesis is catalyzed by the phytochelatin synthase (a special glutamylcysteine dipeptidyl-transpeptidase), a 2–11 dipeptide unit binding to glycine. Homophytochelatins with a similar structure can also be detected, such as peptides containing alanine instead of glycine or the so-called desglycyl peptides containing no glycine. The synthesis of one of the listed phytochelatins is induced by Zn, Cd, Pb, Hg, Sb, Ni, Cu, Ag, Au, Bi, Te, and W.

Certain species can be used as bioindicators in determining the degree of heavy metal accumulation and corresponding phenotypical changes. For instance, algae and mosses can usually accumulate higher amounts of heavy metals than vascular plants (e.g., *Chlorella* enriches Cd, Pb, Hg, Ag, Cu, and Zn, and it can be utilized in industrial sewage purification). Certain moss species concentrate Zn, Cu, Ni, Cd, and Cr without selectivity (1000–10,000 mg/kg), such as the so-called copper moss (*Gymnocolea acutiloba*) living in Cu-containing soil or base rock. Heavy metal excretion may take place through the cell wall, such as Pb and Zn excreted by *Dicranella varia* during the summer drought, which appears as a dust layer on the moss carpet (amounts may reach 40,000–60,000 mg/kg). Some mosses concentrate radionuclids as well. For instance, *Pleurozium schreiberi* was able to bind almost the full amount of nuclear fission products originating from an experimental nuclear explosion. Mosses are also important in the circulation and accumulation of the radioactive isotope Cs-137.

These examples also indicate that vascular plants accumulate much less heavy metal in the free (ionic) state. Generally, they are temporarily dissolved in water via apoplast flow in vascular bundles or adsorbed to other tissue elements. For instance, the leaves and cortex of spruce, pine, locust, and poplar trees can accumulate 1 to 100 mg/kg of heavy metals. These "resistant" or "tolerant" plants are suitable for measuring environment pollution as accumulation indicators. Among the herbaceous plants, an example is the invasive *Solidago canadensis*, which, when located next to motorways, is able to accumulate lead at a level of more than 100 mg/kg.

Balazsy (2000) draws attention to new considerations. *Ambrosia artemisiifolia* is widespread in Europe and its pollen causes allergic reactions. This plant accumulates heavy metals to such a degree at refuse dumps polluted with high levels of heavy metals that a significant amount of chromium, copper, and zinc can be found even in its pollen. According to most available data, the zinc content of the pollen is more significant than that of the vegetative organs of the plant. In certain cases, concentrations in pollen may reach 400 mg/kg of heavy metals when transferred long distances. In addition, by the mediation of pollen, this plant may be increasing the pollution of ecological systems, and high concentrations of heavy metals in pollen are exacerbating allergy symptoms.

CONCLUSION

The relationship between metals and plants is very complicated. Several metals bound to enzymes and vitamins are essential in the life of plants. Others are responsible for photosynthetic and mitochondrial electron transport by redox processes or take part in oxidative and reductive biochemical reactions. Excess metals, especially heavy metals, are toxic to the plant but frequently, when they are bound to special proteins, do not cause any damage. Knowing the amount of metals in plants is of great importance in agriculture, food distribution and preparation, and environment protection. Exact data and correct conclusions can be provided only through the application of chemical analytical methods.

REFERENCES

Balazsy, S., *Dissemination of Metals in the Ecological Systems*, Bessenyei Press, Nyiregyhaza, Hungary, 2000.

Farago, M.E., *Plants and the Chemical Elements,* VCH, Weinheim, Germany, 1994.

Grill, E. et al., Phytochelatins, the heavy-metal-binding petides in plants, are synthesized from glutathione by a specific glutamylcysteine dipeptidyl transpeptidase (polychelatin synthase), *Proc. Natl. Acad. Sci. U.S.A.*, 86, 6838–6842, 1989.

Kubota, H. et al., Phytochelatins (class III metallothioneins) and their desglycol peptides induced by cadmium in normal root culture of *Rubia tinctorum L., Plant Science*, 106, 157–166, 1995.

Lea, P.J. and Leegood, R.C., *Plant Biochemistry and Molecular Biology*, John Wiley & Sons, Chichester, England, 1993.

Salisbury, F.B. and Ross, C.W., *Plant Physiology*, Wadsworth, Belmont, CA, 1992.

Szabo, L., Kevey, B., and Tolgyesi, Gy., Macroelement and microelement accumulation in the plant species of hornbeam-oak communities living on different parent material in the Mecsek mountains. *Bot. Kozlem.*, 72, 1–2, 77–88, 1985.

Tanner, W. and Caspari, T., Membrane transport carriers, *Annu. Rev. Plant Physiol. Plant Mol. Biol.* 47, 595–626, 1996.

Tolgyesi, Gy., *Microelement Content of Plants and Agricultural Aspects of THIS,* Mezogazd, Kiado, Budapest, 1969.

Appendix E: Toxicity of Cyanide

The cyanide ion (CN–) is the toxic agent in cyanide salts, such as sodium cyanide used in electroplating. Because cyanide is a relatively strong base, it reacts easily with many acids to form volatile hydrogen cyanide (HCN). HCN boils at a relatively low temperature (26°C) and thus is a gas at temperatures slightly above room temperature.

Cyanide is often used as a fumigant in storage facilities and cargo transport (e.g., ship holds) because it is toxic to most forms of life and, in gaseous form, can penetrate tiny openings, even insect eggs. The cyanide ion is one of the most rapidly working poisons; lethal doses taken orally act in minutes. Cyanide poisons via asphyxiation, as does carbon monoxide (CO), but the mechanism of cyanide poisoning is different. Instead of preventing the cells from getting oxygen, cyanide interferes with oxidative enzymes, such as *cytochrome oxidase*. Oxidases — enzymes containing a metal, usually iron or copper — catalyze oxidation:

$$\text{Metabolite (H)}_2 + 1/2\ O_2 \rightarrow \text{oxidized substance} + H_2O + E$$

The iron atom in cytochrome oxidase is oxidized from Fe^{2+} to Fe^{3+} to provide electrons for the reduction of O_2. The iron regains electrons from other steps in the process. The cyanide ion forms stable cyanide complexes with metal ions of the oxidase and renders the enzyme incapable of reducing oxygen or oxidizing the metabolite.

$$\text{Cytochrome oxidase (Fe)} + \text{CN} \rightarrow \text{Cytochrome oxidase (Fe)} \ldots \text{CN} \rightarrow \text{complex}$$

In essence, the electrons of the iron ion are "frozen"; they cannot participate in the oxidation-reduction process. Plenty of oxygen gets to the cells, but the mechanism by which the oxygen is used comes to a halt. Hence, the cell dies, and if this process occurs rapidly enough in vital organs, the victim dies. The mechanism of cyanide poisoning is illustrated in Figure E.1.

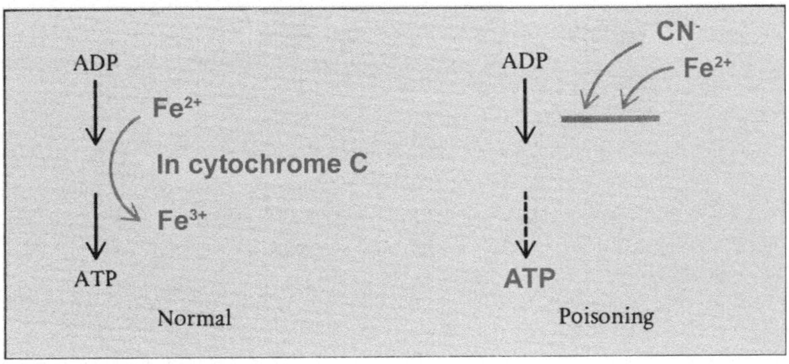

FIGURE E.1 Cyanide poisoning.

Appendix F: Components of Nucleic Acid

Each cell in living organisms contains thousands of protein molecules. All of these molecules are made up of the same 20 amino acids, but in numerous different sequences. Each species has a different set of protein molecules. Protein molecules also vary among individuals within a species, but the range of variation is much less than that among species. Once scientists understood this, the next question was how do the cells know which proteins to synthesize out of the extremely large number of possible amino acid sequences? The answer to this question is *heredity*; that is, an individual gets such information from the parents.

Since the late nineteenth century, biologists suspected that the transmission of hereditary information from one generation to the next takes place in the *nucleus* of the cell, and the structure within the nucleus, called *chromosomes*, has something to do with heredity. The information that determines external characteristics (hair color, eye color, etc.) and internal characteristics (blood group, hereditary diseases, etc.) resides in *genes* located inside the chromosomes. A gene can be defined as a segment of DNA (a segment of nucleotide in DNA) that codes for a functional product.

NUCLEIC ACIDS

Chemical analysis of nuclei showed that they are largely made up of special basic proteins and a type of compound called *nucleic acids*. Nucleic acids are exceedingly large organic molecules containing carbon, hydrogen, nitrogen, and phosphorus.

Nucleic acids are found in all living cells, with the exception of the red blood cells of mammals. There are two kind of nucleic acids found in cells, each with a specific role in the transmission of hereditary information. The two types are *ribonucleic acid* (RNA) and *deoxyribonucleic acid* (DNA). DNA resides in the chromosomes of the nucleus. RNA is not found in the chromosomes, but rather is located elsewhere in the nucleus and even outside the nucleus, in the cytoplasm. The building blocks of nucleic acid chains are *nucleotides*. Nucleotides are composed of three simpler units: a base, a sugar, and phosphorus. Nucleic acids are found in all living cells, with the exception of the red blood cells of mammals.

DEOXYRIBONUCLEIC ACID (DNA)

A molecule of DNA is a chain of many nucleotide units. Each DNA nucleotide is comprised of a nitrogen-containing base, deoxyribose (pentose sugar), and a phosphate group (phosphoric acid molecule). The nitrogen-containing base is *adenine* (A), *guanine* (G), *cytosine* (C), or *thymine* (T). All the nitrogen-containing bases are ring structures containing atoms of carbon, hydrogen, oxygen, and nitrogen. Adenine and guanine are double-ring structures, collectively known as *purines*. Thymine and cytosine are smaller, single-ring structures, called *pyrimidines*. Nucleotides are named according to their nitrogenous base. Thus, a nucleotide containing thymine is called a thymine nucleotide, a nucleotide containing adenine is called an adenine nucleotide, and so on. The term *nucleoside* refers to the combination of a purine or pyrimidine plus a pentose sugar; it does not contain a phosphate group.

Although the chemical composition of the DNA molecule was known before 1900, it was not until 1953 that the organization of the chemical subunits was modeled by James Watson and Francis Crick. Structural characteristics of this model of DNA molecules are summarized below:

1 The molecule consists of two strands with crossbars. The strands twist around each other in the form of a double helix, so that the shape resembles a twisted ladder.
2. The uprights of the DNA ladder, called the backbone, consist of alternating phosphate groups and the deoxyribose (sugar) portions of the nucleotide.
3. The rungs of the ladder contain nitrogenous bases in pairs joined by hydrogen bonds. As shown in Figure F.1, a purine always pairs with a pyrimidine; that is, adenine always pairs with thymine, and cytosine always pairs with guanine.

As discussed previously, cells contain hereditary material called genes, each of which is a segment of a DNA molecule. Genes determine all hereditary traits, and they control all activities that take place within cells. When a cell divides, its hereditary information is passed on to the next generation. This transfer of information is possible because of DNA's unique structure.

Crick (b. 1916) and Watson (b. 1928), working in the Cavendish Laboratory at Cambridge, built scale models of the double helical structure of DNA based on x-ray data from Rosalind Franklin (1920–1958) and Maurice H.F. Wilkins (b. 1916). Knowing distances and angles between atoms, they compared the task to solving a three-dimensional jigsaw puzzle. Watson, Crick, and Wilkins received the Nobel Prize in 1962 for their work relating to the structure of DNA.

Ribonucleic Acid (RNA)

The second principal kind of nucleic acid, RNA differs from DNA in several respects. Whereas DNA is double stranded, RNA is usually single stranded. The five-carbon sugar in the RNA nucleotide is *ribose*, which has one more oxygen atom than deoxyribose, and one of RNA's bases is *uracil*, instead of thymine. At least three different kinds of RNA have been identified in cells, known as messenger RNA, ribosomal RNA, and transfer RNA. Each type of RNA has a specific role in cells.

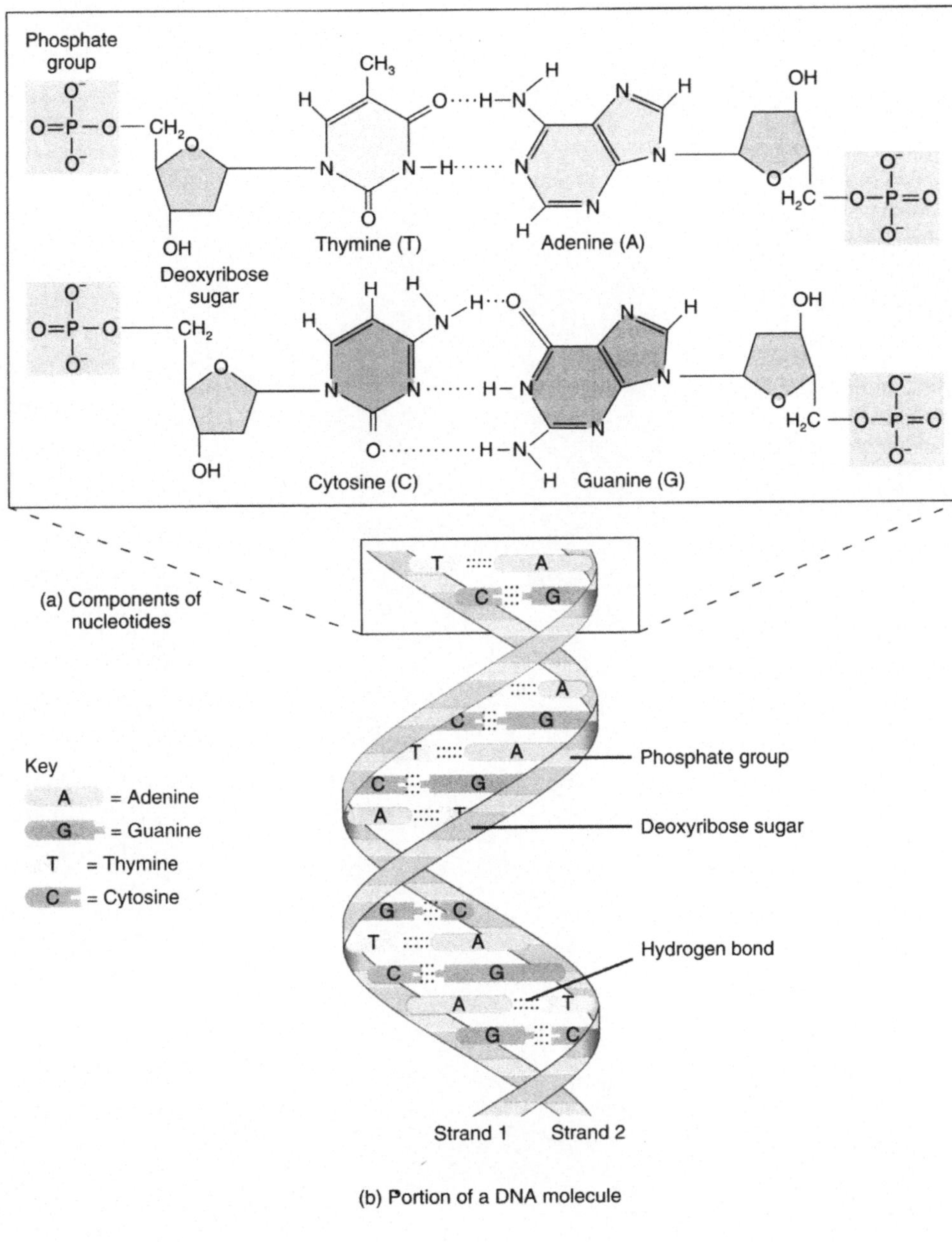

FIGURE F.1 Structure of DNA.

Appendix G: Polarized Light

POLAR AND NONPOLAR COMPOUNDS

All diatomic molecules composed of atoms of various elements are *polar* (e.g., NaCl). In a molecule, a dipole moment occurs when the charge in the chemical bonds is separated; that is, one part of the molecule has a positive charge and the other a negative charge (e.g., water, H^+OH^-). All diatomic molecules built from the same atoms are *nonpolar* (e.g., Cl_2). Nonpolar molecules have no dipole moment. Examples of nonpolar compounds are methane and hexane.

POLARIZATION

Polarization is the process of confining the vibrations of the electric vector, constituting a transverse wave in one direction. In unpolarized radiation, the vector oscillates in all directions perpendicular to the direction of propagation.

POLARIZATION OF LIGHT

An ordinary light beam consists of waves vibrating in all possible planes perpendicular to its path. If the light passes through some substance that permits only one of these components to pass through, the waves in the resulting beam vibrate along the same plane. Such a light beam, described as plane polarized, is illustrated in Figure G.1.

Plane Polarized Light

After reflection or transmission through certain substances (see Polaroid section below), the electric field is confined to one direction and the radiation is referred to as plane polarized light. The plane can be rotated when it passes through certain substances.

Circular Polarized Light

In circular polarized light, the tip of the electric vector describes a circular helix about the direction of propagation with a frequency equal to that of the light. The magnitude of the vector remains constant.

Elliptical Polarized Light

In elliptical polarized light, the vector also rotates about the direction of propagation but the amplitude changes; a projection of the vector on a plane at right angles to the direction of propagation describes an ellipse. Circular and elliptical polarized light are produced using a retardation plate.

POLAROID

In 1808, French physicist Etienne Malus discovered that light can be polarized. One convenient way is to pass ordinary light through a device composed of Iceland spar (crystalline calcium carbonate) called a Nicol prism (invented in 1828 by British physicist William Nicol). A more recently developed polarizing material is Polaroid, invented by an American, E.H. Land. It contains a crystalline

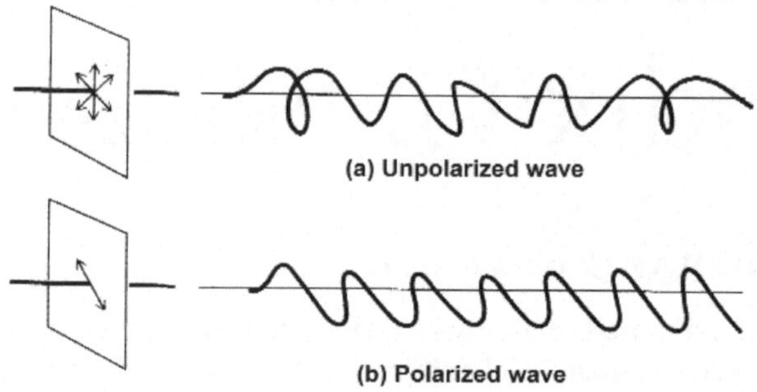

FIGURE G.1 In contrast to ordinary light (a), which vibrates in all planes, polarized light (b) vibrates in only one plane.

organic compound properly oriented and embedded in a transparent plastic. Sunglasses are often made from Polaroid.

Polaroid is a doubly refracting material that plane polarizes unpolarized light passed through it. It consists of a plastic sheet strained in a manner that makes it birefringent by aligning its molecules. Sunglasses incorporating a Polaroid material absorb light that is vibrating horizontally — produced by reflection from horizontal surfaces — and thus reduce glare.

POLAROGRAPHY

Polarography is an electrochemical-based analytical technique. A dropping mercury electrode is used as the cathode along with a large nonpolarizable anode. The dropping mercury electrode consists of a narrow tube through which mercury is slowly passed into a dilute solution of the solution, so as to form small drops at the end of the tube, which fall away. In this way, the cathode has a small surface area and can be kept clean. A variable potential is applied to the cell and a plot of current against potential (*polarogram*) made. As each chemical species is reduced at the cathode (in order of electrode potentials) a step-wise increase in current is obtained. The height of each step is proportional to the concentration of the component. This technique is useful for detecting trace amounts of metals and for investigating solvated complexes.

Appendix H: Stock Metal Solutions

Aluminum (Al)

Dissolve 1.000 g of aluminum metal in dilute HCl via gentle heating. Dilute to 1 liter with reagent-grade water (1000 mg/l).

Antimony (Sb)

Carefully weigh 2.7426 g of antimony potassium tartrate ($K(SbO)C_4H_4O_6.1/2H_2O$), and dissolve in reagent-grade water and dilute to 1 liter (1000 mg/l).

Arsenic (As)

Dissolve 1.320 g of arsenic trioxide (As_2O_3) in 100 ml of reagent-grade water containing 4 g of NaOH. Acidify the solution with 20 ml of concentrated HNO_3 and dilute to 1 liter with reagent-grade water.

Barium (Ba)

Dissolve 1.7787 g of barium chloride ($BaCl_2.2H_2O$) in reagent-grade water and dilute to 1 liter (1000 mg/l).

Beryllium (Be)

Dissolve 11.6586 g of beryllium sulfate ($BeSO_4$) in reagent-grade water containing 2 ml of concentrated HNO_3 and dilute to 1 liter (1000 mg/l).

Cadmium (Cd)

Dissolve 1.000 g of cadmium metal in 20 ml of 1:1 HNO^3 and dilute to 1 liter with reagent-grade water (1000 mg/l).

Calcium (Ca)

Suspend 2.500 g of $CaCO_3$ (dried for 1 h at 180°C) in reagent-grade water and dissolve by adding a minimum of dilute HCl. Dilute to 1 liter with reagent-grade water (1000 mg/l).

Chromium (Cr)

Dissolve 1.923 g of chromium trioxide (CrO_3) in reagent-grade water, acidify with HNO_3, and dilute to 1 liter (1000 mg/l).

Chromium Hexavalent (Cr6+)

Dissolve 0.2829 g of pure dried potassium dichromate ($K_2Cr_2O_7$) and dilute to 1000 ml with reagent-grade water (1 ml = 100 µg Cr).

Cobalt (Co)

Dissolve 1.000 g of cobalt metal in 20 ml of 1:1 HNO_3 and dilute to 1 liter with reagent-grade water, or 4.307 g of cobaltous chloride ($CoCl_2.6H_2O$), and dissolve in reagent-grade water. Add 10 ml of concentrated HNO_3 and dilute to 1 liter with reagent-grade water (1000 mg/l).

Copper (Cu)

Dissolve 1.000 g of electrolytic copper in 5 ml of HNO_3 and dilute to 1 liter with reagent-grade water (1000 mg/l).

Iron (Fe)

Dissolve 1.000 g of analytical-grade iron wire in 10 ml of HNO_3 and reagent-grade water, and dilute to 1 liter (1000 mg/l).

Lead (Pb)

Dissolve 1.599 g of lead nitrate ($Pb(NO_3)_2$) in reagent-grade water, acidify with 10 ml of HNO_3, and dilute to 1 liter (1000 mg/l).

Magnesium (Mg)

Dissolve 1.000 g of Mg metal (analytical reagent grade) in 20 ml of 1:1 HNO_3 and dilute to 1 liter with reagent-grade water (1000 mg/l). Alternatively, dissolve 0.829 g of magnesium oxide (MgO) in 10 ml of concentrated HNO_3, and dilute to 1 liter with reagent-grade water (500 mg/l).

Manganese (Mn)

Dissolve 1.000 g of metal (analytical reagent grade) in 10 ml of HNO_3 and dilute to 1 liter with reagent-grade water (1000 mg/l).

Mercury (Hg)

Dissolve 0.1354 g of mercuric chloride ($HgCl_2$) in 75 ml of reagent-grade water. Add 10 ml of concentrated HNO_3 and adjust the volume to 100 ml with reagent-grade water (1 ml = 1 mg Hg).

Molybdenum (Mo)

Dissolve 1.840 g of ammonium molybdate ($(NH_4)_6Mo_7O_{24}.4H_2O$) in reagent-grade water and dilute to 1 liter (1000 mg/l).

Nickel (Ni)

Dissolve 1.000 g of nickel metal (analytical grade) or 4.953 g of nickel nitrate ($Ni(NO_3).6H_2O$) in 10 ml of concentrated HNO_3, and dilute to 1 liter with reagent-grade water (1000 mg/l).

Osmium (Os)

Procure a certified aqueous standard from a supplier.

Potassium (K)

Dissolve 1.907 g of potassium chloride (KCl, dried at 110°C) and dilute to 1 liter with reagent-grade water (1000 mg/l).

Selenium (Se)

Dissolve 0.3453 g of selenious acid with assay 94.4% H_2SeO_3 and dilute to 200 ml (1000 mg/l).

Silver (Ag)

Dissolve 0.7874 g of anhydrous silver nitrate ($AgNO_3$) in reagent-grade water. Add 5 ml of concentrated HNO_3 and bring to volume in a 500-ml volumetric flask with reagent-grade water (1000 mg/l).

Sodium (Na)

Dissolve 2.542 g of sodium chloride (NaCl) in reagent-grade water, acidify with 10 ml of HNO_3, and dilute to 1 liter with reagent-grade water (1000 mg/l).

Thallium (Tl)

Dissolve 1.303 g of thallium nitrate ($TlNO_3$) in reagent-grade water, acidify with 10 ml concentrated HNO_3, and dilute to 1 liter with reagent-grade water (1000 mg/l).

Tin (Sn)

Dissolve 1.000 g of analytical-grade tin metal in 100 ml of concentrated HCl, and dilute to 1 liter with reagent-grade water (1000 mg/l).

Vanadium (V)

Dissolve 1.7854 g of vanadium pentoxide (V_2O_5) in 10 ml of concentrated HNO_3, and dilute to 1 liter with reagent-grade water (1000 mg/l).

Zinc (Zn)

Dissolve 1.000 g of zinc metal in 10 ml of concentrated HNO_3 and dilute to 1 liter with reagent-grade water (1000 mg/l).

Appendix I: Calculation for Solid Matrices

MOISTURE DETERMINATION

Moisture in a solid is determined by drying a known quantity aliquot of the well-mixed sample at 103 to 105°C in a laboratory oven. After the dried sample is cooled in a desiccator, weigh and calculate its percent moisture by using the following formula:

$$\% \text{ moisture} = [(\text{g of solid} - \text{g of dried solid}) \times 100]/\text{g of solid} \tag{I.1}$$

REPORTING OF RESULTS FOR SOLIDS

For solid matrices, the report is expressed as ppm (milligrams per kilogram) or ppb (micrograms per kilogram). The report should state that the reported value is calculated on the *wet base*, sometimes called *as-is base*, or on the *dry base*.

Wet base means that the original solid sample (soil, sediment, sludge, etc.) contains moisture; therefore, the original weight of the sample incorporates the weight of the moisture. Consequently, the weight of the sample is incorrect. When the moisture content of the sample is known, the analyst can make the necessary corrections. For this calculation, use the following formula:

$$\text{mg/kg on wet base} = [\text{mg/l} \times \text{final volume of the sample after treatment}]/\text{g sample} \tag{I.2}$$

Once the percent moisture of the sample is known, correct the error caused by the moisture content by using the following formula:

$$\text{mg/kg on dry base} = \text{mg/kg on wet base/decimal fraction of dry solid} \tag{I.3}$$

Assume that a 5-g soil sample was weighed and digested for lead (Pb) analysis. After the preparatory process (digestion), the final volume of the "ready-for-analysis sample" was 100 ml. The Pb content of the digestate was found to be 0.56 mg/l (0.56 mg per 1000 ml, or 0.056 mg/100 ml). The 100-ml digestate corresponds to the 5-g soil sample; therefore, the 5-g soil sample contains 0.056 mg Pb. How much Pb will be in a 1000-g (1-kg) sample?

A 1-kg sample will contain 200 times more Pb, or $200 \times 0.056 = 11.2$ mg. Thus, the result is 11.2 mg/Kg (ppm) Pb on wet base. To avoid lengthy calculations, use formula I.2:

$$\text{mg/kg Pb on wet base} = (0.56 \times 100)/5 = 11.2 \text{ mg/kg}$$

The moisture content by weight of the soil sample was 12%; therefore, the dry soil weight is 88%. Using Formula I.3:

$$\text{mg/mg Pb on dry base} = 11.2/0.88 = 12.72 \text{ mg/kg}$$

Appendix J: Plasma

Energy can be obtained by combining light nuclei into a heavier nucleus by nuclear fusion. Such fusion reactions have been observed in the laboratory by means of bombardment using particle accelerators.

FUSION

According to the "big bang" theory of the formation of the universe, 98% of all matter is comprised of helium. The theory postulates that the universe began with an explosion in which matter was formed out of energy, and, at the beginning only the lightest element, hydrogen, existed. Later, as the universe expanded, stars were born when the hydrogen clouds collapsed under gravitational forces. In the cores of these stars, hydrogen nuclei fused together and formed helium.

The transformation of hydrogen nuclei into helium nuclei liberates a large amount of energy in the form of photons — the smallest units of electromagnetic radiation — largely in the following way:

$$H_1^2 + H_1^3 \rightarrow He_2^4 + n_0^1 + \text{energy}$$

This process, called *fusion*, is how the sun generates energy. The reactions occurring in the sun are essentially the same as those in a hydrogen bomb.

Fusion is the combination of very light nuclei. When very light nuclei, such as hydrogen, helium, and lithium are combined or fused to form an element of higher atomic number, energy is released consistent with the greater stability of the elements in this intermediate atomic number range. This energy, which comes from a decrease in mass, is the source of the energy released by the sun and by hydrogen bombs. Nuclear fusion consists of the combining of two or more small nuclei into a larger one, and the corresponding release of energy.

Materials for fusion reactions are available in enormous quantities. Deuterium is a relatively abundant isotope: Of 6500 atoms of hydrogen in seawater, for example, one is a deuterium atom. Thus, the oceans are a potential source of fantastic amounts of deuterium. A single liter of seawater has 1.000×10^{22} atoms of deuterium. A single cubic kilometer of seawater, then, would contain enough deuterium atoms with the potential energy to equal the burning of 1300 billion barrels of crude oil (the approximate total amount of crude oil originally present on the planet).

When a deuterium and a tritium nucleus are transformed into a helium nucleus and a neutron, a large amount of energy is released. Where does this energy come from? The combined mass of the nuclei of two hydrogen isotopes is greater than the combined mass of the helium nucleus plus that of a neutron. When the deuterium and tritium nuclei are converted to helium and a neutron, the extra mass is converted to energy. Based on Albert Einstein's (1879–1955) work, the amount of energy released upon the conversion of mass follows:

$$E = mc^2$$

where

E = energy (J).
m = mass (kg).
c = 3.0×10^8 m/sec.

This equations states that the mass (m) converted (in kilograms), multiplied by the square of the velocity of light (c^2 in m^2/sec^2), is equal to the energy (E) created (in joules).

For example, 1 g of matter completely converted to energy would produce 8.8×10^{13} J, which is enough energy to boil 34 million liters of water initially at 20°C. In summary, a lot of energy can be obtained from a little mass.

PLASMA

Fusion reactions occur rapidly only at temperatures of 100 million °C or more. At these high temperatures, atoms do not exist as such; instead, a plasma forms that consists of unbound nuclei and electrons. Plasma is a gaseous state composed of ions. In plasma, nuclei merge or combine. To achieve these high temperatures, the fusion reaction of a hydrogen bomb is required.

Appendix K: Soxhlet Extraction

A Soxhlet extractor (Figure K.1) can be used to extract solutes from solids. Any volatile solvent can be used. The solvent is vaporized, and in the condensed phase dropped onto the solid substance placed in a thimble. When the liquid level fills the body of the extractor, it automatically siphons back into the flask. This process continues repeatedly.

PROCEDURE SETUP

1. Put the known-weight solid substance in the porous thimble and place it in the inner tube of the Soxhlet extractor.
2. Fill the flask half full of the extracting solvent.
3. Assemble the unit.
4. Turn on the cooling water. Heat.
5. When the extraction is complete, turn off the heat and the cooling water.
6. Dismantle the apparatus, and pour the extraction solvent containing the solute into a beaker. Isolate the extracted component by evaporating the solvent.
7. Determine the component and calculate its concentration according to the analytical method used.

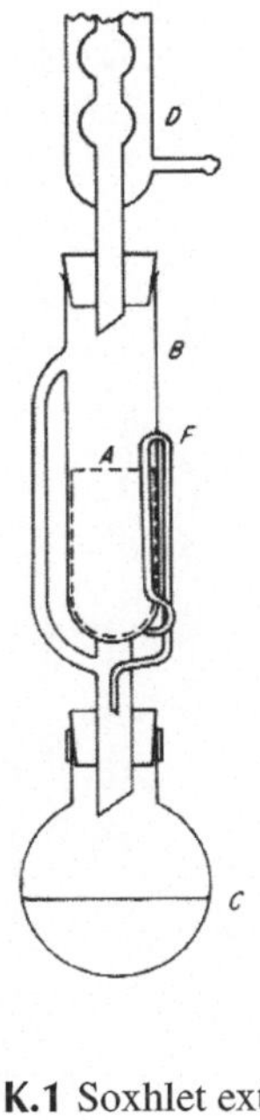

FIGURE K.1 Soxhlet extraction.

Appendix L: SI Units and Conversion Factors

Conversion to Metric Measures

	Symbol	When You Know	Multiply by	To Find	Symbol
Length	in	inches	2.54	centimeters	cm
	ft	feet	30.48	centimeters	cm
	yd	yards	0.9	meters	m
	mi	miles	1.6	kilometers	km
Area	in^2	square inches	6.5	square centimeters	cm^2
	ft^2	square feet	0.09	square meters	m^2
	yd^2	square yards	0.8	square meters	m^2
	mi^2	square miles	2.6	square kilometers	km^2
		acres	0.4	hectares	ha
Mass (weight)	oz	ounces	28	grams	g
	lb	pounds	0.45	kilograms	kg
		short tons (2000 lb)	0.9	tonnes	t
Volume	tsp	teaspoons	5	milliliters	mL
	Tbsp	tablespoons	15	milliliters	mL
	fl oz	fluid ounces	30	milliliters	mL
	c	cups	0.24	liters	L
	pt	pints	0.47	liters	L
	qt	quarts	0.95	liters	L
	gal	gallons	3.8	liters	L
	ft^3	cubic feet	0.03	cubic meters	m^3
	yd^3	cubic yards	0.76	cubic meters	m^3
Temp. °C = 5/9(°F – 32)	°F	Fahrenheit temperature	5/9 (after subtracting 32)	Celsius temperature	°C

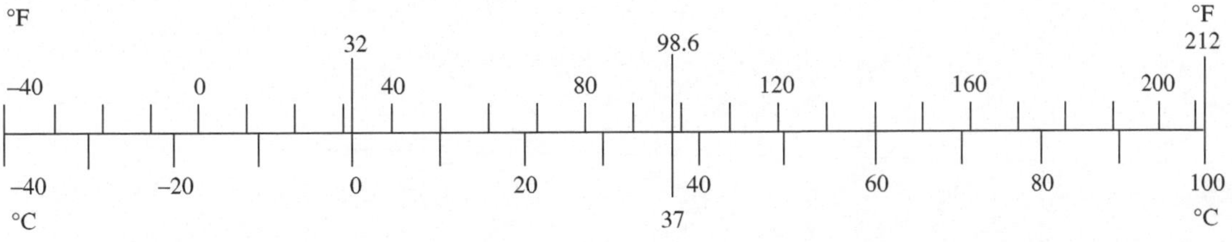

	Conversion from Metric Measures				
	Symbol	When You Know	Multiply By	To Find	Symbol
Length	mm	millimeters	0.04	inches	in
	cm	centimeters	0.4	inches	in
	m	meters	3.3	feet	ft
	m	meters	1.1	yards	yd
	km	kilometers	0.6	miles	mi
Area	cm^2	square centimeters	0.16	square inches	in^2
	m^2	square meters	1.2	square yards	yd^2
	km	square kilometers	0.4	square miles	mi^2
	ha	hectares (10,000 m^2)	2.5	acres	
Mass (Weight)	g	grams	0.035	ounces	oz
	kg	kilograms	2.2	pounds	lb
	t	tonnes (1000 kg)	1.1	short tons	
Volume	mL	milliliters	0.03	fluid ounces	fl oz
	L	liters	2.1	pints	pt
	L	liters	1.06	quarts	qt
	L	liters	0.26	gallons	gal
	m^3	cubic meters	35	cubic feet	ft^3
	m^3	cubic meters	1.3	cubic yards	yd^3
Temp. °F = (9/5°C) + 32	°C	Celsium temperature	9/5 (then add 32)	Fahrenheit temperature	°F

References

Alcamo, L.E., *Fundamentals of Microbiology*, 4th ed., Benjamin Cummings, Menlo Park, CA, 1994.

Ayers, D. et al., *Environmental Science and Technology Handbook*, Government Institutes, Rockville, MD, 1994.

Beaty, R.D., *Concepts, Instrumentation and Techniques in Atomic Absorption Spectrophotometry*, Perkin Elmer, 1988.

Boss, C.B. and Kenneth J. Freeden, *Concepts, Instrumentation, and Techniques in Inductively Coupled Plasma Emission Spectrometry*, Perkin Elmer, 1988.

Brady, J.E. and Holum, J.R., *Chemistry: The Study of Matter and Its Changes*, John Wiley & Sons, New York, 1993.

A Concise Dictionary of Chemistry, 2nd ed., Oxford University Press, Oxford, 1990.

Csuros, M., *Environmental Sampling and Analysis for Technicians*, Lewis Publishers, Boca Raton, FL, 1994.

Csuros, M., *Environmental Sampling and Analysis Lab Manual*, Lewis Publishers, Boca Raton, FL, 1997.

Day, R.A., Jr. and Underwood, A.L., *Quantitative Analysis*, 6th ed., Prentice Hall, New York, 1991.

Driscoll, F.G., *Groundwater and Wells*, 2nd ed., Johnson Division, St. Paul, MN, 1986.

Ebbing, D.D., *General Chemistry*, 4th ed., Houghton Mifflin, Boston, 1993.

Environmental Protection Agency, *Handbook for Sampling and Sample Preservation of Water and Wastewater*, Government Printing Office, Washington, D.C. (EPA 600/4–82–029), 1982.

Environmental Protection Agency, *Methods for Chemical Analysis of Water and Wastes,* rev., Government Printing Office, Washington, D.C. (EPA-600/4–79–020), March 1983.

Environmental Protection Agency, *Standard Methods for the Examination of Water and Wastewater*, 18th ed., Government Printing Office, Washington, D.C. (APHA-AWWA-WPCF), 1992.

Environmental Protection Agency, *Test Methods for Evaluating Solid Waste*, 3rd ed., Government Printing Office, Washington, D.C. (EPA SW 846), 1986.

Friedman, B., *Environmental Ecology*, Academic Press, New York, 1989.

Fritz, J.S. and Schenk, G.H., *Quantitative Analytical Chemistry*, 5th ed., Allyn & Bacon, Boston, 1987.

Furr, A.K., *CRC Handbook of Laboratory Safety*, 4th ed., CRC Press, Boca Raton, FL, 1995.

Greenfield, S., Jones, I.L.I., and Berry, C.T., High pressure plasma spectroscopic emission sources, *Analyst*, 89, 713–720, 1964.

Joesten, M.D., *World of Chemistry Essentials*, Saunders College, Fort Worth, TX, 1993.

Keenan, J., *Quality Assurance in Chemical Measurements*, Lewis Publishers, Boca Raton, FL, 1988.

Kenkel, J., *Analytical Chemistry for Technicians,* 2nd ed., Lewis Publishers, Boca Raton, FL, 1990.

Malachowski, M.J. and Goldberg, A.F., *Health Effects of Toxic Substances*, Government Institutes, Rockville, MD, 1995.

Martini, F., *Fundamentals of Anatomy and Physiology*, 2nd ed., Prentice Hall, New York, 1992.

Sullivan, T.F.P., *Environmental Law Handbook*, 16th ed., Government Institutes, Rockville, MD, 2001.

Wolfe, D.H., *Introduction to College Chemistry,* 2nd ed., McGraw-Hill, New York, 1988.

Index

N

T

U

V

W

X

Y

Z